LabVIEW 8™ Student Edition

LabVIEW 8™ Student Edition

Robert H. Bishop
The University of Texas at Austin

PEARSON

Prentice
Hall

Upper Saddle River, NJ 07458

Library of Congress Cataloging-in-Publication Data on file.

Vice President and Editorial Director, ECS: *Marcia J. Horton*
Executive Managing Editor: *Vince O'Brien*
Managing Editor: *David A. George*
Production Editor: *Wendy Kopf*
Director of Creative Services: *Paul Belfanti*
Art Director and Cover Manager: *Paul Gourhan*
Cover Design: *National Instruments, Inc.*
Art Editor: *Xiaohong Zhu*
Manufacturing Manager: *Alexis Heydt-Long*
Manufacturing Buyer: *Lisa McDowell*
Marketing Manager: *Tim Galligan*

© 2007 by Pearson Education, Inc.
Pearson Prentice Hall
Pearson Education, Inc.
Upper Saddle River, NJ 07458

Printed in the United States of America

10 9 8 7 6 5 4 3 2

ISBN 0-13-199918-4

Pearson Education Ltd.
Pearson Education Australia Pty.
Pearson Education Singapore, Pte. Ltd.
Pearson Education North Asia Ltd.
Pearson Education Canada, Inc.
Pearson Educación de Mexico, S.A. de C.V.
Pearson Education—Japan
Pearson Education Malaysia, Pte. Ltd.
Pearson Education Inc., *Upper Saddle River, New Jersey*

To my parents, W. Robert Bishop
and Anna Maria DiPietro Bishop

CONTENTS

4   Editing and Debugging Virtual Instruments   147

8   Charts and Graphs   342

A   Instrument Control   569

B   Other LabVIEW Applications   597

PREFACE

Learning with LabVIEW 8™ is the textbook that accompanies the *LabVIEW 8 Student Edition* from National Instruments, Inc. This textbook, as well as the Lab-VIEW software, has undergone a significant revision from the previous edition to include all the latest features and technologies of LabVIEW 8. One exciting new feature, as well as significant new addition to the textbook is MathScript, a text-based math oriented computing environment built right into LabVIEW 8. As you read through the book and work through the examples, we hope you will agree that this book is more of a personal tour guide than a software manual.

The LabVIEW graphical development environment was built specifically for applications in engineering and science, with built-in functionality designed to reduce development time for design and simulation in signal processing, control, communcations, electronics and more. The *LabVIEW 8 Student Edition* delivers all the capabilities of the full version of LabVIEW, widely considered the industry standard for design, test, measurement, automation, and control applications. With LabVIEW, students can design graphical programming solutions to their homework problems and laboratory experiments—an ideal tool for science and engineering applications—that is also fun to use! The *LabVIEW 8 Student Edition* affords students the opportunity for self-paced learning and independent project development.

The goal of this book is to help students learn to use LabVIEW on their own. With that goal in mind, this book is very art-intensive with over 400 figures in all. That means that there are numerous screen captures in each section taken from a typical LabVIEW session. The figures contain additional labels and pointers added to the LabVIEW screen captures to help students understand what they are seeing on their computer screens as they follow along in the book.

The most effective way to use *Learning with LabVIEW* is to have a concurrent LabVIEW session in progress on your computer and to follow along with the steps in the book. A directory of virtual instruments has been developed by the author exclusively for use by students using *Learning with LabVIEW* and is available on **www.prenhall.com/bishop**. These virtual instruments complement the material in the book. In most situations, the students are asked to develop the virtual instrument themselves following instructions given in the book, and then comparing their solutions with the solutions provided by the author to obtain immediate feedback. In other cases, students are asked to run a specified virtual instrument as a way to demonstrate an important LabVIEW concept.

THE *LABVIEW 8 STUDENT EDITION*

With the renewed emphasis in higher education on hands-on laboratory experience, many educational institutions are beginning to improve their laboratory facilities with the goal of increasing student exposure to practical problems. Educators continue to be pushed by industry to produce college graduates with experience in acquiring and analyzing data, constructing computer-based simulations of physical systems, and multi-purpose computer programming. To that end, LabVIEW offers a powerful, efficient, and easy-to-use development environment that provides educators the tools to teach their students a wide range of topics with just one open, industry-standard tool. Using LabVIEW you can create your own virtual instruments (VIs). An effective way to improve laboratory instrumentation is by modifying and improving a LabVIEW computer program (that is, a virtual instrument that emulates a standard instrument), rather than to retrofit a laboratory with new hardware equipment. In LabVIEW, the software is the instrument.

THE *LABVIEW 8 STUDENT EDITION*

The *LabVIEW 8 Student Edition* software package is a powerful and flexible instrumentation, analysis, and control software platform for PCs running Microsoft Windows or Apple Macintosh OS X. The student edition is designed to give students early exposure to the many uses of graphical programming. LabVIEW not only helps reinforce basic scientific, mathematical, and engineering principles, but it encourages students to explore advanced topics as well. Students can run LabVIEW programs designed to teach a specific topic, or they can use their skills to develop their own applications. LabVIEW provides a real-world, hands-on experience that complements the entire learning process.

WHAT'S NEW WITH THE LABVIEW 8 STUDENT EDITION?

The demand for LabVIEW in colleges and universities has led to the development of *LabVIEW 8 Student Edition* based on the industry version of LabVIEW 8. This is a new and significant software revision that delivers all of the graphical programming capabilities of the full edition. With the student edition, students can design graphical programming solutions for their classroom problems and laboratory experiments on their personal computers. The *LabVIEW 8 Student Edition* features include the following:

- Express VIs that bring interactive, configuration-based application design for acquiring, analyzing, and presenting data.
- Interactive measurement assistants and a redesigned NI-DAQ driver make creating data acquisition and instrument control applications easier than ever.
- Access to the new interactive MathScript environment providing a command line computation capability and the MathScript Node for integrating textual scripts programmatically.
- Full LabVIEW advanced analysis capability, including 13 analysis Express VIs and over 400 native analysis and signal processing functions.
- Full compatibility with all National Instruments data acquisition and instrument control hardware.
- Support for all data types used in the LabVIEW Full Development System.

ORGANIZATION OF LEARNING WITH LABVIEW

This textbook serves as a LabVIEW resource for students. The pace of instruction is intended for both undergraduate and graduate students. The book is comprised of 11 chapters and should be read sequentially when first learning LabVIEW. For more experienced students, the book can be used as a reference book by using the index to find the desired topics. The 11 chapters are as follows:

CHAPTER 1: LabVIEW Basics—This chapter introduces the LabVIEW environment and helps orient students when they open a virtual instrument. Concepts such as windows, toolbars, menus, and palettes are discussed.

CHAPTER 2: Virtual Instruments—The components of a virtual instrument are introduced in this chapter: front panel, block diagram, and icon/connector pair. This chapter also introduces the concept of controls (inputs) and indicators (outputs) and how to wire objects together in the block diagram. Express VIs are introduced in the chapter.

CHAPTER 3: MathScript—This chapter introduces the new interactive MathScript environment, which combines a mathematics-oriented text-based language with the intuitive graphical dataflow programming of LabVIEW. Both the interactive MathScript environment for command line computation and the MAthScript Node for integrating textual scripts within the LabVIEW block diagram are discussed.

CHAPTER 4: Editing and Debugging Virtual Instruments—Resizing, coloring, and labeling objects are just some of the editing techniques introduced in this chapter. Students can find errors using execution highlighting, probes, single-stepping, and breakpoints, just to name a few of the available debugging tools.

CHAPTER 5: Sub VIs—This chapter emphasizes the importance of reusing code and illustrates how to create a VI icon/connector. It also shows parallels between LabVIEW and text-based programming languages.

CHAPTER 6: Structures—This chapter presents loops, case structures, and flat sequence structures that govern the execution flow in a VI. The Formula Node is introduced as a way to implement complex mathematical equations.

CHAPTER 7: Arrays and Clusters—This chapter shows how data can be grouped, either with elements of the same type (arrays) or elements of a different type (clusters). This chapter also illustrates how to create and manipulate arrays and clusters.

CHAPTER 8: Charts and Graphs—This chapter shows how to display and customize the appearance of single and multiple charts and graphs.

CHAPTER 9: Data Acquisition—The basic characteristics of analog and digital signals are discussed in this chapter, as well as the factors students need to consider when acquiring and generating these signals. This chapter introduces students to the Measurement and Automation Explorer (MAX) and the DAQ Assistant.

CHAPTER 10: Strings and File I/O—This chapter shows how to create and manipulate strings on the front panel and block diagram. This chapter also explains how to write data to and read data from files.

CHAPTER 11: Analysis—LabVIEW can be used in a variety of ways to support analysis of signals and systems. Several important analysis topics are discussed in this chapter, including how to use LabVIEW for signal generation, signal processing, linear algebra, curve fitting, formula display on the front panel, differential equations, finding roots (zero finder), and integration and differentiation.

APPENDIX A: Instrument Control—The components of an instrument control system using a GPIB or serial interface are presented in this appendix. Students are introduced to the notion of instrument drivers and of using the Measurement and Automation Explorer (MAX) to detect and install instrument drivers. The Instrument I/O Assistant is introduced.

APPENDIX B: Other LabVIEW Applications—This appendix briefly discusses other features of LabVIEW, including sound card programming, the LabVIEW Project, Shared Variables, LabVIEW Toolkits and integrating LabVIEW with Multisim.

The important pedagogical elements in each chapter include the following:

1. A brief table of contents and a short preview of what to expect in the chapter.
2. A list of chapter goals to help focus the chapter discussions.

3. Margin icons that focus attention on a helpful hint or on a cautionary note.

Helpful hint   Cautionary note

4. An end-of-chapter summary and list of key terms.

KEY TERMS

5. Sections entitled **Building Blocks** near the end of each chapter present the continuous development and modification of a virtual instrument for calculating and generating a pulse-width modulated signal. The student is expected to construct the VIs based on the instructions given in the sections. The same VI is used as the starting point and then improved in each subsequent chapter as a means for the student to practice with the newly introduced chapter concepts.

BUILDING BLOCK

6. Many worked examples are included in each chapter including several new examples introduced in this edition. In most cases, students construct the VIs discussed in the examples by following a series of instructions given in the text. In the early chapters, the instructions for building the VIs are quite specific, but in the later chapters, students are expected to construct the VIs without precise step-by-step instructions. Of course, in all chapters, working versions of the VIs are provided for all examples in the Learning directory included as part of the *LabVIEW 8 Student Edition*. Here is a sample of the worked examples:

   - Temperature system demonstration.
   - Solving a set of linear differential equations.

- Building your first virtual instrument.
- Computing area, diameter, and radius of a circle.
- Computing and graphing the time value of money.
- Studying chaos using the logistic difference equation.
- Acquiring data.
- Writing ASCII data to a file.

7. A section entitled *Relaxed Reading* that describes how LabVIEW is being utilized to solve interesting real-world problems. The material is intended to give students a break from the technical aspects of learning LabVIEW and to stimulate thinking about how LabVIEW can be used in various other situations.

8. End-of-chapter exercises, problems, and design problems reinforce the main topics of the chapter and provide practice with LabVIEW.

ORIGINAL SOURCE MATERIALS

Learning with LabVIEW 8 was developed with the aid of several important references. Each reference is a manual published by National Instruments and one manual is actually used in a hands-on classroom environment taught by National Instruments instructors in their own classroom. Specifically, the following materials were used as primary sources:

- *LabVIEW Basics I*, Course Software Version 8.0, Copyright © 2005.
- *LabVIEW User Manual*, Copyright © 2005.

By design, there is a strong correlation between the material contained in the NI LabVIEW manuals and the material presented in this book. The information contained in the manuals was reduced in scope and refined to make it more accessible to students learning LabVIEW on their own.

LABVIEW 8 STUDENT EDITION SOFTWARE

It is assumed that the reader has a working knowledge of either the Windows or the Mac OS X operating system. If your computer experience is limited, you may first want to spend some time familiarizing yourself with your computer in order to understand the operation of your Mac or PC. You should know how to access pull-down menus, open and save files, install software from a CD, and use a mouse. You will find previous computer programming experience helpful—but not necessary.

A set of virtual instruments has been developed by the author for this book. You will need to obtain the Learning directory from the companion website to this book at Prentice Hall:

http://www.prenhall.com/bishop

For more information, you may also want to visit the NI Student Edition website at

http://www.ni.com/labviewse

All of the VI examples in this book were tested by the author on a Dell PC running Windows XP Professional. Obviously, it is not possible to verify each VI on all the available Windows and Macintosh platforms that are compatible with LabVIEW so if you encounter platform-specific difficulties, please let us know.

If you would like information on upgrading to the LabVIEW 8 Professional Version, please write to

National Instruments
att.: Academic Sales
11500 North Mopac Expressway
Austin, TX 78759

or visit the National Instruments website: **http://www.ni.com**

LIMITED WARRANTY

The software and the documentation are provided "as is," without warranty of any kind, and no other warranties, either expressed or implied, are made with respect to the software. National Instruments does not warrant, guarantee, or make any representations regarding the use, or the results of the use, of the software or the documentation in terms of correctness, accuracy, reliability, or otherwise and does not warrant that the operation of the software will be uninterrupted or error-free. This software is not designed with components and testing for a level of reliability suitable for use in the diagnosis and treatment of humans or as critical components in any life-support systems whose failure to perform can reasonably be expected to cause significant injury to a human. National Instruments expressly disclaims any warranties not stated herein. Neither National Instruments nor Pearson Education shall be liable for any direct or indirect damages. The entire liability of National Instruments and its dealers, distributors, agents, or employees are set forth above. To the maximum extent permitted by applicable law, in no event shall National Instruments or its suppliers be liable for any damages, including any special, direct, indirect, incidental, exemplary, or consequential damages, expenses, lost profits, lost savings, business interruption, lost business information, or any other damages arising out of the use, or inability to use, the software or the documentation even if National Instruments has been advised of the possibility of such damages.

ACKNOWLEDGMENTS

Thanks to all the folks at National Instruments for their assistance and input during the development of *Learning with LabVIEW 8*. A very special thanks to Jim Cahow and Erik Luther of NI for providing day-to-day support during the final months of the project. Finally, I wish to express my appreciation to Lynda Bishop for assisting me with the manuscript preparation, for providing valuable comments on the text, and for handling my personal day-to-day activities associated with the entire production.

KEEP IN TOUCH!

The author and the staff at Pearson Prentice Hall and at National Instruments would like to establish an open line of communication with the users of the *LabVIEW 8 Student Edition*. We encourage students to e-mail the author with comments and suggestions for this and future editions.

Keep in touch!

ROBERT H. BISHOP
rhbishop@mail.utexas.edu
JIM CAHOW
National Instruments Academic Resources Manager
jim.cahow@ni.com

CHAPTER 1

LabVIEW Basics

Welcome to the *LabVIEW Student Edition*! **LabVIEW** is a powerful and complex programming environment. Once you have mastered the various concepts introduced in this book you will have the ability to develop applications in a graphical programming language and to develop virtual instruments for design, control, and test engineering. This introductory chapter provides a basic overview of LabVIEW and its components.

GOALS

1. Installation of the *LabVIEW Student Edition*.

2. Familiarization with the basic components of LabVIEW.

3. Introduction to front panels and block diagrams, shortcut and pull-down menus, palettes, VI libraries, and online help.

1.1   SYSTEM CONFIGURATION REQUIREMENTS

The *LabVIEW Student Edition* is distributed on CD-ROM and contains versions for Windows and Macintosh.

Windows XP/2000/XP × 64

Processor:	Pentium III/Celeron 866 MHz or equivalent
RAM:	256 MB minimum; 512 MB recommended
Screen Resolution:	1024 × 768 pixels
Operating System:	Windows XP/2000
Disk Space:	900 MB

Macintosh OS X 10.3 or later

Processor:	G3 minimum; G4 recomended
RAM:	256 MB minimum; 512 MB recommended
Screen Resolution:	1024 × 768 pixels
Operating System:	Mac OS X (10.3 or later)
Disk Space:	500 MB

1.2   INSTALLING THE *LABVIEW STUDENT EDITION*

 Disable any automatic virus detection programs before you install. Some virus detection programs interfere with the installation program.

Windows

1. Log on as an administrator or as a user with administrator privileges.
2. Insert the LabVIEW 8.0 installation CD and follow the instructions that appear on the screen.
3. After installation, check your hard disk for viruses and enable any virus detection programs you disabled.

LabVIEW relies on a licensing activation process to obtain an activation code. The Activation Wizard is a part of the NI License Manager that takes you through the steps of enabling software to run on a machine. If you have questions regarding activation, visit the NI website at http://www.ni.com/labviewse/.

Macintosh

1. Insert the LabVIEW 8.0 installation CD.

2. Run the appropriate installation program. The following types of installations of LabVIEW 8.0 are available:

   - **Easy Installation**—Installs all LabVIEW 8.0 files, including LabVIEW, VI libraries, GPIB drivers, VISA drivers, examples, and help files. The LabVIEW Professional Development System also installs source code control (SCC) tools.

   - **Custom Installation**—If you select this option, you select the files to install. To select a custom installation, select **Custom Install** from the **Easy Install** pull-down menu. You must select the **LabVIEW 8.0 Core** component to install the core set of LabVIEW files necessary for running LabVIEW 8.0.

3. Follow the instructions that appear on the screen.

4. After installation, check your hard disk for viruses and enable any virus detection programs you disabled.

 The LabVIEW installation folder might contain the following items:

- *cintools—Tools for calling C object code from LabVIEW*

- *AppLibs—(LabVIEW Professional Development System). Contains stub for stand-alone applications.*

Refer to LabVIEW Help for a complete list of the folders installed in LabVIEW directory structure.

1.3   THE LABVIEW ENVIRONMENT

LabVIEW is short for **Lab**oratory **V**irtual **I**nstrument **E**ngineering **W**orkbench. It is a powerful and flexible graphical development environment created by the folks at National Instruments—a company that creates hardware and software products that leverage computer technology to help engineers and scientists take measurements, control processes, and analyze and store data. National Instruments was founded over twenty-eight years ago in Austin, Texas by James Truchard (known as Dr. T), Jeffrey Kodosky, and William Nowlin. At the time, all three men were working on sonar applications for the U.S. Navy at the Applied Research Laboratories at The University of Texas at Austin. Searching for a way to connect test equipment to DEC PDP-11 computers, Dr. T decided to develop an interface bus. He recruited Jeff and Bill to join him in his endeavor, and together they successfully developed LabVIEW and the notion of a "virtual instrument." In the process they managed to infuse their new company—National Instruments—with an entrepreneurial spirit that still pervades the company today.

Engineers and scientists in research, development, production, test, and service industries as diverse as automotive, semiconductor, aerospace, electronics, chemical, telecommunications, and pharmaceutical have used and continue to use LabVIEW to support their work. LabVIEW is a major player in the area of testing and measurements, industrial automation, and data analysis. For example, scientists at NASA's Jet Propulsion Laboratory used LabVIEW to analyze and display Mars Pathfinder Sojourner rover engineering data, including the position and temperature of the rover, how much power remained in the rover's battery, and generally to monitor Sojourner's overall health. This book is intended to help you learn to use LabVIEW as a programming tool and to serve as an introduction to the power of graphical programming and the myriad applications to which it can be applied.

LabVIEW programs are called **Virtual Instruments**, or VIs for short. LabVIEW is different from text-based programming languages (such as Fortran and C) in that LabVIEW uses a graphical programming language, known as the G programming language, to create programs relying on graphic symbols to describe programming actions. LabVIEW uses a terminology familiar to scientists and engineers, and the graphical icons used to construct programs in G are easily identified by visual inspection. You can learn LabVIEW even if you have little programming experience, but you will find knowledge of programming fundamentals helpful. If you have never programmed before (or maybe you have programming experience but have forgotten a few things) you may want to review the basic concepts of programming before diving into the G programming language.

LabVIEW provides an extensive library of virtual instruments and functions to help you in your programming. An important new development introduced in LabVIEW 8.0 is the MathScript environment. The MathScript environment is discussed in detail in Chapter 3. The advent of MathScript heralds the era of the text-based command line environment complementing the LabVIEW graphical programming environment. Users will now have the capability to make quick calculations or computations at a LabVIEW command line prompt. MathScript also enables users to integrate their scripts with LabVIEW block diagrams, easily mixing graphical programming and powerful user interfaces with text-based scripts. LabVIEW also contains application-specific libraries for data acquisition (discussed in Chapter 9), file input/output (discussed in Chapter 10), and data analysis (discussed in Chapter 11). It includes conventional program debugging tools with which you can set breakpoints, single-step through the program, and animate the execution so you can observe the flow of data. Editing and debugging VIs is the topic of Chapter 4.

LabVIEW has a set of VIs for data presentation on various types of charts and graphs. Chapter 8 discusses the process of presenting data on charts and graphs.

The LabVIEW system consists of LabVIEW application executable files and many associated files and folders. LabVIEW uses files and directories to store information necessary to create your VIs. Some of the more important files and directories are:

1. The **LabVIEW** executable. Use this to launch LabVIEW.

2. The **vi.lib** directory. This directory contains libraries of VIs such as data acquisition, instrument control, and analysis VIs; it must be in the same directory as the LabVIEW executable. Do not change the name of the **vi.lib** directory, because LabVIEW looks for this directory when it launches. If you change the name, you cannot use many of the controls and library functions.

3. The **examples** directory. This directory contains many sample VIs that demonstrate the functionality of LabVIEW.

4. The **user.lib** directory. This directory is where you can save VIs you have created, and they will appear in the LabVIEW **Functions** palette.

5. The **instr.lib** directory. This directory is where your instrument driver libraries are placed if you want them to appear in the **Functions** palette.

6. The **Learning** directory. This file contains a library of VIs that you will use with the *Learning with LabVIEW* book.

The files in the **Learning** *directory must be downloaded from the site* http://www.prenhall.com/bishop—*the* **Learning** *directory VIs are not found on the installation CD! You can access the Prentice Hall site through the Internet using any standard Web browser.*

1.4   THE GETTING STARTED SCREEN

When you launch LabVIEW by double-clicking on its icon, the Getting Started screen appears as in Figure 1.1. The Getting Started screen contains a navigation dialog box that includes introductory material and common commands. The dialog box includes a menu that allows users to quickly create new VIs, select among the most recently opened LabVIEW files, find examples, and search the LabVIEW Help. On the left-hand side of the Getting Started screen are *file* functions, such as creating a new VI or for opening an existing VI. On the right-hand side are *resource* functions. Information and resources are reachable from the Getting Started screen to help you learn about LabVIEW using online manuals and tutorials, with convenient access to Internet links to resources on the National Instruments website.

FIGURE 1.1
The Getting Started screen.

On the left-hand side of the Getting Started screen you can:

- Click on **New** to open the New Dialog Box to create a blank VI or one based on a VI template. For quick access, it is also possible to select a new VI or a VI from a template directly on the Getting Started screen.

- Click on **Open** to open an existing VI by browsing to find the desired file, or select a VI from the list of recently used VIs.

On the right-hand side of the screen you can:

- Access manuals and helpful information about LabVIEW, if desired, by making a selection in the **New to LabVIEW?** or **Upgrading LabVIEW?** list. For example, selecting **Getting Started with LabVIEW** will open a pdf document published by National Instruments, Inc. of the same name.

- Click **Web Resources** to connect to the Internet assets, such as the discussion groups, and to see what courses are available for LabVIEW training.

- Select **Examples** or **Find Examples** to begin a tour of the plethora of example VIs through the Example Finder dialog box. At the bottom of the right-hand side is the access point for LabVIEW examples.

Throughout this book, use the left mouse button (if you have one) unless we specifically tell you to use the right one.

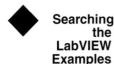

Searching the LabVIEW Examples

In this exercise you will search through the list of example VIs and demonstrations that are included with the *LabVIEW Student Edition*. Open the LabVIEW application and get to the Getting Started screen. The search begins at the LabVIEW Getting Started screen by clicking on **Find Examples**, as shown in Figure 1.2. The NI Example Finder screen displays the numerous examples available with LabVIEW, as illustrated in Figure 1.3. The examples can be browsed by Task or by Directory Structure. In Figure 1.3, the examples are sorted by Task.

To reach the desired example—in this case, we are searching for the Function Waveform Generation—select **Analyzing and Processing Signals**, and then **Signal and Noise Generation**, as shown in Figure 1.3. Selecting **Function Waveform Generation** opens up the associated virtual instrument (VI) (more on VIs in Chapter 2). Just for fun, you can start the VI running and see what happens. Start the VI by clicking on the **Run** button, as shown in Figure 1.4. Stop the VI by clicking on the **Abort Execution** button. Try it! ◆

FIGURE 1.2
LabVIEW examples

FIGURE 1.3
The NI Example Finder

FIGURE 1.4
The Function Waveform Generation front panel

1.5   PANEL AND DIAGRAM WINDOWS

An untitled front panel window appears when you select **Blank VI** from the Getting Started screen. The front panel window is the interface to your VI code and is one of the two LabVIEW windows that comprise a virtual instrument. The other window—the block diagram window—contains program code that exists in a graphical form (such as icons, wires, etc.).

Front panels and block diagrams consist of graphical objects that are the G programming elements. Front panels contain various types of controls and indicators (that is, inputs and outputs, respectively). Block diagrams contain terminals corresponding to front panel controls and indicators, as well as constants, functions, subVIs, structures, and wires that carry data from one object to another. Structures are program control elements (such as For Loops and While Loops).

Figure 1.5 shows a front panel and its associated block diagram. You can find the virtual instrument First VI.vi shown in Figure 1.5 in the Chapter 1 folder within the directory Learning. That VI can be located by choosing **Open** on the Getting Started screen and navigating to the Chapter1 folder in the Learning directory and then selecting First VI.vi. Once you have the VI front panel open, find the **Run** button on the panel toolbar and click on it. Your VI is now running. You can turn the knob and vary the different inputs and watch the output changes reflected in the graph. Give it a try! If you have difficulty getting things to work, then just press ahead with the material in the next sections and come back to this VI when you are ready.

1.5.1   Front Panel Toolbar

A toolbar of command buttons and status indicators that you use for controlling VIs is located on both the front panel and block diagram windows. The front panel toolbar and the block diagram toolbar are different, although they do each contain some of the same buttons and indicators. The toolbar that appears at the top of the front panel window is shown in Figure 1.6.

While the VI is executing, the **Abort Execution** button appears. Although clicking on the abort button terminates the execution of the VI, as a general rule you should avoid terminating the program execution this way and either let the VI execute to completion or incorporate a programmatic execution control (that is, an on-off switch or button) to terminate the VI from the front panel.

The **Broken Run** button replaces the **Run** button when the VI cannot compile and run due to coding errors. If you encounter a problem running your VI, just click on the **Broken Run** button, and a window will automatically appear on the desktop that lists all the detected program errors. And then, if you double-click on one of the specific errors in the list, you will be taken automatically to the location in the block diagram (that is, to the place in the code) where the error exists. This is a great debugging feature! More discussion on the issue of debugging VIs can be found in Chapter 4.

Panel palette

Free label

Owned label

Knob control

Numeric control

Horizontal slider

FRONT PANEL

Icon pane

Graph legend

Graph

Round LED

Push button

Numeric constant
Knob control terminal

Horizontal slider terminal

BLOCK DIAGRAM

Graph terminal

For Loop structure

While Loop structure

Wire data path   SubVI

FIGURE 1.5
A front panel and the associated block diagram

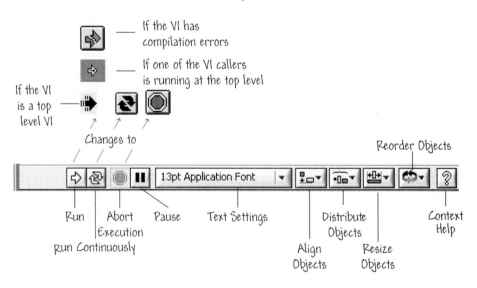

If the VI has compilation errors

If one of the VI callers is running at the top level

If the VI is a top level VI

Changes to

Run

Abort Execution

Run Continuously

Pause

Text Settings

Align Objects

Distribute Objects

Resize Objects

Reorder Objects

Context Help

FIGURE 1.6
The front panel toolbar

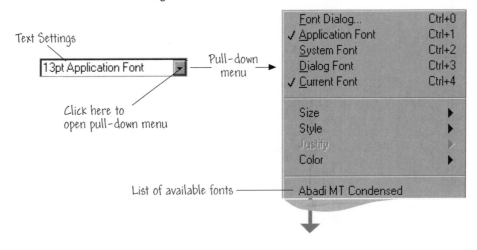

Text Settings

Click here to
open pull-down menu

Pull-down
menu

List of available fonts

FIGURE 1.7
The **Text Settings** pull-down menu

Clicking on the **Run Continuously** button leads to a continuous execution of the VI. Clicking on this button again disables the continuous execution—the VI stops when it completes normally. The behavior of the VI and the state of the toolbar during continuous run is the same as during a single run started with the **Run** button.

The **Pause/Continue** button pauses VI execution. To continue program execution after pausing, press the button again, and the VI resumes execution.

The **Text Settings** pull-down menu, shown in Figure 1.7, sets font options—font type, size, style, and color.

The **Align Objects** pull-down menu sets the preferred alignment of the various objects on either the front panel or the block diagram. After selecting the desired objects for alignment, you can set the preferred alignment for two or more objects. For example, you can align objects by their left edges or by their top edges. The various alignment options are illustrated in the **Align Objects** pull-down menu shown in Figure 1.8. Aligning objects is very useful in organizing the

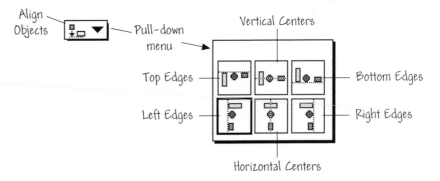

Align
Objects

Pull-down
menu

Vertical Centers

Top Edges

Bottom Edges

Left Edges

Right Edges

Horizontal Centers

FIGURE 1.8
The **Align Objects** pull-down menu

FIGURE 1.9
The **Distribute Objects** pull-down menu

VI front panel (and the block diagram, for that matter). On the surface, it may appear that aligning the front panel objects, while making things "neat and pretty," does not contribute to the goal of a functioning VI. As you gain experience with constructing VIs, you will find that they are easier to debug and verify if the interface (front panel) and the code (block diagram) are organized to allow for easy visual inspection.

The **Distribute Objects** options pull-down menu, shown in Figure 1.9, sets the preferred distribution for two or more objects. For example, you can evenly space selected objects, or you can remove all the space between the objects.

The **Resize Objects** objects pull-down menu, shown in Figure 1.10, is used to resize multiple front panel to the same size. For example, this feature allows you to select multiple objects and resize them all to the same height as the object within the selected group with the maximum height.

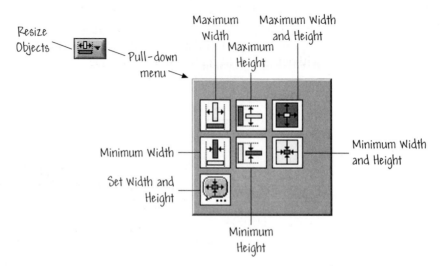

FIGURE 1.10
The **Resize Objects** pull-down menu.

Clicking on the **Context Help** button displays the **Context Help** window. When activated, this feature shows information about LabVIEW objects, such as functions, VIs, constants, structures, and palettes, as you move the cursor over the object.

1.5.2   Block Diagram Toolbar

The block diagram toolbar contains many of the same buttons as the front panel toolbar. Five additional program debugging features are available on the block diagram toolbar and are enabled via the buttons, as shown in Figure 1.11.

Clicking on the **Highlight Execution** button enables execution highlighting. As the program executes, you will be able to see the data flow through the code on the block diagram. This is extremely helpful for debugging and verifying proper execution. In the execution highlighting mode, the button changes to a brightly lit light bulb.

Clicking on the **Retain Wire Values** button allows you to create probes after the VI runs. In other words, after the VI executes successfully, you can place a probe on the block diagram and the most recent value of the data that flowed through the wire at that location will be displayed in the probe. You may find this capability useful when debugging complex block diagrams.

LabVIEW debugging capabilities allow you to single-step through the VI node to node. A **node** is an execution element, such as a For Loop or subVI. You will learn more about the different types of nodes as you proceed through the book, but for the time being you can think of a node as a section of the computer code that you want to observe executing. Each node blinks to show it is ready to execute.

The **Step Into** button allows you to step into a node. Once you have stepped into the node, you can single-step through the node.

The **Step Over** button steps over a node. You are in effect executing the node without single-stepping through the node.

FIGURE 1.11
The block diagram toolbar

The **Step Out** button allows you to step out of a node. By stepping out of a node, you can complete the single-stepping through the node and go to the next node.

The **Warning** indicator only appears when there is a potential problem with your block diagram. The appearance of the warning does not prevent you from executing the VI. You can enable the Warning indicator on the Options. . .≫ Debugging menu in the **Tools** pull-down menu.

1.6   SHORTCUT MENUS

LabVIEW has two types of menus—**pull-down** menus and **shortcut** menus. We will focus on shortcut menus in this section and on pull-down menus in the next section. Our discussions here in Chapter 1 are top-level; we reserve the detailed discussions on each menu item for later chapters as they are used.

To access a shortcut menu, position the cursor on the desired object on the front panel or block diagram and click the right mouse button on a PC-compatible or hold down the <command> key and then click the mouse button on the Mac. In most cases, a shortcut menu will appear since most LabVIEW objects have shortcut menus with options and commands. You will also find that you can **right-click** on the empty front panel and block diagram space, giving you access to the **Controls** and **Functions** palettes and other important palettes. The options presented to you on shortcut menus depend on the selected object—right-clicking on a numeric control will open a different shortcut menu than right-clicking on a For Loop. When you construct a program in G, you will use shortcut menus extensively!

Many shortcut and pull-down menus contain submenus, as shown in Figures 1.12 and 1.13.

FIGURE 1.12
An example of a pull-down menu expanding into a submenu

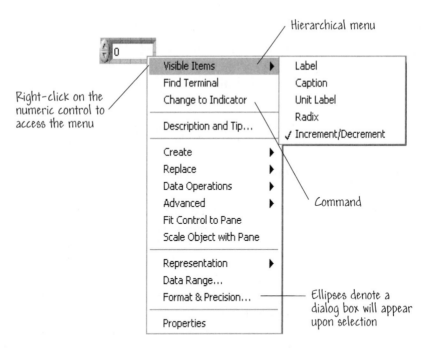

FIGURE 1.13
An example of a shortcut menu

On a PC-compatible, right-click on the object to open a shortcut menu. On a Mac, press <control> and simultaneously click on the object.

Menu items that expand into submenus are called **hierarchical** menus and are denoted by a right arrowhead on the menu. Hierarchical menus will present you with different types of options and commands. One typical option is the so-called mutually exclusive option. This means that if you select the option (by clicking on it), a check mark will appear next, indicating the option is selected; otherwise the option is not selected.

Right-clicking on different areas of an object may lead to different shortcut menus. If you right-click and do not see the anticipated menu selection, then right-click somewhere else on the object.

Another type of menu item opens dialog boxes containing options for you to use to modify and configure your program elements. Menu items leading to dialog boxes are denoted by ellipses (. . .). Menu items without right arrowheads or ellipses are generally commands that execute immediately upon selection. **Create Constant** is an example of a command that appears in many shortcut menus. In some instances, commands are replaced in the menu by their inverse commands when selected. For example, after you choose **Change to Indicator**, the menu selection is replaced by **Change to Control**.

1.7   PULL-DOWN MENUS

The menu bar at the top of the LabVIEW screen, shown in Figure 1.14, contains the important pull-down menus. In this section we will introduce the pull-down menus: **File**, **Edit**, **View**, **Project**, **Operate**, **Tools**, **Window**, and **Help**.

 *In **Windows**, the menus display only the most recently used items by default. Click the arrows at the bottom of a menu to display all items. You can display all menu items by default by selecting Tools≫Options and selecting **Environment** from the top pull-down menu.*

The level of discussion on pull-down windows given here is consistent with our previous discussions. The main goal is to introduce the menus.

FIGURE 1.14
The menu bar

 Some menu items are unavailable while a VI is in Run mode.

1.7.1   File Menu

The **File** pull-down menu, shown in Figure 1.15, contains commands associated with file manipulations. For example, you can create new VIs or open existing ones from the **File** menu. You use options in the **File** menu primarily to open, close, save, and print VIs. As you observe each pull-down menu, notice that selected commands and options have shortcuts listed beside them. These shortcuts are keystroke sequences that can be used to choose the desired option without pulling down the menu. For example, you can open a new VI by pressing Ctrl-O on the keyboard, or you can access the **File** pull-down menu and select **Open**.

FIGURE 1.15
Pull-down menus—**File**

1.7.2   Edit Menu

The **Edit** menu, shown in Figure 1.16, is used to modify front panel and block diagram objects of a VI. The **Undo** and **Redo** options are very useful when you are editing because they allow you to undo an action after it is performed, and once you undo an action you can redo it. By default, the maximum number of undo steps per VI is eight—you can increase or decrease this number if desired.

1.7.3   View Menu

The **View** menu, shown in Figure 1.17, contains items that display LabVIEW environment information. For example, you can easily access the Getting Started screen and display the **Tools**, **Controls**, and **Functions** palettes.

 *Some menu items are available only on specific operating systems and with specific LabVIEW development systems. For example, the **Navigation** window is available only in the LabVIEW Full and Professional Development Systems.*

File	Edit	View	Project	Operate	Tools	Window	Help		
Undo Delete			Ctrl+Z			Undo an action immediately after it is performed
Redo			Ctrl+Shift+Z			Redo an action after an undo
Cut			Ctrl+X			Remove the selected object and place it on the clipboard
Copy			Ctrl+C			Copy the selected object and place it on the clipboard
Paste			Ctrl+V			Place a copy of the active clipboard comments nto the active window
Delete						Delete the selected object
Select All			Ctrl+A			Select all objects in the active window
Make Current Values Default						Set current values as defaults
Reinitialize Values to Default						Set all controls and indicators to default values
Customize Control...						Modify the currently selected front panel control object
Import Picture from File...						Import a graphic file into the LabVIEW clipboard
Set Tabbing Order...						Set the order of front panel objects
Remove Broken Wires			Ctrl+B			Delete all faulty wiring connections
Create SubVI						Convert selected block diagram objects to subVIs
Disable Alignment Grid on Panel			Ctrl+#			Enable/disable alignment grid
Align Items			Ctrl+Shift+A			Align selected objects in active window
Distribute Items			Ctrl+D			Distribute active objects in active window
VI Revision History			Ctrl+Y			Document changes to the current VI
Run-Time Menu...						Create custom run-time menus for the current VI
Find and Replace...			Ctrl+F			Search for VIs, other objects, or text
Show Search Results			Ctrl+Shift+F			Show the search result and access each result in the list

FIGURE 1.16
Pull-down menus—**Edit**

FIGURE 1.17
Pull-down menus—**View**

Selecting **Error List** and **VI Hierarchy** opens display windows that list the errors in the current VI and permits the user to view the subVIs and other nodes that make up the VIs in memory, respectively. **Browse Relationships** contains items for viewing aspects of the current VI and its hierarchy. These functions are helpful when developing and debugging complex VIs.

Class Browser and **Toolbars** are helpful for advanced users of LabVIEW, and not discussed in detail in this book.

1.7.4   Project Menu

The **Project** menu, shown in Figure 1.18, contains items used for basic LabVIEW project operations, such as opening, closing, and saving projects; building specifications; and editing deployments. Projects are used to group together LabVIEW (and non-LabVIEW) files and to deploy or download files to targets, such as real-time (RT) or FPGA targets. The use of projects is most likely to be of interest to advanced students and users of LabVIEW. This book does not present topics associated with projects in any detail.

The use of projects in LabVIEW is a powerful new development with Lab-VIEW 8. It is, however, a topic for more advanced programmers. For more information about projects, please refer to the online help.

FIGURE 1.18
Pull-down menus—**Project**

Execute the VI	**Run** Ctrl+R
Stops execution of the VI	Stop Ctrl+.
Open a node and pause	Step Into Ctrl+Down
Execute a node and pause at the next node	Step Over Ctrl+Right
Finish executing the current node and pause	Step Out Ctrl+Up
Display the **Find** dialog box with breakpoints preselected	Breakpoints...
Pause execution when VI is called	Suspend when Called
Print the VI front panel	Print at Completion
Log front panel data to a file	Log at Completion
Display data logging options	Data Logging
Toggle between run and edit modes	Change to Run Mode Ctrl+M
Connect to a front panel running remotely	Connect to Remote Panel...
Select the stand-alone application or shared library to debug	Debug Application or Shared Library...

FIGURE 1.19
Pull-down menus—**Operate**

1.7.5 Operate Menu

The **Operate** menu, shown in Figure 1.19, can be used to run or stop your VI execution, change the default values of your VI, and switch between the run mode and edit mode.

1.7.6 Tools Menu

The **Tools** menu, shown in Figure 1.20, is used to communicate with instruments and data acquisition boards, compare VIs, build applications, enable the Web server, and access other options of LabVIEW.

An important link is to the main National Instruments website, where you can obtain general information about the company and its products. If you use LabVIEW to control external instruments, you will be interested in the **Instrument Driver Network** link which connects you to over 600 LabVIEW-ready instrument drivers. This can be found in the Tools≫Instrumentation hierarchical menu. Refer to Appendix A for more information on instrument drivers.

1.7.7 Window Menu

The **Window** menu, shown in Figure 1.21, is used for a variety of activities. You can toggle between the front panel and block diagram, and you can "tile" both windows so you can see them at the same time (one above the other or side-by-side). All the open VIs are listed in the menu (at the bottom), and you can switch between the open VIs. Also, if you want to show the palettes on the desktop, you can use the **Window** menu to select either (or both) palettes (more on palettes in the next section).

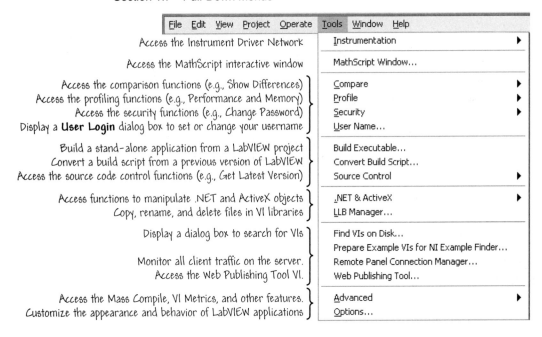

FIGURE 1.20
Pull-down menus—**Tools**

FIGURE 1.21
Pull-down menus—**Windows**

1.7.8   Help Menu

The **Help** menu, shown in Figure 1.22, provides access to the extensive LabVIEW online help facilities. You can view information about front panel or block diagram objects, activate the online reference utilities, and view information about your LabVIEW version number and computer memory. The direct pathways to the internet are provided by **Web Resources** and **Student Edition Web Resources**, which connect you directly to the main source of information concerning the *LabVIEW Student Edition* available on the web. You should make a point to surf this website—this is where the latest and greatest information, news, and updates on the both the software *LabVIEW Student Edition* and on the book *Learning with LabVIEW* will be posted.

Display the Context Help window to provide basic reference information ⎫
Lock the current contents of the Context Help window ⎭

Display the LabVIEW help ⎫
Access the complete reference information for any current errors in your VI ⎬
Access VI reference information from the electronic documentation ⎭

Access the Search Examples section of the LabVIEW help electronic documentation ⎫
Display the Instrumentation Driver Finder ⎬
Access internet links to online National Instruments resources. ⎭

Access patent information about LabVIEW ⎫
Show your LabVIEW version and memory information ⎭

File Edit View Project Operate Tools Window **Help**

Show Context Help	Ctrl+H
Lock Context Help	Ctrl+Shift+L
Search the LabVIEW Help...	Ctrl+?
Explain Error...	
Help for This VI	
Find Examples...	
Find Instrument Drivers...	
Web Resources...	
Patents...	
About LabVIEW...	

FIGURE 1.22
Pull-down menus—**Help**

1.8   PALETTES

Palettes are graphical panels that contain various tools and objects used to create and operate VIs. You can move the palettes anywhere on the desktop that you want—preferably off to one side so that they do not block objects on either the front panel or block diagram. It is sometimes said that the palettes float. The three main palettes are the **Tools**, **Controls**, and **Functions** palettes.

1.8.1   Tools Palette

A **tool** is a special operating mode of the mouse cursor. You use tools to perform specific editing functions, similar to how you would use them in a standard paint program. You can create, modify, and debug VIs using the tools located in the floating **Tools** palette, shown in Figure 1.23. If the **Tools** palette is not visible, select **Tools Palette** from the **View** pull-down menu to display the palette. After you select a tool from this palette, the mouse cursor changes to the appropriate shape.

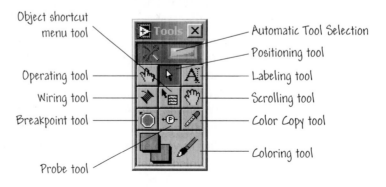

Object shortcut menu tool
Operating tool
Wiring tool
Breakpoint tool
Probe tool
Automatic Tool Selection
Positioning tool
Labeling tool
Scrolling tool
Color Copy tool
Coloring tool

FIGURE 1.23
The **Tools** palette

When the **Automatic Tool Selection** button (located at the top center of the **Tools** palette, as shown in Figure 1.23) is enabled, LabVIEW automatically selects the corresponding tool from the **Tools** palette as you move the cursor over objects on either the front panel or the block diagram. If necessary, you can disable the automatic tool selection feature by clicking on the **Automatic Tool Selection** button to toggle the state. The green light on the button will turn off to indicate that the automatic tool selection is off. You also can manually select a tool on the **Tools** palette to disable the automatic tool selection feature. This allows you to select a tool manually by clicking the tool you want on the **Tools** palette.

One way to access the online help is to place any tool found in the **Tools** palette over the object of interest in the block diagram window. If online help exists for that object, it will appear in a separate Help window. This process requires that you first select **Show Context Help** from the **Help** pull-down menu.

*On a **Windows** platform, a shortcut to accessing the **Tools** palette is to press the <Shift> and the right mouse button on the front panel or the block diagram. On a **Macintosh** platform, you can access the **Tools** palette by pressing <control–shift> and the mouse button on the front panel or block diagram.*

1.8.2   Controls Palette

The **Controls** palette, shown in Figure 1.24, consists of top-level icons representing subpalettes that contain a full range of available objects that you can use to create front panels. Controls simulate input devices and provide a pipeline to move data to the block diagram. Examples of controls include knobs, push buttons, and dials. The controls are located on subpalettes based on the types of controls and the desired appearance, including **Modern**, **System**, **Classic**, **Express**, and others. The **Modern** and **Classic** views present the same controls but with different appearances, with the **Classic** appearance dating back several versions of LabVIEW. The **Express** palette presents the more commonly utilized controls in a more compact viewing pane.

Front panel objects generally have a high-color appearance, thus, most likely the **Modern** view will provide the most appealing appearance. It is suggested to use the controls on the **Classic** palette to create VIs for 256-color and 16-color monitor settings. The **System** controls are designed for use in dialog boxes and will not be utilized a great deal by new users of LabVIEW.

*The palette view can be altered in the **Tools≫Advanced≫Edit Palette Set** dialog box.*

You can access the subpalettes by clicking on the desired top-level icon. As illustrated in Figure 1.24, the subpalette **Numeric**, when opened, will reveal

FIGURE 1.24
The **Controls** palette

the various numeric controls (by control we mean "input") and indicators (by indicator we mean "output") that you will utilize on your front panel as a way to move data into and out of the program code (on the block diagram). The topic of numeric controls and indicators is covered in Chapter 2. Each item on the **Controls** palette is discussed in more detail in the chapter in which it is first utilized.

If the **Controls** palette is not visible, you can open the palette by selecting **Controls Palette** from the **View** pull-down menu. You can also access the **Controls** palette by right-clicking on an open area in the front panel window.

*The **Controls** palette is available only when the front panel window is active.*

1.8.3   Functions Palette

The **Functions** palette, shown in Figure 1.25, works in the same general way as the **Controls** palette. It consists of top-level icons representing subpalettes, which contain a full range of available objects that you can use in creating the block diagram. The **Functions** palette contains functions (the vital elements of VIs), as well as many VIs that ship with LabVIEW. The **Express** palette presents key express VIs, such as signal simulation. You access the subpalettes by clicking on

FIGURE 1.25
The **Functions** palette

the desired top-level icon. Many of the G program elements are accessed through the **Functions** palette. For example, the subpalette **Structures** (see Figure 1.25) contains For Loops, While Loops, and Formula Nodes—all of which are common elements of VIs.

If the **Functions** palette is not visible, you can open the palette by selecting **Functions Palette** from the **View** pull-down menu. You can also access the **Functions** palette by right-clicking on an open area in the block diagram window.

*The **Functions** palette is available only when the block diagram window is active.*

1.9   OPENING, LOADING, AND SAVING VIS

When you click the **New** link on the LabVIEW Getting Started screen (see Figure 1.1), the **New** dialog box appears, as shown in Figure 1.26. Selecting the Blank VI from the **Create New** list opens a blank VI front panel and block diagram.

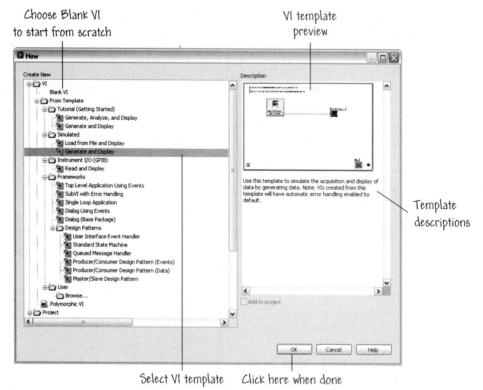

FIGURE 1.26
New dialog box

Remember that from the VI pull-down menu, you can select File≫New *to display the* **New** *dialog box (refer to Section 1.7.1). You can then open a blank by selecting* **Blank VI**.

The **New** dialog box can be employed to help you create LabVIEW applications by utilizing VI templates. If you prefer not to start with a blank VI to build your VI from scratch, you can start with a VI template. This may simplify your programming task. When you select a template in the **Create New** list, previews of the VI and a description of the template appear in the **Description** section. Figure 1.26 shows the Generate and Display VI template.

Making a selection from the **VI from Template** opens a front panel and block diagram with components you need to build different types of VIs. For example, the DAQ VI template opens a front panel and a block diagram with the components you need to measure or generate signals using the DAQ Assistant Express VI and NI-DAQmx. The Instrument I/O VI template opens a front panel and block diagram with the components you need to communicate with an external instrument attached to the computer through a port, such as a serial or GPIB-enabled device. The **Browse for Template** button displays the **Browse** dialog box so you can navigate to a VI or VI template.

You load a VI into memory by choosing the **Open** option from the **File** menu. When you choose that option, a dialog box similar to the one in Figure 1.27 appears. VI directories and VI libraries appear in the dialog box next to a representative symbol. VI libraries contain multiple VIs in a compressed format.

FIGURE 1.27
Locating the desired VI

You can open a VI library or directory by clicking on its icon and then on **Open**, or by double-clicking on the icon. The dialog box opens VI libraries as if they were directories. Once the directory or library is opened, you can locate your VI and load it into memory by clicking on it and then on **OK**, or by double-clicking on the VI icon.

If LabVIEW cannot immediately locate a subVI (think of this as a subroutine) called by the VI, it begins searching through all directories specified by the VI Search Path (Tools≫Options≫Paths). A status dialog box will appear (sometimes the status box disappears so fast that you cannot see it was even there!) as the VI loads. The *Searching* field in the status box lists directories or VIs as LabVIEW searches through them. The *Loading* field lists the subVIs of your VI as they are loaded into memory. If a subVI cannot be found, you can have LabVIEW ignore the subVI by clicking on **Ignore SubVI** in the status box, or you can click on **Browse** to search for the missing subVI.

You can save your VI to a regular directory or VI library by selecting **Save**, **Save As**, **Save All**, or **Save with Options** from the **File** menu. You also can transfer VIs from one platform to another (for example, from LabVIEW for Macintosh to LabVIEW for Windows). LabVIEW automatically translates and recompiles the VIs on the new platform. Because VIs are files, you can use any file transfer method or utility to move your VIs between platforms. Porting VIs over networks using FTP protocol, Z- or X-Modem protocol, and other similar utilities eliminates the need for additional file translation software. If you port your VIs via magnetic media (such as floppy disks), you will need to use a file transfer utility (such as MacDisk or Transfer Pro).

1.10   LABVIEW HELP OPTIONS

The two common help options that you will use as you learn about LabVIEW programming are the **Context** and the **LabVIEW Help**. Both help options can be accessed in the **Help** pull-down menu.

1.10.1   Context Help Window

To display the help window, choose **Show Context Help** from the **Help** pull-down menu. If you have already placed objects on the front panel or block diagram, you can find out more about those objects by placing the cursor on the front panel or block diagram. This process causes the **Context Help Window** to appear showing the icon associated with the selected object and displaying the wires attached to each terminal. As you will discover, some icon terminals must be wired and others are optional. To help you locate terminals that require wiring, in the help window required terminals are labeled in bold, recommended

FIGURE 1.28
A simple context help window

connections in plain text, and optional connections are gray. The example in Figure 1.28 displays a help window in the so-called **Simple Context Help** mode.

On the lower left-hand side of the help window is a button to switch between the simple and detailed context help modes. The simple context emphasizes the important terminal connections—de-emphasized terminals are shown by wire stubs. The detailed help displays all terminals, as illustrated in Figure 1.29.

On the lower left-hand side of the help window is a lock icon that locks the current contents of the help window, so that moving the tool over another function or icon does not change the display. To unlock the window, click again on the lock icon at the bottom of the help window.

The **More Help** icon is the question mark located in the lower left-hand portion of the context help window. This provides a link to the description of the object in the online reference documentation, which features detailed descriptions of most block diagram objects.

1.10.2   LabVIEW Help

The LabVIEW online reference contains detailed descriptions of most block diagram objects. This information is accessible either by clicking on the More Help icon in the Context Help window, choosing **Contents and Index** from the **Help** menu, or clicking on the sentence **Click here for more help** in the Context Help window.

FIGURE 1.29
A detailed Context Help window

   BUILDING BLOCK

1.11   BUILDING BLOCKS: PULSE WIDTH MODULATION

At the end of each chapter in this book, you will find a Building Blocks section. The purpose of this section is to give you the opportunity to apply the main principles and techniques learned in the chapter. You will apply the knowledge gained in the chapter to improve a VI of your design that has continuously evolved from chapter to chapter. This notion of a continuous design project employs the interesting application of **pulse width modulation**.

Pulse width modulation is a method used to digitally encode an analog signal level through modulation of the duty cycle. Pulse width modulation is used in a variety of applications, from regulating power transfer to a load to conveying information over a communications channel. Through the use of this technique, analog circuits can be controlled digitally, thus reducing system costs and power consumption.

Figure 1.30(a) depicts a signal with a 50% duty cycle. The duty cycle of a signal is defined to be the ratio of pulse width to total cycle time. As illustrated in

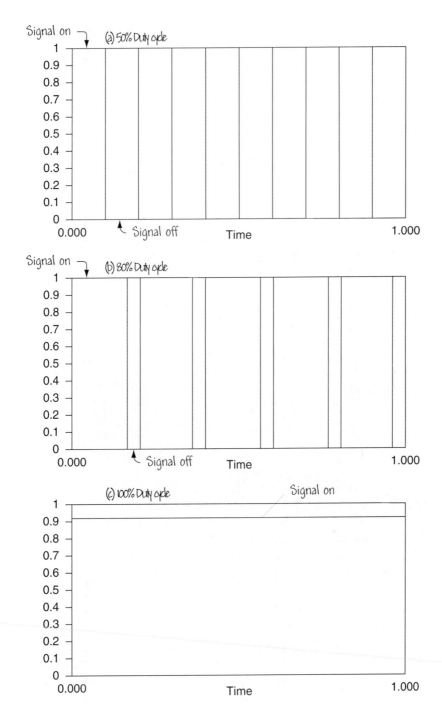

FIGURE 1.30
Signal duty cycle: (a) 50%, (b) 80%, and (c) 100%.

Fig. 1.30(a), the output is "on" for 50% of each cycle. The duty cycle is increased to 80% in Figure 1.30(b). The length of the cycle remains the same, but the longer pulse length, or "on" time results in an increased duty cycle. The signal in Figure 1.30(c) is always "on," so the duty cycle is 100%. Note that adjusting the ratio of "on" time to total time allows you to create an analog value from a digital signal.

   In this first building block, you will open and run a VI that embodies the characteristics of a VI that you will develop in the subsequent chapters. The VI you should open is called **PWM with Square Wave VI.vi** and is included in the **Building Blocks** folder of the **Learning** directory. The VI is shown in Fig. 1.31. Run the VI and observe the change resulting from varying the period and the duty cycle. Keep the period constant and adjust the duty cycle. Observe that the pulse length, or amount of time that the signal is asserted, changes, but the signal still toggles at the same rate. Now keep the duty cycle constant and adjust the period. You should see that the ratio of on time to off time remains the same, but the

FIGURE 1.31
The PWM with Square Wave VI front panel and block diagram.

overall signal frequency changes. Open the **Context Help** window and hold your cursor over the VI icon, located in the upper right hand corner on the front panel, to see a description of the VI and a diagram showing all of its inputs and outputs.

1.12   RELAXED READING: REMOTE CONTROLLED EXCAVATION WITH LABVIEW

A project was recently undertaken to develop an excavator to remove chemical and explosive hazards. Safety concerns suggested that the hazardous materials be removed using a remote-controlled excavator at a remote location to minimize the exposure of chemical and explosive hazards to the equipment operator. In addition, the design of the excavator should include six independent movements, vision capabilities, operating parameters, and remote ignition and emergency stop links.

The CASE CX-160 excavator shown in Figure 1.32 operates by conventional means with two dual-axis joysticks and two single-axis foot pedals. Each of these axes actuate through two stages of hydraulic spool valves. The excavator has six independent movements: arm, boom, bucket, swing, and left and right track. To control these movements, FieldPoint modules were placed at the operator station and in the excavator. Programs were then loaded on the FieldPoint modules that convert joystick inputs to excavator movements. A FieldPoint distributed I/O system operates at 10 Hz and controls the six independent movements at the remote location. Electrical power was available in 12 and 24 volts with a maximum current of 45 amps. An independent emergency shutdown circuit was employed to fulfill the safety requirements.

FIGURE 1.32
CASE CX-160 Excavator

To ensure stability and overall safety, the robust excavator controllers had to be remotely controlled in real time. The FieldPoint modules, in conjunction with LabVIEW and LabVIEW Real-Time, provided intelligent operation, as well as the failsafe conditions that the excavator required. An electronic sensing and acquisition system was selected that consisted of various FieldPoint distributed I/O modules that link together using FieldPoint network modules. FieldPoint distributed I/O quickly facilitated client-based systems through simple integration with LabVIEW. FieldPoint modules communicated through Ethernet and serial data transfer protocols, while data transmitted between modules through wireless serial radio modems. The basic excavator operation reads analog voltages from control input devices at the remote station. The analog voltage then travels across the network for implementation on the excavator FieldPoint modules. The wireless distributed I/O system is depicted in Figure 1.33.

At the remote operator station, the operator can remotely toggle the power of vision systems and work lights. Four joystick buttons provide pan-and-tilt capabilities for the excavator cameras. To ensure the vehicle is running at capacity, the operator must monitor vehicle parameters, such as fuel level, water and oil temperature, and warning messages in real time. These parameters are normally displayed on an on-board liquid crystal display next to the joystick. This data is reported via an available RS-232 port located on-board the excavator's computer. The status of the excavator is processed and sent back across the wireless network

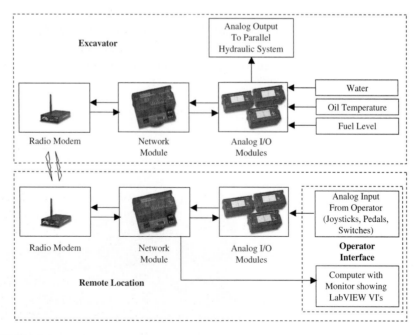

FIGURE 1.33
Wireless Distributed I/O System

PROBLEMS

P1.1 In this problem, you will open and run an existing VI. On the Getting Started screen, click **Find Examples** at the bottom right. In the **Example Finder**, open the Waveform Min Max example.vi, located in **Fundamentals≫Waveforms**.

(a) Run the VI by clicking on the **Run** button.

(b) Vary the amplitude and frequency by typing in new values and pressing **Enter**, or by clicking the arrows on the left side of the controls. Observe how the outputs change for the new input values.

(c) Under the **Window** pull-down menu, select **Show Block Diagram**. Open the **Context Help** window and hold the cursor over the objects on the block diagram to see terminal connections and a brief description of the functionality of various block diagram objects.

FIGURE P1.1
The front panel for the Waveform Min Max example VI

P1.2 In this problem we want to open an existing VI from the Learning directory. You can open the VI by either selecting **Open** from the Getting Started screen, or if you are already in LabVIEW, you can use the **File** pull-down menu (see Figure 1.15) and select **Open**. In both cases, you must navigate through your local file structure to find the desired VI. Find, open, and run Running Dog.vi located in the **Learning** directory. Navigate to the **Instructional VIs** folder and open **CompSci.llb**.

This VI is only available on the Windows platform. If you are on a Macintosh platform, locate, open, and run Control Mixer Process.vi *located in the library* controlmix.llb *found in the subfolder* Instructional VIs *within the* Learning *directory.*

P1.3 You can construct games using LabVIEW! In this problem, you will download a LabVIEW game of your choice from the LabVIEW Zone. Go to the Getting Started screen and select **Labviewzone** from the **Web Resources** area. Once you are in the LabVIEW Zone, click on **Code Sharing**, then select **Games**.

At this location, you can select a game to download. Go ahead and download and play a game. Document the following: Which game did you download? What was the purpose of the game? Investigate the block diagram of the game to see the code.

You can go directly to the LabVIEW games by visiting the website http://www.ni.com/devzone/lvzone/games.htm.

P1.4 Find the following objects. Write down the palette on which each can be found.

(a) Round LED

(b) Add function

(c) Or function

(d) String control

P1.5 The three types of nodes in LabVIEW are functions, subVI nodes, and structures. Search the NI Example Finder and find examples of VIs that employ each type of node. Explain how you determined the classification of each node you chose. Find two examples of each type of node.

DESIGN PROBLEMS

D1.1 On the Getting Started screen, select **Find Examples** to open the NI Example Finder. Locate **Express Comparison.vi** found in the subfolder **Mathematics** within the **Analyzing and Processing Signals** folder. The front panel should resemble the front panel in Figure D1.1.

(a) Run the VI by selecting **Run** on the **Operate** menu, or by clicking the **Run** button.

Move slides to vary input values

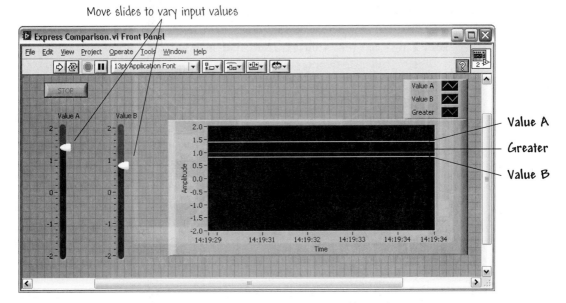

FIGURE D1.1
The front panel for the Express Comparison VI

(b) Move the slides on the controls on the front panel and observe how the corresponding values on the chart change.

(c) Verify that the signal Greater is high (that is, it takes a value of 1) when Value A is greater than Value B.

(d) Open the block diagram and double-click the Express Comparison VI, which is currently configured to perform the Greater Than function. Change the settings in the configuration window so that the Express VI will instead perform the Less Than function. Note that the name on the chart of the result of the Express Comparison VI automatically changes from Greater to Less.

(e) Run the VI again and verify that the signal Less is high when Value A is less than Value B. Push **Stop** to end.

Save the VI as Less Than Comparison.vi in the Users Stuff folder in the Learning directory.

D1.2 In this design problem, you will take the first step to construct your own VI. On the Getting Started screen, select **New** and open a **Blank VI**. Go to the block diagram and place a Compound Arithmetic function, found in the **Programming≫ Numeric** subpalette. By default, this function returns the sum of two inputs.

(a) Using LabVIEW **Context Help**, find out how to configure the function to compute other operations besides addition.

(b) Using what you learned in Step (a), configure the function to multiply two inputs.

(c) Using LabVIEW **Context Help**, find out how to change the Compound Arithmetic function to compute operations on more than two inputs.

(d) Using what you learned in Step (c), configure the function to multiply three inputs together.

Virtual Instruments

Virtual instruments (VI) are the building blocks of LabVIEW programming. We will see in this chapter that VIs have three main components: the front panel, the block diagram, and the icon and connector pair. We will revisit the front panel and block diagram concepts first introduced in Chapter 1. An introduction to wiring the elements together on the block diagram is presented, although many of the debugging issues associated with wires are left to Chapter 4. The important notion of data flow programming is also discussed in this chapter. Finally—you will have the opportunity to build your first VI!

GOALS

1. Gain experience by running more worked examples.

2. Understand the three basic components of a virtual instrument.

3. Begin the study of programming in G.

4. Understand the notion of data flow programming.

5. Build your first virtual instrument.

2.1   WHAT ARE VIRTUAL INSTRUMENTS?

LabVIEW programs are called virtual instruments (VIs) because they often have the look and feel of physical systems or instruments. The illustration in Figure 2.1 shows an example of a front panel. A VI and its components are analogous to main programs and subroutines from text programming languages like C and Fortran. VIs have both an interactive user interface—known as the front panel—and the source code—represented in graphical form on the block diagram. LabVIEW provides mechanisms that allow data to pass easily between the front panel and the block diagram.

The block diagram is a pictorial representation of the program code. The block diagram associated with the front panel in Figure 2.1 is shown in Figure 2.2. The block diagram consists of executable icons (called nodes) connected (or **wired**) together. We will discuss wiring later in this chapter. The important concept to remember is that, in the G programming language, the block diagram is the source code.

The art of successful programming in G is an exercise in **modular programming**. After dividing a given task into a series of simpler subtasks (in G these subtasks are called subVIs and are analogous to subroutines), you then construct a virtual instrument to accomplish each subtask. Chapter 5 focuses on building subVIs. The resulting subtasks (remember, these are called subVIs)

FIGURE 2.1
A virtual instrument front panel

FIGURE 2.2
The virtual instrument block diagram associated with the front panel in Figure 2.1

are then assembled on a top-level block diagram to form the complete program. Modularity means that you can execute each subVI independently, thus making debugging and verification easier. Furthermore, if your subVIs are general purpose programs, you can use them in other programs.

VIs (and subVIs) have three main parts: the front panel, the block diagram, and the icon/connector. The front panel is the interactive user interface of a VI—a window through which the user interacts with the code. When you run a VI, you must have the front panel open so you can pass inputs to the executing program and receive outputs (such as data for graphical display). The front panel is indispensable for viewing the program outputs. It is possible, as we will discuss in Chapter 10, to write data out to a file for subsequent analysis, but generally you will use the front panel to view the program outputs. The front panel contains knobs, push buttons, graphs, and many other controls (the term *controls* is interchangeable with *inputs*) and indicators (the term *indicators* is interchangeable with *outputs*).

The block diagram is the source code for the VI. The source code is "written" in the G programming language. We use the term *written* loosely, since in fact the code is made up of graphical icons, wires, and such, rather than traditional "lines of code." The block diagram is actually the executable code. The **icons** of a block diagram represent lower-level VIs, built-in functions, and program control structures. These icons are wired together to allow the data flow. As you will learn later in this chapter, the execution of a G program is governed by the data flow and not by a linear execution of lines of code. This concept is known as **data flow programming**.

The **icons** and **connectors** specify the pathways for data to flow into and out of VIs. The icon is the graphical representation of the VI in the block diagram and the connector defines the inputs and outputs. All VIs have an icon and a connector. As previously mentioned, VIs are hierarchical and modular. You can use them as top-level (or calling) programs or as subprograms (or subVIs) within other programs. The icon and connector are shown in Figure 2.3 for the subVI Demo Tek FG 5010.vi. This subVI can be found in the center of the Frequency Response.vi diagram in Figure 2.2. It simulates a Textronix FG 5010 function generator.

FIGURE 2.3
The icon and connector of the Demo Tek FG 5010.vi subVI shown in Figure 2.2

2.2 SEVERAL WORKED EXAMPLES

Before you construct your own VI, we will open several existing LabVIEW programs and run them to see how LabVIEW works. The first VI example—Moonlanding.vi—can be found in the suite of examples provided as part of LabVIEW. The second VI example—ODE Example.vi—illustrates how Lab-VIEW can be used to simulate linear systems. In this example, the motion of a mass-spring-cart system is simulated, and you can observe the effects of changing any of the system parameters on the resulting motion of the cart.

Moonlanding Demo

In this example, you will open and run the virtual instrument called Moonlanding.vi. At the Getting Started screen, select **Find Examples** and when the NI Example Finder appears, select the **Browse according to: Directory Structure** button on the upper left-hand side. Click on the express folder and upon opening find Moonlanding.vi and double click on the VI. The front panel window appears and should resemble the one shown in Figure 2.4. The front panel contains a tank, a slide control, a dial, a chart, and several numeric indicators.

Run button —

Adjust the
power to
control the
descent

Watch the
fuel level!

FIGURE 2.4
Moonlanding front panel

Run the VI by clicking on the **Run** button. The front panel toolbar changes as the VI switches from edit mode to run mode. For example, once the VI begins executing, the **Stop** button will change appearance on the front panel toolbar (it changes from a shaded symbol to a red stop sign). Also, the **Run** button changes appearance to indicate that the VI is running.

In this VI, you get to practice landing on the Moon. Initially the altitude is 10,000 meters. For a successful landing you need to have a velocity of 5 m/s or less when the altitude is zero (that is, when you have landed). You control the engine power by adjusting the Power setting in the slide control. While the VI is running

use the **Operating** tool to change the power setting in the vertical slide control. The amount of fuel remaining is displayed in the fuel tank indicator, as well as in the numeric indicator. The speed of the lander is displayed in the indicator dial and in a numeric indicator. This illustrates two methods for displaying data to the user.

From 10,000 m to 1000 m you are in the Normal Flight regime. Once you reach 1000 m you enter the Final Approach. The dial indicators for the Final Approach begin operation once 1000 m is reached. The flight path is updated in the Altitude Chart.

Switch to the block diagram by choosing **Show Block Diagram** from the **Window** pull-down menu. The block diagram shown in Figure 2.5 is the underlying code for the VI. At this point in the learning process, you may not understand all of the block diagram elements depicted in Figure 2.5—but you will eventually! Note that this VI employs Express VIs as key elements of the code.

Express VIs

FIGURE 2.5
Moonlanding demonstration block diagram—the code

When you are finished experimenting with the **Moonlanding.vi**, close the VI and subVI by selecting **Close** from the **File** pull-down menu on each open front panel. Remember—do not save any changes!

 *Selecting **Close** from the **File** pull-down menu of a block diagram closes the block diagram window only. Selecting **Close** on a front panel window closes both the front panel and the block diagram.* ◆

 A Linear System of Equations Example

In this example we use LabVIEW to solve a set of linear, constant coefficient, ordinary differential equations. Many physical systems can be modeled mathematically as a set of linear, constant coefficient differential equations of the form

$$\dot{\mathbf{x}} = \mathbf{A}\mathbf{x}$$

with initial conditions $\mathbf{x}(0) = \mathbf{x}_o$, where the matrix $\mathbf{A}$ is a constant matrix.

Suppose we want to model the motion of a mass-spring-damper system, as shown in Figure 2.6. Let m represent the mass of the cart, k represent the spring constant, and b the damping coefficient. The position of the cart is denoted by y, and the velocity is the time derivative of the position, that is, the velocity is $\dot{y}$.

FIGURE 2.6
A simple mass-spring-damper system

Equating the sum of forces to the mass times acceleration (using Newton's Second Law) we obtain the equation of motion:

$$m\ddot{y} + b\dot{y} + ky = 0,$$

with the initial conditions $y(0) = y_o$ and $\dot{y}(0) = \dot{y}_o$. The motion of the cart is described by the solution of the second-order linear differential equation above. For this simple system we can obtain the solution analytically. For more complex systems it is usually necessary to obtain the solution numerically using the computer. In this exercise we seek to obtain the solution numerically using LabVIEW.

It is sometimes convenient for obtaining the numerical solution to rewrite the second-order linear differential equation as two first-order differential equations. We first define the state vector of the system as

$$\mathbf{x} = \begin{pmatrix} y \\ \dot{y} \end{pmatrix},$$

where the components of the state vector are given by

$$x_1 = y \quad \text{and} \quad x_2 = \dot{y}.$$

Using the definitions of the state vector and the equation of motion, we obtain

$$\dot{x}_1 = x_2$$
$$\dot{x}_2 = -\frac{k}{m}x_1 - \frac{b}{m}x_2.$$

Writing in matrix notation yields

$$\dot{\mathbf{x}} = \mathbf{A}\mathbf{x},$$

where

$$\mathbf{A} = \begin{bmatrix} 0 & 1 \\ -k/m & -b/m \end{bmatrix}.$$

In this example, let

$$\frac{k}{m} = 2 \quad \text{and} \quad \frac{b}{m} = 4.$$

Choose the initial conditions as

$$x_1 = 10 \quad \text{and} \quad x_2 = 0.$$

Now we are ready to compute the solution of the system of ordinary differential equations numerically with LabVIEW.

- Select **Open VI** from the Getting Started screen or use the **File** menu to open the VI.
- Open the ODE Example.vi located in the folder Chapter 2 in the directory Learning.

The front panel depicted in Figure 2.7 will appear. Verify that the initial conditions $\mathbf{x}_o$ are set correctly to $x_1 = 10$ and $x_2 = 0$. Run the VI by clicking on the **Run** button. About how long does it take for the cart position to come to rest? Change the initial position to $x_1 = 20$. Now how long does it take for the cart to come to rest? What is the maximum value of the cart velocity?

FIGURE 2.7
Linear system of equations front panel

LabVIEW does not accept values in numeric displays until you press the **Enter** *button on the toolbar or click the mouse in an open area of the window.*

Open and examine the block diagram by choosing **Show Block Diagram** from the **Window** menu. The code is shown in Figure 2.8. If you need to solve a set of linear ordinary differential equations, you can start with this VI and modify it as necessary. The idea of starting with a VI that solves a related problem is a good approach in the early stages of learning LabVIEW. Access the **Help** pull-down menu and select **Show Context Help**. Move the cursor over various objects on the block diagram and read what the online help has to say.

Most VIs are hierarchical and modular. After creating a VI, you can (with a little work configuring the icon and connector) use the VI as a subVI (similar to a subroutine) in the block diagram of another VI. By creating subVIs, you can construct modular block diagrams that make your VIs easier to debug. The ODE Example.vi uses a subVI named ODE Linear System Numeric.vi. Open the subVI by double-clicking on the appropriate subVI icon (see Figure 2.8). The icon for the subVI is labeled. The front panel shown in Figure 2.9 should appear.

The icon and connector provide the graphical representation and parameter definitions required to use a VI as a subVI in the block diagrams of other VIs. The icon and connector are located in the upper right corner of the VI front panel (see Figure 2.9). The icon is a graphical representation of the VI when used as

Sub VI

FIGURE 2.8
Linear system of equations block diagram

FIGURE 2.9
ODE Linear System Numeric subVI

a component in a G program, that is, when used in the block diagram of other VIs. An icon can be a pictorial representation or a textual description of the VI, or a combination of both. The icon for the ODE Linear System Numeric subVI includes both text and graphics.

Every VI (and subVI) has a connector. The connector is a set of terminals that correspond to its controls and indicators. When you show the connector for the first time, LabVIEW will suggest a connector pattern that has one terminal for each control or indicator on the front panel—you can choose a different pattern. In Chapter 5, you will learn how to associate front panel controls and indicators with connector terminals. The connector terminals determine where you must wire the inputs and outputs on the icon. These terminals are analogous to parameters of a subroutine. You might wonder where the icon is located relative to the connector. It is at the same location—the icon sits on top of the connector pattern. The icon and connector of the ODE Linear System Numeric subVI are shown in Figure 2.10.

FIGURE 2.10
The icon and connector of the ODE Linear System Numeric subVI

FIGURE 2.11
The default icon

Every VI has a default icon, which is displayed in the icon panel in the upper right corner of the front panel and block diagram windows as shown in Figure 2.11. You will learn how to edit the VI icon in Chapter 5. You may want to personalize VI icons so that they transmit information by visual inspection about the contents of the underlying VI.

When you are finished experimenting, close the VI by selecting **Close** from the **File** menu. ◆

2.3 THE FRONT PANEL

The front panel of a VI is a combination of controls and indicators. Controls simulate the types of input devices you might find on a conventional instrument, such as knobs and switches, and provide a mechanism to move input from the front panel to the underlying block diagram. On the other hand, indicators provide a mechanism to display data originating in the block diagram back on the front panel. Indicators include various kinds of graphs and charts (more on this topic in Chapter 8), as well as numeric, Boolean, and string indicators. Thus, when we use the term *controls* we mean "inputs," and when we say *indicators* we mean "outputs."

You place controls and indicators on the front panel by selecting and "dropping" them from the **Controls** palette. Once you select a control (or indicator) from the palette and release the mouse button, the cursor will change to a "hand" icon, which you then use to carry the object to the desired location on the front panel and "drop" it by clicking on the mouse button again. Once an object is on the front panel, you can easily adjust its size, shape, and position (see Chapter 4). If the **Controls** palette is not visible, right-click on an open area of the front panel window to display the **Controls** palette.

2.3.1 Numeric Controls and Indicators

You access the numeric controls and indicators from the **Numeric** subpalette located under the **Modern** subpalette of the **Controls** palette, as shown in Figure 2.12. As the figure shows, there are quite a large number of available numeric controls and indicators. The two most commonly used numeric objects are the numeric control and the numeric indicator. When you construct your first

(a) Commonly used numeric controls

(b) All numeric controls

FIGURE 2.12
Numeric controls and indicators

VI later in this chapter, you will get the chance to practice dropping numeric controls and indicators on the front panel. Once a numeric control is on the front panel, you click on the increment buttons (that is, the up and down arrows on the left-hand side of the control) with the **Operating** tool to enter or change the displayed numerical values. Alternatively, you can double-click on the current value of the numeric control with the **Auto** tool, which will highlight the value, and you can then enter a different value.

FIGURE 2.13
Navigating the **Functions** and **Controls** palettes

*You can tack down the **Numeric** palette (and most other palettes) to the screen so they are visible at all times by clicking on the thumbtack on the top left corner of the palette.*

You can use the tree structure (see Figure 2.12) on the **Controls** and **Functions** palettes to navigate and search for controls, VIs, and functions. When you left-click a subpalette icon, the palette expands to display the subpalette you selected. If you right-click a subpalette icon, the subpalette appears, but the **Controls** (or **Functions**) palette remains in view. An example of a subpalette icon on the **Controls** palette is the **Numeric** subpalette. The **Controls** and **Functions** palettes contain two navigation buttons (as illustrated in Figure 2.13):

- **Search**—Changes the palette to search mode. In search mode, you can perform text-based searches to locate controls, VIs, or functions in the palettes.

- **Options**—Allows you to configure the appearance of the palettes.

Suppose you need to search for a function that provides a random number between 0 and 1. On the **Functions** palette, select the **Search** button and in the dialog box that appears, enter the word "random," as illustrated in Figure 2.14. In the results list of the search, select **Random Number (0-1)«Express Numeric»**. The **Express Numeric** palette will appear and the Random Num function is indeed located, as shown in Figure 2.14.

2.3.2   Boolean Controls and Indicators

You access the Boolean controls and indicators from the **Boolean** subpalette located under the **Modern** subpalette of the **Controls** palette, as shown in Figure 2.15. As with the numeric controls and indicators, there are quite a large number of available Boolean controls and indicators. Boolean controls and indicators simulate switches, buttons, and LEDs and are used for entering and displaying Boolean (True-False) values. For example, you might use a Boolean LED in a temperature monitoring system as a warning signal that a measured temperature has exceeded some predetermined safety limit. The measured temperature is too high, so an LED indicator on the front panel turns from green to

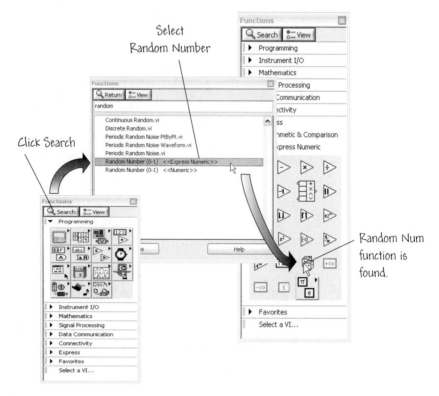

FIGURE 2.14
Searching the **Functions** palette for a Random Number VI

red! Before continuing, take a few moments to familiarize yourself with the types of available numeric and Boolean controls and indicators shown in Figures 2.12 and 2.15, so that when you begin to construct your own VIs, you will have a feel for what is available on the palettes.

2.3.3   Configuring Controls and Indicators

Right-clicking on a numeric control displays the shortcut menu, as shown in Figure 2.16. You can change the defaults for controls and indicators using options from the shortcut menus. For example, under the submenu **Representation** you will find that you can choose from 14 representations of the numeric control or indicator, including 32-bit single precision, 64-bit double precision, signed integer 8-bit, and more. The representation indicates how much memory is used to represent a number. This choice is important if you are displaying large amounts of data and you want to conserve computer memory. Another useful item on the shortcut menu is the capability to switch from control and indicator using **Change to Indicator** and vice versa. We will discuss each item in the shortcut menu on an as-needed basis.

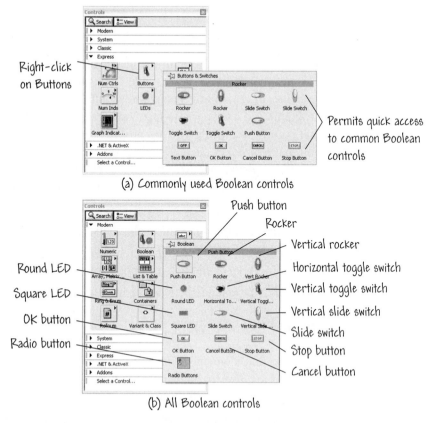

(a) Commonly used Boolean controls

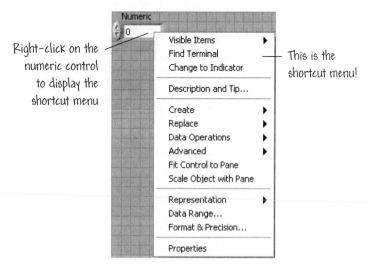

(b) All Boolean controls

FIGURE 2.15
Boolean controls and indicators

FIGURE 2.16
Configuring a numeric control using the shortcut menu

2.4   THE BLOCK DIAGRAM

The graphical objects comprising the block diagram together make up what is usually called the source code. The block diagram (visually resembling a computer program flowchart) corresponds to the lines of text found in text-based programming languages. In fact, the block diagram is the actual executable code that is compiled while you are programming, allowing instant feeedback when you make mistakes such as wiring incompatible data types. The block diagram is built by wiring together objects that perform specific functions.

The components of a block diagram (a VI is depicted in Figure 2.17) belong to one of three classes of objects:

- **Nodes**: Program execution elements.
- **Terminals**: Ports through which data passes between the block diagram and the front panel and between nodes of the block diagram.
- **Wires**: Data paths between terminals.

2.4.1   VIs and Express VIs

An important element of LabVIEW is the so-called Express VI. These VIs are provided to allow for quick construction of VIs designed to accomplish common tasks, such as acquisition, analysis, file reading, and writing. Express VIs are

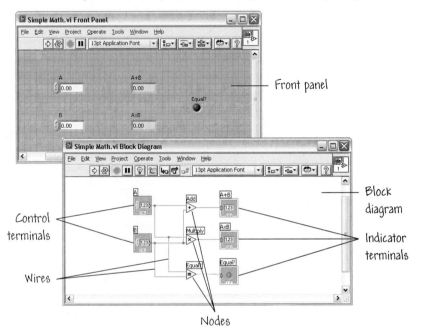

FIGURE 2.17
A typical VI illustrating nodes, terminals, and wires

nodes (see Section 2.4.2 for more information on nodes) that require minimal wiring because they are configured with dialog boxes. The positive implications of minimizing the required wiring will become more evident as we proceed through the forthcoming chapters.

LabVIEW views a VI placed on the block diagram to be a subVI (more on subVIs in Chapter 5). A subVI is itself a VI that can be used as an element on a block diagram. In general, a block diagram can have VIs and Express VIs as elements. One difference in their use is that when you double-click a subVI, its front panel and block diagram appear. When you double-click an Express VI a dialog box appears in which you can configure the VI to meet your needs. The idea is that the Express VI allows you to quickly configure a VI by interacting with a dialog box rather than reconfiguring the code in a subVI block diagram. When an Express VI is available for the task you need, it will be better, in general, to consider the use of the Express VI.

VIs and Express VIs are distinguishable on the block diagram through the use of colored icons. By default, icons for Express VIs appear on the block diagram as expandable nodes (see Section 2.4.2) with icons surrounded by a blue field, while icons for VIs have white backgrounds. The other common element on the block diagram is the function (or primitive), and you can easily identify functions since their icons have pale yellow backgrounds, meaning you cannot open the code behind the function.

2.4.2 Nodes

Nodes are analogous to statements, functions, and subroutines in text-based programming languages. There are three node types—**functions**, **subVI nodes**, and **structures**. Functions are the built-in nodes for performing elementary operations such as adding numbers, file I/O, or string formatting. Functions are the fundamental operating element of a block diagram. The Add and Multiply functions in Figure 2.17 represent one type of node. SubVI nodes are VIs that you design and later call from the diagram of another VI. You can also create subVIs from Express VIs. Structures—such as For Loops and While Loops—control the program flow.

Express VIs and subVIs can be displayed as either icons or as expandable nodes. By default, most subVIs appear as icons that are not expandable. On the other hand, most Express VIs appear as expandable icons. Figure 2.18 depicts the various possibilities for displaying subVIs and Express VIs.

Expandable nodes appear as icons surrounded by a colored field. SubVIs appear with a yellow field, and Express VIs appear with a blue field. By default, subVIs appear as icons on the block diagram, and Express VIs appear as expandable nodes. To display a subVI or Express VI as an expandable node, right-click the subVI or Express VI and select **View As Icon** from the shortcut menu.

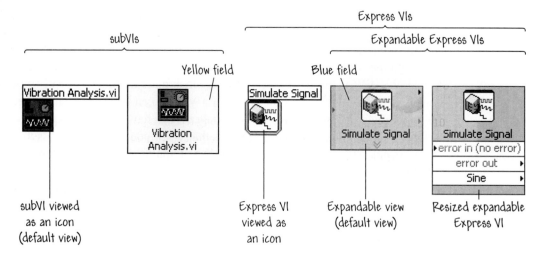

FIGURE 2.18
Expandable Icons versus Icons

You can resize the expandable node to make wiring even easier. When you place the **Positioning** tool over an expandable node, resizing handles will appear at the top and bottom of the node. Placing the cursor over a resizing handle will transform the cursor to the resizing cursor, which can be used to drag the lower border of the node down to display input and output terminals. In Figure 2.18, the Simulate Signal Express VI is resized to display the inputs "error in (no error)" and outputs "error out" and "Sine." For Express VIs with several inputs and outputs, the inputs and order in which they appear can be selected by clicking on the input and selecting the input or output you want in that position. Resizing the expandable node takes a larger amount of space on the block diagram which can then clutter a complex program, but it does aid in the readability of your code. Therefore, use icons if you want to conserve space on the block diagram and use expandable nodes to make wiring easier and to aid in documenting block diagrams.

2.4.3 Terminals

Terminals are analogous to parameters and constants in text-based programming languages. There are different types of terminals—control and indicator terminals, node terminals, constants, and specialized terminals that you will find on various structures. In plain words, a terminal is any point to which you can attach a wire to pass data.

For example, in the case of control and indicator terminals, numeric data entered into a numeric control passes to the block diagram via the control terminals when the VI executes. When the VI finishes executing, the numeric output data passes from the block diagram to the front panel through the indicator terminals. Data flows in only one direction—from a "source" terminal to one or more "destination" terminals. In particular, controls are source terminals and indicators are destination terminals. The data flow direction is from the control terminal to the indicator terminal, and not vice versa. Clearly, controls and indicators are not interchangeable.

Control and indicator terminals belong to front panel controls and indicators and are automatically created or deleted when you create or delete the corresponding front panel control or indicator. The block diagram of the VI in Figure 2.17 shows terminals belonging to five front panel controls and indicators. The Add and Multiply functions shown in the figure also have node terminals through which the input and output data flow into and out of the functions.

You can configure front panel controls or indicators to appear as icon terminals or data type terminals on the block diagram. The terminals shown in Figure 2.18 are all icon terminals. The data type terminals were the standard representation of terminals through LabVIEW 6.1, but with LabVIEW 7.0 and higher, by default, front panel objects appear as icon terminals. To display a terminal as a data type on the block diagram, right-click the terminal and select **View As Icon** from the shortcut menu to remove the checkmark.

Control terminals have thick *borders and indicator terminals have* thin *borders. It is important to distinguish between thick and thin borders since they are not functionally equivalent. Additionally, small black arrows point out of (to the right) controls and into (from the left) indicators to depict data flow.*

Data types indicate what objects, inputs, and outputs you can wire together. You cannot wire together objects with incompatible data types. For example, a switch has a green border (Boolean data type) so you can wire a switch to any input with a green label on an Express VI. Similarly, a knob has an orange border (Numerical data type) so you can wire a knob to any input with an orange label. However, you cannot wire an orange knob (Numeric) to an input with a green label (Boolean).

Wires are the same color as the terminal.

Associated with the Express VI is the **dynamic data type**. The dynamic data type stores information generated or acquired by an Express VI and appears on the block diagram as a dark blue terminal. You can wire the dynamic data type to any indicator or input that accepts numeric, waveform, or Boolean data. The objective is to wire the dynamic data type to an indicator (graph, chart, or numeric) that can best present the data.

Most subVIs and functions do not accept the dynamic data type directly. To use a subVI or function to process the dynamic data type data, you must convert the dynamic data type using the Convert from Dynamic Data Express VI to change the dynamic data type to numeric, Boolean, waveform, or array data types. As illustrated in Figure 2.19, the Convert to Dynamic Data Express VI and the Convert from Dynamic Data Express VI are found on the **Signal Manipulation** palette. When you place the Convert from Dynamic Data Express VI on

FIGURE 2.19
The Dynamic data type

the block diagram, a dialog box appears and displays options that let you specify how you want to format the data that the Convert from Dynamic Data Express VI returns. In Figure 2.19 the Convert to Dynamic Data Express VI is configured to transfer double numeric data from a knob to the dynamic data type used by the Simulate Signal Express VI.

When you wire a dynamic data type to an array indicator, LabVIEW automatically places the Convert from Dynamic Data Express VI on the block diagram. Double-click the Convert from Dynamic Data Express VI to open the dialog box to control how the data appears in the array.

Transforming data for use in Express VIs occurs in a similar fashion. You use the Convert to Dynamic Data Express VI to convert numeric, Boolean, waveform, and array data types to the dynamic data type for use with Express VIs. When you place the Convert to Dynamic Data Express VI on the block diagram, a dialog box appears that allows you to select the kind of data to convert to the dynamic data type.

2.4.4   Wiring

Wires are data paths between terminals and are analogous to variables in conventional languages. How then can we represent different data types on the block diagram? Since the block diagram consists of graphical objects, it seems appropriate to utilize different wire patterns (shape, line style, color, etc.) to represent different data types. In fact, each wire possesses a unique pattern depending on the type of data (numeric, Boolean, string, etc.) that flows through the wire. Each data type appears in a different color for emphasis. To determine the data types on a given wire, match up the colors and styles with the wire types as shown in Table 2.1.

The hot spot of the **Wiring** tool is the tip of the unwound wiring segment as seen in Figure 2.20. To wire from one terminal to another, click the hot spot of the **Wiring** tool on the first terminal (you can start wiring at either terminal!), move the tool to the second terminal, and click on the second terminal. There is no need to hold down the mouse button while moving the **Wiring** tool from one

TABLE 2.1   Common wire types

	Scalar	1D array	2D array	Color
Numeric	————	▬▬▬▬	▭▭▭▭	Orange (floating point) & Blue (integer)
Boolean	············	▬▬▬▬	▭▭▭▭	Green
String	∿∿∿∿∿	⋙⋙⋙⋙	▓▓▓▓	Pink

terminal to the other. The wiring process is illustrated in Figure 2.21. When you wire two terminals, notice that moving the **Wiring** tool over one of the terminals causes that terminal to blink. This is an indication that clicking the mouse button will make the wire connection.

Hot spot

FIGURE 2.20
The hot spot on the **Wiring** tool

The VI shown in Figure 2.21 is easy to construct. It consists of one numeric control and one numeric indicator wired together. Open a new VI and try to build the VI! A working version can be found in the Chapter 2 folder in the Learning directory—it is called Wiring Demo.vi. The function of the VI is to set a value for Input control on the front panel and to display the same input at the numeric indicator Output indicator.

To delete a wire as you are wiring:

- **Windows**—Click the right mouse button or click on the origination terminal.
- **Macintosh**—Hold down <option> and click, or click on the origination terminal.

FIGURE 2.21
Wiring terminals

When wiring two terminals together, you may want to bend the wire to avoid running the wire under other objects. This is accomplished during the wiring process by clicking the mouse button to tack the wire down at the desired location of the bend, and moving the mouse in a perpendicular direction to continue the wiring to the terminal. Another way to change the direction of a wire while wiring is to press the space bar while moving the **Wiring** tool.

Automatic wiring is also available. Instead of tacking down the wire, simply connect the two terminals and LabVIEW will choose the best path. If you have an existing wire that you would like to fix, right-click on the wire and choose **Clean Up Wire** on the pull-down menu.

Windows: All wiring is performed using the left *mouse button.*

Tip strips make it easier to identify function and node terminals for wiring. When you move the **Wiring** tool over a terminal, a tip strip appears, as illustrated in Figure 2.22. Tip strips are small text banners that display the name of each terminal. When you place the **Wiring** tool over a node, each input and output will show as a wire stub—a dot at the end of the wire stub indicates an input. Tip strips should help you wire the terminals correctly.

It is possible to have objects wired automatically. A feature of LabVIEW is the capability to automatically wire objects when you first drop them on the diagram. After you select a node from the **Functions** palette, move that node close to another node to which you want to wire the first node. Terminals containing similar datatypes with similar names will automatically connect. You can disable the automatic wiring feature by pressing the space bar. You can adjust the auto wiring settings from the Tools≫Options≫Block Diagram window.

Since it is important to correctly wire the terminals on functions, LabVIEW provides an easy way to show the icon connector to make the wiring task easier. This is accomplished by right-clicking on the function and choosing Visible Items≫Terminals from the shortcut menu, as illustrated in Figure 2.23. To return to the icon, right-click on the function and deselect Visible Items≫Terminals.

FIGURE 2.22
Tip strip

FIGURE 2.23
Showing terminals

2.5 BUILDING YOUR FIRST VI

In this section you will create your first virtual instrument to perform the following functions:

- Add two input numbers and display the result.
- Multiply the same two input numbers and display the result.
- Compare the two input numbers and turn on an LED if the numbers are equal.

Begin by considering the front panel shown in Figure 2.24. It has two numeric control inputs for the numbers A and B, two numeric indicator outputs to display the results $A + B$ and $A \times B$, respectively, and a round LED that will turn on when the input numbers A and B are identical.

FIGURE 2.24
The front panel for your first VI

As with the development of most sophisticated computer programs, constructing VIs is an art, and you will develop your own style as you gain experience with programming in G. With that in mind, you should consider the following steps as only one possible path to building a working VI that carries out the desired calculations and displays the results.

1. Open a new front panel by choosing **New VI** from the **File** menu.

2. **Create the numeric controls and indicators**. The two front panel controls are used to enter the numbers, and the two indicators are used to display the results of the addition and multiplication of the input numbers.

    (a) Select **Numeric Controls** from the **Controls≫Express** palette. If the **Controls** palette is not visible, right-click in an open area of the front panel to gain access to the palette.

    (b) Drop the control on the front panel, as illustrated in Figure 2.25. Drag the control to the desired location and then click the mouse button to complete the drop.

    (c) Type the letter A inside the label box (which appears above the control) and press the **Enter** button on the front panel toolbar. If you do not type the control label before starting other programming actions (such as dropping the other control on the front panel), the label box will remain labeled with the default label Numeric. If the control or indicator does

FIGURE 2.25
Placing the controls and indicators on the front panel

Click here on **Enter** after typing in the label text "A"

Change label to A

FIGURE 2.26
Labeling the numeric control and indicators on the front panel

not have a label, you can right-click on the control and select **Label** from the **Visible Items** menu. The label box appears, and you can then edit the text using the **Labeling** tool (see Figure 2.26).

(d) Repeat the above process to create the second numeric control and the two numeric indicators. You can arrange the controls and indicators in any manner that you choose—although a neat and orderly arrangement is preferable. Add the labels to each control and indicator using Figure 2.24 as a guide.

3. **Create the Boolean LED**. This indicator will turn on if the two input numbers are identical, or remain off if they do not match.

(a) Select **Round LED** from the **Modern≫Boolean** subpalette of the **Controls** palette. Place the indicator on the front panel, drag it to the desired location, and then click the mouse button to complete the process.

(b) Type Equal? inside the label box and click anywhere outside the label when finished, or click on the **Enter** button.

Each time you create a new control or indicator, LabVIEW automatically creates the corresponding terminal in the block diagram. When viewed as icons, the terminals are graphical representations of the controls or indicators.

The Block Diagram

1. Switch your center of activity to the block diagram by selecting **Show Block Diagram** from the **Window** pull-down menu. The completed block diagram is shown in Figure 2.27. It may be helpful to display the front panel and block diagram simultaneously using either the **Tile Left and Right** or the **Tile Up and Down** options found in the **Window** pull-down menu. For this example, the up and down option works better in the sense that all the block diagram and front panel objects can be displayed on the screen without having to use the scrollbars.

FIGURE 2.27
The block diagram window for your first VI

2. Now we want to place the addition and multiplication functions on the block diagram. Select the Add function from the **Programming≫Numeric** sub-palette of the **Functions** palette. If the **Functions** palette is not visible, right-click on an open area of the block diagram to gain access to the palette. Drop the Add function on the block diagram in approximately the same position as shown in Figure 2.27. The label for the Add function can be displayed using the shortcut menu and selecting Visible Items≫Label. This is illustrated in Figure 2.28. Following the same procedure, place the Multiply function on the block diagram and display the label.

FIGURE 2.28
Right-click to select the Visible Items≫Label option

3. Select the Equal? function from the **Express≫Arithemetic & Comparison ≫Express Comparison** subpalette of the **Functions** palette and place it on the block diagram, as shown in Figure 2.29. The Equal? function compares two numbers and returns True if they are equal or False if they are not. To get more information on this function, you can activate the online help by choosing **Show Context Help** from the **Help** menu. Then placing the cursor over the Equal? function (or any of the other functions on the block diagram) leads to the display of the online help information.

FIGURE 2.29
Adding the Equal? function to the VI

4. Using the **Wiring** tool, wire the terminals as shown in Figure 2.27. As seen in Figure 2.30, to wire from one terminal to another, click the **Wiring** tool on the first terminal, move the tool to the second terminal, and click on the second terminal. Remember that it does not matter on which terminal you initiate the wiring. To aid in wiring, right-click on the three functions and choose Visible Items≫Terminals. Having the terminals shown explicitly helps to wire more quickly and accurately. Once the wiring is finished for a given function, it is best to return to the icon by right-clicking on the function and choosing **Visible Items**, and deselecting **Terminals**.

5. Switch back to the front panel window by clicking anywhere on it or by choosing **Show Front Panel** from the **Window** menu.

Show Terminals: Here we have
two inputs and one output

Using the **Wiring** tool,
wire each element as shown

FIGURE 2.30
Wiring from one terminal to another

6. Save the VI as **Simple Math.vi**. Select **Save** from the **File** menu and make sure to save the VI in the **Users Stuff** folder within the **Learning** directory.

*In case you cannot get your VI to run properly, a working version of the VI (called **Simple Math.vi**) is located in the **Chapter 2** folder within the **Learning** directory.*

7. **Enter input data.** Enter numbers in the numeric controls utilizing the **Auto** tool by double-clicking in the numeric control box and typing in a number. The default values for *A* and *B* are 0 and 0, respectively. You can run the VI using these default values as a first try! When you use the default values, the LED should light up since *A* = *B*.

8. Run the VI by clicking on the **Run** button.

9. Experiment with different input numbers—make *A* and *B* identical and verify that the LED does indeed turn on.

10. When you are finished experimenting, close the VI by selecting **Close** from the **File** menu.

2.6 DATA FLOW PROGRAMMING

The principle that governs VI execution is known as data flow. Unlike most sequential programming languages, the executable elements of a VI execute only when they have received all required input data—in other words, data flows out of the executable element only after the code is finished executing. The notion of data flow contrasts with the control flow method of executing a conventional

FIGURE 2.31
Block diagram that adds two numbers and then computes the sine of the result

program, in which instructions execute sequentially in the order specified by the programmer. Another way to say the same thing is that the flow of traditional sequential code is instruction driven, while the data flow of a VI is data driven.

Consider the VI block diagram, shown in Figure 2.31, that adds two numbers and then computes the sine of the result. In this case, the block diagram executes from left to right, not because the objects are placed in that order, but because one of the inputs of the Sine & Cosine function is not valid until the Add function has added the numbers together and passed the data to the Sine & Cosine function. Remember that a node (in this case, the Sine function) executes only when data is available at all of its input terminals, and it supplies data to its output terminals only when it finishes execution. Open DataFlowA.vi located in the Chapter 2 folder of the Learning directory, click on execution highlighting, and then run the VI.

Consider the example in Figure 2.32. Which code segment would execute first—the one on the left or the one on the right? You cannot determine the answer just by looking at the segments. The one on the left does not necessarily execute first. In a situation where one code segment must execute before another, and there is no type of dependency between the functions, you must use a Sequence structure to force the order of execution. To observe the data flow on the code in Figure 2.32, open DataFlowB.vi located in the folder Chapter 2 in the Learning directory. Before running the VI, click on the **Highlight Execution** button and then watch the flow of the execution.

When ready, click here   Click here to begin
to begin execution   execution highlighting

FIGURE 2.32
Which code executes first?

2.7 BUILDING A VI USING EXPRESS VIS

In this section, you will create your first VI using Express VIs. The objective is to construct a VI that generates a sawtooth signal at an amplitude that we prescribe on the front panel and displays the sawtooth signal graphically on the front panel. LabVIEW provides a VI template containing information that will help you in building this VI.

At the LabVIEW Getting Started screen (see Figure 1.1), select **VI from Template** under the **New** menu to display the **New** dialog box. Select **Generate and Display** from the **Simulated** tree in the **Create New** list (see Section 1.9 and Figure 1.26). This template VI generates and displays a signal.

1. Click the **OK** button to open the template. You also can double-click the name of the template VI in the **Create New** list to open the template.

2. Examine the front panel of the VI, as shown in Figure 2.33. The front panel appears with a gray background and includes a waveform graph and a stop button. The title bar of the front panel indicates that this window is the front panel for the Generate and Display [GenerateDisplay1.vi] VI.

If the front panel is not visible, you can display the front panel by selecting Window≫Show Front Panel. *Press the* <Ctrl-E> *keys to switch from the front panel to the block diagram or from the block diagram to the front panel.*

FIGURE 2.33
Building a VI using the Simulate Signal Express VI

3. Examine the block diagram of the VI shown in Figure 2.33. The block di-
    agram appears with a white background and includes the Simulate Signal
    Express VI, a waveform graph icon, a stop button icon, and a While Loop
    (more on loops in Chapter 6). The title bar of the block diagram indicates
    that this window is the block diagram for the Generate and Display [Gener-
    ateDisplay1.vi] VI.

*If the block diagram is not visible, you can display the block diagram
by selecting Window≫Show Block Diagram.*

4. On the front panel toolbar, click the **Run** button and verify that a sine wave
    appears on the graph.

5. Stop the VI by clicking the Stop button located at the bottom right-hand side of the front panel.

*Although the **Abort Execution** button may seem, at first glance, to operate like a stop button, the **Abort Execution** button does not always properly close the VI. It is recommended to always stop your VIs using the Stop button on the front panel. Use the **Abort Execution** button only when errors prevent you from terminating the application using the Stop button.*

We can now add other elements to the VI template to begin the process of constructing the VI. First, we will add a control to the front panel to use to vary the sawtooth signal amplitude. In our VI, the control will supply the sawtooth amplitude data to the block diagram.

6. On the front panel, right-click in an open area to display the **Controls** palette.

7. Move the cursor over the icons on the **Controls** palette to locate the **Numeric** subpalette under the **Modern** palette (it should be located in the leftmost area of the first row as shown in Figure 1.24). Notice that when you move the cursor over icons on the **Controls** palette, the name of that subpalette appears in the gray space above all the icons on the palette. When you idle the cursor over any icon on any palette, the full name of the subpalette, control, or indicator appears.

8. Click the **Numeric** icon to access the **Numeric** palette.

9. Since many physical instruments have knobs to vary the operational parameters of the instrument, we will select the **Knob** control on the **Numeric** palette and place it on the front panel to the left of the waveform graph, as shown in Figure 2.34. This knob will provide control over the amplitude of the sawtooth signal once it is properly wired.

10. Select File≫Save As and save this VI as Acquiring a Signal.vi in the Users Stuff folder in the Learning directory. Notice that the VI name appears in the closed brackets in the title bar (see Figure 2.34).

Save the VIs you edit or create using this book, in the Users Stuff folder in the Learning directory.

Examine the block diagram of the VI as it currently is configured. Notice that the block diagram has a blue icon labeled Simulate Signal. This icon represents the Simulate Signal Express VI, which simulates a sine wave by default. To meet our objectives, we must reconfigure the Express VI to simulate a sawtooth signal.

11. Display the block diagram by selecting Window≫Show Block Diagram. The Simulate Signal Express VI depicted in Figure 2.35 simulates a signal

FIGURE 2.34
Adding a **Knob** to the Generate and Display template

based on the configuration that you specify. We must interact with the Express VI dialog box to output a sawtooth signal, since the default is a sine wave signal.

12. Right-click the Simulate Signal Express VI and select **Properties** from the shortcut menu to display the **Configure Simulate Signal** dialog box, as shown in Figure 2.35.

13. Select **Sawtooth** from the **Signal Type** pull-down menu. Notice that the waveform on the graph in the **Result Preview** section changes to a sawtooth wave.

14. Click the **OK** button to apply the current configuration and close the **Configure Simulate Signal** dialog box.

15. Now we want to expand the Simulate Signal Express VI to show the inputs and outputs to make the wiring easier. Move the cursor over the down arrow at the bottom of the Simulate Signal Express VI as illustrated in Figure 2.36.

16. When a double-headed arrow appears, click and drag the border of the Express VI down to show one additional input. By default, **error out** appears. Click on **error out** and select **Amplitude** from the pull-down menu. Because the **Amplitude** input appears on the block diagram, you can configure the amplitude of the sawtooth wave on the block diagram.

In Figure 2.35, observe how the signal parameter **Amplitude** is an option in the **Configure Simulate Signal** dialog box. When inputs, such as **Amplitude**, appear on the block diagram and in the configuration dialog box, you can configure the inputs in either location.

FIGURE 2.35
The **Configure Simulate Signal** dialog box

Now we can complete the VI by finishing up the wiring. To use the **Knob** control to change the amplitude of the signal, the **Knob** must be wired to the **Amplitude** input on the Simulate Signal Express VI.

17. Move the cursor over the **Knob** terminal until the **Positioning** tool appears (see Figure 1.23 for review of tools). The **Positioning** tool, represented by an arrow, is used to select, position, and resize objects.

18. Click the **Knob** terminal to select it, then drag the terminal to the left of the Simulate Signal Express VI. Make sure the **Knob** terminal is inside the While Loop. Deselect the **Knob** terminal by clicking a blank space on the block diagram.

The cursor does not switch to another tool while an object is selected.

19. Move the cursor over the arrow on the right-hand side of the **Knob** terminal. This will result in the cursor becoming the **Wiring** tool. Now we can use the **Wiring** tool to wire the **Knob** to the Express VI.

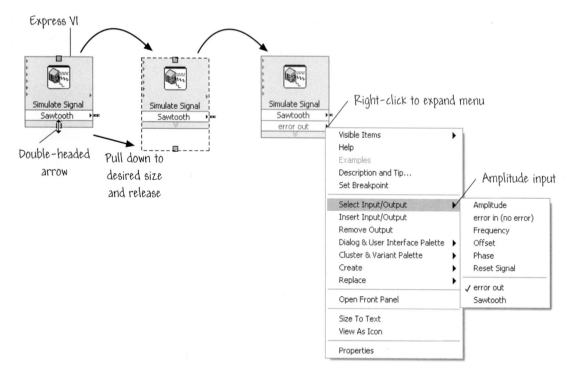

FIGURE 2.36
Input and Output of the Simulate Signal Express VI

20. When the **Wiring** tool appears, click the arrow and then click the **Amplitude** input of the Simulate Signal Express VI to wire the two objects together. When the wire appears and connects the two objects, then data can flow along the wire from the **Knob** to the Express VI. The final block diagram is shown in Figure 2.37.

21. Select File≫Save to save this VI.

Now that the VI is ready for execution, we can see if we have successfully achieved the goal of generating a signal and displaying it graphically on the front panel.

22. Return to the front panel by selecting Window≫Show Front Panel and click the **Run** button.

23. Move the cursor over the knob control. Notice how the cursor becomes the **Operating** tool. The **Operating** tool can be used to change the value of the sawtooth amplitude.

24. Using the **Operating** tool, turn the knob to adjust the amplitude of the sawtooth wave. Notice how the amplitude of the sawtooth wave changes as you turn the knob. Also notice that the y-axis on the graph autoscales to

Knob is wired to the
Express VI at the
Amplitude input

FIGURE 2.37
The block diagram of the completed VI to generate and display a signal

account for the change in amplitude. To indicate that the VI is running, the
Run button changes to a darkened arrow. You cannot edit the front panel or
block diagram while the VI runs.

25. Click the Stop button to stop the VI when you are finished experimenting.

Your VI development is now finished. The utility of using the provided VI
templates is evident in this example. We were able to start from a point much
closer to the desired final VI using the Generate and Display VI than starting
with a blank VI. Also, the utility of the Express VIs is demonstrated with the
Simulate Signal VI. This VI was easily configured to simulate a sawtooth signal
and provided a quick and easy solution to our problem. Clearly, it would have
taken more effort to construct the VI from scratch.

BUILDING BLOCK

2.8   BUILDING BLOCKS: PULSE WIDTH MODULATION

The ultimate goal in developing the VIs in the Building Blocks as we progress from chapter to chapter is to create a pulse width modulation VI to output a periodic pulse whose length and magnitude are prescribed by controls on the front panel. Recall from your experimentation with the VI in the Chapter 1 Building Block section that an adjustment to either the period or the duty cycle impacts the pulse length or on time. Increasing the period will increase the total on time as well as the total off time. Increasing the duty cycle will increase the on time and decrease the off time such that the total cycle time remains constant. In this Building Block, we will construct a VI that will be used when the output signal is high to determine whether the output should toggle to low or remain high on the next clock cycle. This VI will be used in subsequent Building Block VIs. In Chapter 5 you will learn how to modify this VI to be used within another VI as a subVI.

Open a new VI and save it as **Falling Edge.vi**. Define the output of the VI as a Boolean indicator called **Falling Edge**. This indicator will be True when a "falling edge" is to occur upon the next clock cycle. The inputs to this VI are **Clock**, **Period**, and **Duty Cycle**. The input **Clock** is an integer value denoting the current value of the clock. The time units of the clock are not relevant at this time. The input **Period** is an integer value greater than or equal to 100 specifying the desired cycle duration. The input **Duty Cycle** is a value between 0 and 1 describing the ratio of on time to total cycle time.

The resolution of the clock limits the precision with which we can specify **Period** and **Duty Cycle**. In this VI, we will limit the **Period** to a minimum value of 100 so that duty cycle can be expressed to the nearest hundredth. For example, if we set **Duty Cycle** = 0.25 and set **Period** = 100, the clock will increment 25 times before the signal toggles. If we had instead specified **Duty Cycle** as 0.333, then the clock would need to increment 33.3 times before the output would need to change. Since this is not an integer value, this is not a feasible set of inputs. A similar problem would be encountered for any period of less than 100. Assume for now that the user will always provide valid inputs for **Duty Cycle** and **Period**.

Figure 2.38 shows a completed version of **Falling Edge.vi**. The output of the VI should be True whenever the percentage of the period specified by **Duty Cycle** has elapsed. In order to determine whether this event has occurred, subtract the pulse length (determined by the multiplication of the duty cycle and the period) from the current value of the clock and check to see if this value is evenly divisible by the period length. If this division produces an integer, then **Falling Edge** is True and the output should go low on the next clock cycle. To check if the result

Numeric controls

Boolean indicator

Quotient & Remainder

Found on the **Programming» Numeric** palette

FIGURE 2.38
The Falling Edge VI

of the division is an even multiple of the period, we use the modulus VI and then test for a remainder of 0.

Experiment with different inputs to the VI and verify that it produces the expected outputs. Try running the VI in **Highlight Execution** with Duty Cycle = 0.5 and Period = 100. Set the Clock to 0. What happens? The Falling Edge should not be detected. Now try the input Clock = 50. Leaving **Highlight Execution** on, what happens this time? You should see that the Falling Edge is detected.

When you are done experimenting with your new VI, save it as Falling Edge.vi in the Users Stuff folder in the Learning directory. You will use this VI as a Building Block in later chapters—so make sure to save your work!

A working version of Falling Edge.vi can be found in the Building Blocks folder of the Learning directory.

You will need to construct a similar VI which determines when a Rising Edge is to occur for use in the final pulse width modulation VI. This task is fully specified in an end-of-chapter problem.

2.9   RELAXED READING: LABVIEW AUTOMATES BRAIN WAVE EXPERIMENTS

Brain potentials are divided into spontaneous and event-related categories. The spontaneous potentials result from the regular brain activity (EEG potentials). The event-related potentials (ERP) result from external brain excitation (events) and are divided into evoked and anticipatory. Evoked potentials appear *after* excitation as a reflex of the brain. Anticipatory potentials appear *before* the corresponding event and represent an expectation of an excitation.

The most prominent example of the expectation-related potential is the contingent negative variation (CNV) potential extracted from a CNV experiment. In the CNV experiment, two brain stimuli, denoted by S_1 and S_2, are applied to the subject with a constant interval between stimuli. S_1 is a warning stimulus, and S_2 is an imperative stimulus to which the subject must react. The procedure is repeated tens of times, during which an ERP is produced in the EEG trace between the stimuli and shapes itself toward a specific CNV wave. The ERP after 10 to 20 trials can clearly show both the evoked (short) potential due to S_1 as well as the anticipatory (late, expectancy) potential together with the preparatory potential prior and due to S_2.

The dynamic CNV experiment is an extension of the CNV experiment involving switching S_2 on and off. The switching occurs automatically after fulfilling certain conditions in the experiment's environment, thus forcing a cyclic process of building and degrading of the CNV wave. The subject is not informed about the nature of both stimuli, so the expectation of appearance (absence) of S_2 during the experiment completely corresponds to the learning process. The CNV wave can be qualified by one of its parameters, such as amplitude or slope. After the experiment, a statistical curve of the qualifying parameter is drawn across the trials. This statistical curve is denoted as the electroexpectogram (EXG) and directly presents the subject's cognitive capabilities.

The AEP Research Tool was written in LabVIEW to perform the data acquisition, signal processing, analysis of the data traces, and reporting. The hardware used in the experiment consisted of a bandpass amplifier for mV ranges, an AT-MIO-E board, a sound card for applying the stimuli, and a push-button switch with a TTL interface. The system acquires brain wave activity data on two differential analog channels.

The stimulus S_1 is an audio 1 kHz warning beep and S_2 is a longer 2 kHz imperative beep. It is essential that the subject be aware neither of the nature nor of the number of the stimuli. The data acquisition lasts for 7s. The stimulus S_1 is issued at 1s into the acquisition phase; S_2 is issued at 3s, if applied by the algorithm. The number of trials in the experiment is set to maximum 100 successful. The gap between two consecutive trials varies from 12 to 15s. During the experiment, the subject learns about the number, nature, and order of the stimuli, thus demonstrating the process of learning by shaping the ERP wave toward the

FIGURE 2.39
Main DAQ Front Panel

expected CNV. The subject has to react upon hearing S_2 by pressing the button and immediately stopping it.

The main front panel of the AEP Research Tool is shown in Figure 2.39. The front panel displays the acquired EEG signal, the extracted CNV potential and its linearized model, as well as the required measurements and calculated values. The audio stimulation is performed through a sound card. The reaction time is measured by an onboard counter, started by a digital output from the card issuing a pulse at the same moment as the start of S_2 and stopped by the user pressing the button or the time-out pulse, applied again by the same digital line.

The EXG curve represents a cognitive wave obtained from the human brain showing the oscillatory change of the expectancy status in the human brain and is a manifestation of the expectation process and the learning process taking place in the human brain. The AEP Research Tool successfully performs the expected task and produces an excellent nine-page report containing all required statistics. Experiments during the test phase indicated that different categories of subjects (healthy adults, groups of people with distinctive neurological disturbances, and children) may produce quite different electroexpectograms, but remain very similar within the groups, because some are not able to raise their CNV waves above the thresholds. The design of this experiment illustrates the possibilities of LabVIEW-based systems in the medical field.

For more information, please contact:

Roman Golubovski
roman.golubovski@iname.com

2.10 SUMMARY

Virtual instruments (VIs) are the building blocks of LabVIEW. The graphical programming language is known as the G programming language. VIs have three main components: the front panel, the block diagram, and the icon and connector pair. VIs follow a data flow programming convention in which each executable node of the program executes only after all the necessary inputs have been received. Correspondingly, output from each executable node is produced only when the node has finished executing.

KEY TERMS

Boolean controls and indicators: Front panel objects used to manipulate and display or input and output Boolean (True or False) data.

Connector: Part of the VI or function node that contains its input and output terminals, through which data passes to and from the node.

Connector pane: Region in the upper right corner of the front panel that displays the VI terminal pattern. It underlies the icon pane.

Data flow programming: Programming system consisting of executable nodes in which nodes execute only when they have received all required input data and produce output automatically when they have executed.

G programming language: Graphical programming language used in LabVIEW.

Icon: Graphical representation of a node on a block diagram.

Icon pane: Region in the upper right corner of the front panel and block diagram that displays the VI icon.

Input terminals: Terminals that emit data. Sometimes called source terminals.

Modular programming: Programming that uses interchangeable computer routines.

Numeric controls and indicators: Front panel objects used to manipulate and display numeric data.

Output terminals: Terminals that absorb data. Sometimes called destination terminals.

String controls and indicators: Front panel objects used to manipulate and display or input and output text.

Terminals: Objects or regions on a node through which data passes.

Tip strips: Small yellow text banners that identify the terminal name and make it easier to identify function and node terminals for wiring.

Wire: Data path between nodes.

Wiring tool: Tool used to define data paths between source and sink terminals.

EXERCISES

E2.1 In this first exercise you get to play a drawing game. Open Spirograph.vi located in Chapter 2 of the Learning directory. The front panel should look like the one shown in Figure E2.1. Did you ever use a spirograph drawing kit as a kid? This VI is the computer version of that children's game. To play the drawing game, run the VI and experiment by varying the controls. After clicking on the **Run** button to start the VI execution, click on the Begin drawing push button on the front panel. Once the drawing starts, the Stop drawing label appears in the push button. The VI execution is halted by clicking on the Stop drawing button.

E2.2 Construct a VI that uses a round push-button control to turn on a square light indicator whenever the push button is depressed. The front panel is very simple and should look something like the one shown in Figure E2.2.

E2.3 Open Temperature System Demo.vi, located in the tempsys.llb found in the subfolder apps within the examples folder in the LabVIEW 8 directory. The front panel and block diagram are shown in Figure E2.3. Using the **Context Help** (Ctrl-H), determine the inputs and outputs of the Temperature Status subVI. Sketch the subVI icon and connector showing the inputs and outputs.

E2.4 Open the VI you created in Section 2.5 called Simple Math.vi.

(a) Locate all of the controls in this VI, specifying their label and type (numeric, Boolean, or string).

FIGURE E2.1
The Spirograph.vi front panel

FIGURE E2.2
A front panel using a round push-button control and a square light indicator

(b) Locate all of the indicators in this VI, specifying their label and type (numeric, Boolean, or string).

(c) Locate and identify all of the nodes in this VI.

(d) Locate and identify all of the terminals in this VI.

E2.5   Create a VI which compares the inputs on two numeric controls and turns on an LED if these values are equal. Use the Equal? function on the **Programming≫Comparison** palette. The front panel should resemble Figure E2.5.

FIGURE E2.3
The Temperature System Demo VI

FIGURE E2.5
The Compare Numbers VI front panel

E2.6 Design a VI which determines whether a number input on a floating point numeric control is an integer. Place an LED on the front panel that lights whenever the input is an integer. One method of accomplishing this goal is seen in the Building Blocks exercise for this chapter. However, there are multiple solutions.

P2.1   Open the VI created in Section 2.7, Acquiring a Signal.vi. Modify this VI so that it meets the following specifications:

- Two sine waves are generated using the Simulate Signal Express VI and displayed on corresponding graphs on the front panel.
- The amplitude of each sine wave can be adjusted to any value using numeric controls on the front panel.
- Using the Comparison express VI on the **Comparison** palette, compare the two signals and light an LED if the two signals are equal.

Save the VI as Compare Signals.vi in the Users Stuff folder in the Learning directory. The front panel should resemble the front panel seen in Figure P2.1.

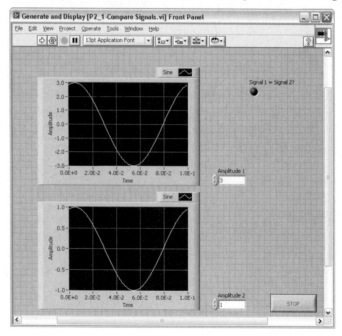

FIGURE P2.1
The front panel for the Compare Signals VI

P2.2   Construct a VI that performs the following tasks:

- Takes two floating-point numbers as inputs on the front panel: X and Y.
- Subtracts Y from X and displays the result on the front panel.
- Divides X by Y and displays the result on the front panel.
- If the input $Y = 0$, a front panel LED lights up to indicate division by zero.

Name the VI Subtract and Divide.vi and save it in the Users Stuff folder in the Learning directory.

P2.3 Construct a VI that uses a vertical slide control for input and a meter indicator for output display. A front panel and block diagram that can be used as a guide are shown in Figure P2.3. Referring to the block diagram, you see a pair of dice, which is the icon for a Random Number function. You will find the Random Number function on the **Programming≫Numeric** palette. When running the VI, any input you provide via the vertical slide will be reflected on the meter indicator. The Random Number function adds "noise" to the input so that the meter output will not be exactly the same as the input. Run the VI in **Run Continuously** mode and vary the slide input.

FIGURE P2.3
Using a vertical slide control and a meter indicator

P2.4 Create a VI that has a numeric control to input a number x and uses the Add and Multiply functions to calculate $3x^2 + 2x + 5.0$. Display the output using a numeric indicator.

P2.5 Use a VI template and modify Express VIs to create a program that generates a Triangle Wave with a frequency of 125 Hz and added noise.

P2.6   Construct a VI in which the user guesses an integer between 1 and 5 using a numeric control. Generate a random integer in this range, display it with a numeric indicator, and compare it to the input. If the numbers are equal, light an LED and display a message using the One Button Dialog function to announce that the user guessed correctly. If the numbers are not equal, display a message notifying the user that the guess was not correct.

Save the VI as Guessing Game.vi in the Users Stuff folder in the Learning directory. The front panel should resemble the front panel seen in Figure P2.6.

FIGURE P2.6
The front panel for the Guessing Game VI

P2.7   Construct a VI that accepts an input in feet and converts this value to its equivalent in both meters and miles, displaying each on an indicator on the front panel. Recall that 1 mile = 5,280 feet and 1 meter = 3.281 feet.

Using the VI, confirm that 10,000 feet is equivalent to 1.89394 miles and 3047.85 meters. Save the VI as Convert.vi in the Users Stuff folder in the Learning directory.

P2.8   Design a VI that computes both the sum and the difference of two numeric inputs, labeled x and y. It should then light one of three LEDs, depending on whether or not the sum is greater than, equal to, or less than the difference between the two inputs. The front panel should resemble the front panel seen in Figure P2.8.

Using the VI, test the following three conditions:

1. $x = 4$ and $y = 1$

2. $x = 3$ and $y = 0$

3. $x = 2$ and $y = -3$

Save the VI as Comparison.vi in the Users Stuff folder in the Learning directory.

FIGURE P2.8
The front panel for the Comparison VI

DESIGN PROBLEMS

D2.1 In the Building Blocks section at the end of this chapter, you created a VI called **Falling Edge.vi** to be used in determining where falling edges should occur on the output of the pulse width modulation VI. Now create a VI called **Rising Edge.vi** that returns a value of True on its Boolean output, **Rising Edge**, whenever a full output cycle has completed and a rising edge is to occur on the next clock cycle. Unlike the Falling Edge VI, this VI will only have two inputs: **Period** and **Clock**. The input **Duty Cycle** is not necessary in this VI since a change in the duty cycle of a signal changes only the location of the falling edge and not the location of the rising edge.

Save the VI as **Rising Edge.vi** in the **Building Blocks** folder in the **Learning** directory. The front panel should resemble the front panel seen in Figure D2.1.

D2.2 In this design problem you will create a VI that simulates a vending machine. The vending machine sells three items:

1. Candy bars for $0.80 each,
2. Potato chips for $0.60 a bag, and
3. Chewing gum for $0.40.

The vending machine accepts only five dollar bills, one dollar bills, quarters, dimes, and nickels.

Inputs on the front panel should include a numeric control for the user to enter the amount of money inserted into the vending machine and three more integer numeric controls that designate how many of each item the user wishes to purchase from the vending machine. Your VI should check to see if the amount of money input is greater than or equal to the total cost of the selected purchase. If there is not enough money, display a message notifying the customer that more

FIGURE D2.1
The front panel for the Rising Edge VI

money is needed to meet the total, using the Display Message to User Express VI. Then light an LED indicator on the front panel and display the amount needed on a numeric indicator. If enough money is inserted into the vending machine based on the user selection, output the change the user will receive, showing the quantity of dollar bills, quarters, dimes, and nickels to be dispensed by the vending machine.

CHAPTER 3

MathScript

MathScript is a new feature that adds math-oriented, textual programming to LabVIEW, both at the command-prompt level and within virtual instruments. MathScript is a high-level, text-based computing language with an easily accessible syntax and ample functionality to address computing tasks related to signal processing, analysis, and mathematics. In this chapter, we discuss the two pathways to MathScript: the MathScript Interactive Window and the MathScript Node. The essentials of creating user-defined functions and scripts, of saving and loading data files, and of using the MathScript Node are presented. With the MathScript Interactive Window, students can interact with LabVIEW through a command prompt and with MathScript Nodes scripts students can integrate MathScript with virtual instruments.

GOALS

1. Learn to use the MathScript Interactive Window.
2. Learn to define functions and create scripts.
3. Learn to use MathScript Nodes to integrate scripts with virtual instruments.

3.1   WHAT IS MATHSCRIPT?

The virtual instrument (VI) introduced in Chapter 2 is the central programming element of LabVIEW graphical programs. The VI is the computational workhorse. With VIs and subVIs (introduced in subsequent chapters) you can build applications to perform a variety of tasks, including simulation and design of complex systems, acquiring data and prototyping systems with real hardware, and deploying and controlling complete systems. But what if you wanted to perform a one-time mathematical operation and did not want to build a VI for the task? What if you had a sequence of computations that you wanted to perform to test an idea or to assist with your VI development? One can think of many instances when it would be very helpful to have easy, integrated access to a math-oriented, text-based computing environment.

MathScript is a new feature allowing you to access a text-based math-oriented language with a command-prompt from within the LabVIEW development environment. MathScript adds math-oriented, textual programming to LabVIEW, both at the command prompt level and within VIs and subVIs, through the use of MathScript Nodes.

Now, working within the LabVIEW environment, you can choose a textual approach (with command-line prompt and scripts), a graphical approach (with VIs), or a combination of the two (with MathScript Nodes within the VI). You can choose between the graphical environment and the command-line prompt in the most effective way for you to address your technical computing needs. MathScript does not require additional third-party software to compile and execute. When you install LabVIEW, you will automatically have access to MathScript.

LabVIEW 8 MathScript supports only Windows XP/2000.

MathScript is a high-level, text-based language with an easily accessible syntax and ample functionality to address tasks related to signal processing, analysis, and mathematics. MathScript includes more than five hundred built-in functions. There are linear algebra functions, curve-fitting functions, digital filters, functions for solving differential equations, and probability and statistics functions. And since MathScript employs a commonly used syntax, it follows that you can work with many of your previously developed mathematical computation scripts, or any of those openly available in engineering textbooks or on the internet.

The fundamental math-oriented data types in MathScript are matrices with built-in operators for generating data and accessing elements. In addition, you can extend MathScript with custom user-defined functions. An overview of the features of MathScript is presented in Table 3.1.

TABLE 3.1   Overview of the Features of MathScript

MathScript Feature	Description
Powerful textual math	MathScript includes more than 500 functions for math, signal processing, and analysis; functions cover areas such as linear algebra, curve fitting, digital filters, differential equations, probability/statistics, and much more
Math-oriented data types	MathScript uses matrices as fundamental data types, with built-in operators for generating data, accessing elements, and other operations
Compatible	MathScript is generally compatible with the m-file script syntax used by The MathWorks, Inc. MATLAB® software, Comsol, Inc. COMSOL Script™ software, and others
Extensible	You can extend MathScript by defining your own custom functions
Part of LabVIEW	MathScript does not require additional third-party software to compile and execute

*You can find more information on MathScript at **www.ni.com/mathscript**, including lists of built-in MathScript functions and links to online examples.*

3.2   ACCESSING THE MATHSCRIPT INTERACTIVE WINDOW

The **MathScript Interactive Window** provides a user interface in which you can enter and execute MathScript commands and see the result after the commands complete. You can enter commands one-at-a-time through a command line or as a group through a simple text editor. You can access the interactive window from the Getting Started screen or any VI by selecting Tools≫MathScript Window, as illustrated in Figure 3.1. The MathScript Interactive Window is a user interface composed of a **Command Window** (the user command inputs), an **Output Window** echoing the inputs and showing the resulting outputs, a **Script Editor** window (for loading, saving, compiling, and running scripts), a **Variables** window (showing variables, dimensions, and type), and a **Command History** window providing a historical account of commands and interactions with MathScript. A new MathScript Interactive Window is shown in Figure 3.2 with the various components highlighted.

The programmatic interface of MathScript is the MathScript Node, a structure that provides the capability to run scripts from within a VI. The MathScript Node is discussed in Section 3.7.

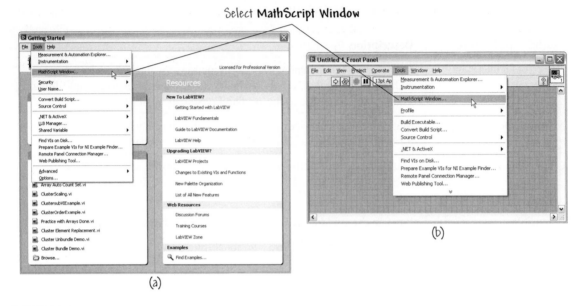

FIGURE 3.1
Accessing the MathScript Interactive Window from (a) the Getting Started screen, or (b) the **Tools** pull-down menu on the front panel or block diagram

As you work, the Output Window updates to show your inputs and the subsequent results. The Command History window tracks your commands. The History view is very useful because there you can access and reuse your previously executed commands by double-clicking a previous command to execute it again. You can also navigate through the history of previous commands (which

FIGURE 3.2
The basic components of the MathScript Interactive Window

will appear in the Command Window) by using the <↑> and <↓> keys. In the Script Editor window, you can enter and execute groups of commands and then save the commands in a file (called a script) for use in a later LabVIEW session. Taken together, the various elements of the MathScript Interactive Window constitute an interactive command-line tool that can be accessed without leaving LabVIEW.

3.2.1   The Command History Window

The commands you entered in previous sessions using the MathScript Interactive Window will reappear in subsequent sessions. In the Command History window you will find a header that shows the day and time when you entered the commands. This feature allows you to easily discern when the commands were entered. If the Command History window gets too full and you want to clear it out, you can right-click the Command History window and select **Clear History** from the shortcut menu. This process is illustrated in Figure 3.3.

*In a manner similar to the clearing of the Command History window, you can clear the Output Window. To accomplish this task, right-click the Output Window and select **Clear** from the shortcut menu, as shown in Figure 3.3. You can also use the MathScript function clc to clear the Output Window by typing in clc in the Command Window.*

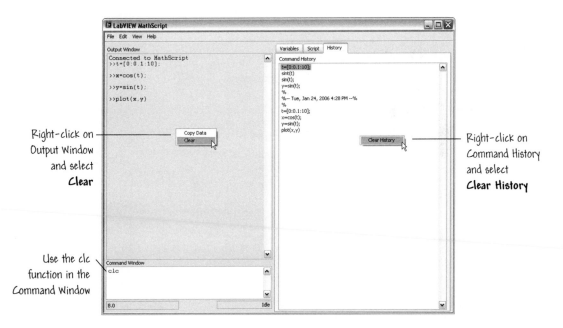

FIGURE 3.3
Clearing the Command History Window and the Output Window

3.2.2   Copying Output Window Data

You can copy data from the Output Window and paste it in the Script Editor window or a text editor. Right-click the Output Window and select **Copy Data** from the shortcut menu to copy the contents of the Output Window to the clipboard.

You also can highlight text in the Output Window and select Edit≫Copy or press the <Ctrl-C> keys to copy the selected text to the clipboard.

3.2.3   Viewing Data in a Variety of Formats

An interesting feature of the MathScript Interactive Window is the ability to view your variables in a variety of formats, as shown in Figure 3.4. Table 3.2 shows the available data types in MathScript. Depending on the variable type of your data, the available viewing formats include: numeric, string, graph, XY graph, sound, surface, and picture. You can also edit a variable in the Preview Pane when the display type is **Numeric** or **String**. Selecting **Sound** plays the data as a sound, but works for one-dimensional variables only. The remaining display types show the data as graphs of one sort or another: **Graph** displays the data on a waveform graph, **XY Graph** displays the data on an XY graph, **Surface** displays the data on a 3D surface graph, and **Picture** displays the data on an intensity graph. Chapter 8 provides additional information on plotting in LabVIEW.

FIGURE 3.4
Showing the data types in various formats

TABLE 3.2   Overview of LabVIEW Data Types and Relationship to Syntax

LabVIEW Data Type	MathScript Language Syntax	MathScript Node Syntax
Double-precision, floating-point numeric	Real	Scalar≫DBL
Complex double-precision, floating-point numeric	Complex	Scalar≫CDB
1D array of double-precision, floating-point numeric	1-D Array of Real	1D-Array≫DBL 1D
1D array of complex double-precision, floating-point numeric	1-D Array of Complex	1D-Array≫CDB 1D
Multidimensional array of double-precision, floating-point numeric	2-D Array of Real	Matrix≫Real Matrix (2D only)
Multidimensional array of complex double-precision, floating-point numeric	2-D Array of Complex	Matrix≫Complex Matrix (2D only)
String	String	Scalar≫String
Path	Path	N/A
1D array of string	N/A	1D-Array≫String 1D

*The **Graphical first?** button specifies whether to display first the numerical or graphical representation of variables in the Preview Pane. You can view only the numerical representation of scalar variables.*

MathScript Variable Types

In this example you will use the MathScript Interactive Window to construct a plot of the cosine function. You will also learn how to view the variables in a variety of formats.

Open a new MathScript Interactive Window from the Getting Started screen by selecting Tools≫MathScript Window, as illustrated in Figure 3.1. In the Command Window enter the time from $t = 0$ seconds to $t = 10$ seconds in increments of 0.1 seconds, as follows:

> t=[0:0.1:10]; .

Then, compute the $y = \cos(t)$ as follows:

> y=cos(t); .

FIGURE 3.5
(a) Entering the time, computing $y = \cos(t)$, and viewing the variable y in numerical form (b) Viewing the variable y in graphical form

Notice that in the Variables window the two variables t and y appear, as illustrated in Figure 3.5(a). Select the variable y and note that in the Preview Pane the variable appears in the numeric format. Now, in the pull-down menu above the Preview Pane, select **Graph**. The data will now be shown in graphical form, as illustrated in Figure 3.5(b).

*The graph can be undocked from the Preview Pane for resizing and customization. To undock the graph, right-click on the graph and select **Undock Window**. The window can now be resized and the plot can be customized interactively and printed.*

As an alternative to using the Preview Pane, you can also obtain a plot of y versus t programmatically using the plot command:

$$\boxed{\text{plot(t,y)}}\ .$$

The process is illustrated in Figure 3.6. A new window appears that presents the graph of y versus t. Following the same procedure, see if you can obtain a plot of $y = \cos(\omega t)$ where $\omega = 4$ rad/sec. ◆

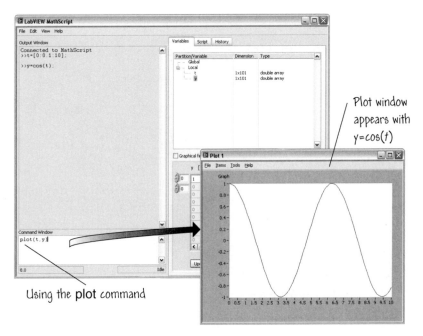

Plot window appears with y=cos(t)

Using the **plot** command

FIGURE 3.6
Obtaining a plot of the cosine function using the plot command

3.3 MATHSCRIPT HELP

You can display several types of help content for MathScript by calling different help commands from the Command Window. Table 3.3 lists the help commands you can call and the type of help these commands display in the Output Window.

TABLE 3.3   Help Commands for MathScript

Command	Description of Help Provided
help	Provides an overview of the MathScript window
help classes	Provides a list of all classes of MathScript functions and topics as well as a short description of each class
help *class*	Provides a list of the names and short descriptions of all functions in a particular MathScript class Example: help basic
help *function*	Provides reference help for a particular MathScript function or topic, including its name, syntax, description, inputs and outputs, examples to type in the Command Window, and related functions or topics Example: help abs

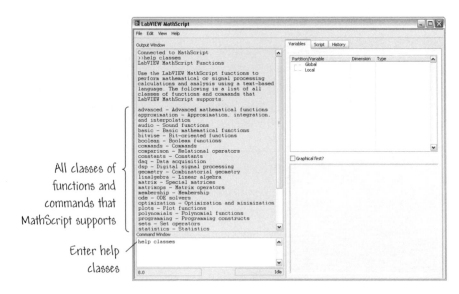

All classes of functions and commands that MathScript supports

Enter help classes

FIGURE 3.7
Accessing the help for MathScript classes, members, and functions

As illustrated in Figure 3.7, entering help classes in the Command Window results in an output showing all classes of functions and commands that Math-Script supports. Examples of the classes of functions are basic and matrixops. Entering help basic in the Command Window results in a list of the members of the basic class, including abs (i.e., the absolute value), conj (i.e., the complex conjugate function), and exp (i.e., the exponential function). Then, entering help abs in the Command Window will result in an output that contains a description of the abs function, including examples of its usage and related topics.

Students usually use a combination of the help commands shown in Table 3.3 to find functions. For example, if you are looking for a function to perform a filter, the common approach would be to first enter help classes *to view the available classes of functions. From that list of functions, you would then enter* help dsp *to get a list of DSP functions, including many filter functions. From there, you can choose the functions you want to learn more about and enter* help filter, *for example, to access the help for that function.*

3.4   SYNTAX

The syntax associated with MathScript is straightforward. Most students with some experience programming a text-based language will be comfortable with the programming constructs in MathScript. If you need help getting started with MathScript, you can access help by selecting Help≫Search the LabVIEW Help from the MathScript Interactive Window (or alternatively on the block diagram or front panel of your VI) and typing mathscript in the search window.

Eleven basic MathScript syntax guidelines are:

1. **Scalar operations:** MathScript is ideally suited for quick mathematical operations, such as addition, subtraction, multiplication, and division. For example, consider the addition of two scalar numbers, 16 and 3. This is a simple operation that you might perform on a calculator. This can be accomplished using the MathScript command:

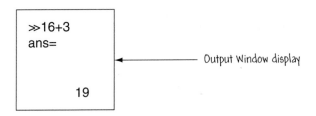

```
≫16+3
ans=


          19
```

Output Window display

In MathScript, if you perform any calculation or function without assigning the result to a variable, the default variable ans is used. If you want to assign the value of the addition of two scalars to the variable x, enter the following command:

```
≫x=16+3
x=


          19
```

In the same manner, you can add two scalar variables y and z by entering the following commands:

```
≫y=16;
≫z=3;
≫y+z
ans=
          19

≫disp(y);
          16
```

Semicolon eliminates output display

No semicolon leads to output display

Using disp leads to output even with the semicolon

Notice that in the previous example a semicolon was used for the first two lines, and no output was displayed. In MathScript, if you end a command line with a semicolon, the MathScript Interactive Window does not display the output for that command.

*Some functions display output even if you end the command line with a semicolon. For example, the **disp** function displays an output even if followed by a semicolon.*

You use the symbol '-' for subtraction, the symbol '/' for division, and the symbol '*' for multiplication, as illustrated below:

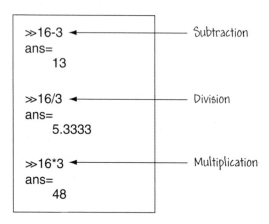

```
≫16-3   ←——————  Subtraction
ans=
    13

≫16/3   ←——————  Division
ans=
    5.3333

≫16*3   ←——————  Multiplication
ans=
    48
```

2. **Creating matrices and vectors:** To create row or column vectors and matrices, use white space or commas to separate elements, and use semicolons to separate rows. Consider for example, the matrix **A** (a column vector),

$$\mathbf{A} = \begin{bmatrix} 1 \\ 2 \\ 3 \end{bmatrix}.$$

In MathScript syntax, you would form the column vector as

A=[1;2;3] .

Similarly, consider the matrix **B** (a row vector)

$$\mathbf{B} = \begin{bmatrix} 1 & -2 & 7 \end{bmatrix}.$$

In MathScript syntax, you would form the row vector as

B=[1,-2,7] or B=[1 -2 7] .

As a final example, consider the matrix **C** (a 3 × 3 matrix):

$$C = \begin{bmatrix} -1 & 2 & 0 \\ 4 & 10 & -2 \\ 1 & 0 & 6 \end{bmatrix}.$$

In MathScript syntax, you would form the matrix as

C=[-1 2 0; 4 10 -2; 1 0 6] or C=[-1, 2, 0; 4, 10, -2; 1, 0, 6] .

3. **Creating vectors using the colon operator:** There are several ways to create a one-dimensional array of equally distributed elements. For example, you will often need to create a vector of elements representing time. To create a one-dimensional array equally distributed and incremented by 1, use the MathScript syntax,

```
≫t=1:10
t=
        1   2   3   4   5   6   7   8   9   10
```

.

To create a one-dimensional array equally distributed and incremented by 0.5, use the MathScript syntax,

```
≫t=1:0.5:10
t=
        1   1.5   2   2.5   3   3.5   4   4.5   5   5.5
  6   6.5   7   7.5   8   8.5   9   9.5   10
```

.

4. **Accessing individual elements of a vector or matrix:** You will often want to access specific elements or subsets of a vector or matrix. Consider the 3 × 3 matrix **C**:

$$C = \begin{bmatrix} -1 & 2 & 0 \\ 4 & 10 & -2 \\ 1 & 0 & 6 \end{bmatrix}.$$

In MathScript syntax, you can access the element in the second row and third column of the matrix **C**, as follows:

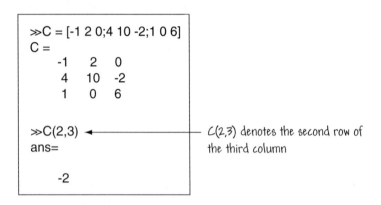

```
≫C = [-1 2 0;4 10 -2;1 0 6]
C =
        -1    2    0
         4   10   -2
         1    0    6

≫C(2,3)
ans=

        -2
```

C(2,3) denotes the second row of the third column

You can assign this value to a new variable by entering the following command:

```
≫F=C(2,3)
F=

        -2
```

You can also access an entire row or an entire column of a matrix using the colon operator. In MathScript syntax, if you want to access the entire second row of matrix **C**, enter the following command:

```
≫C(2,:)
ans=

        4    10    -2
```

The second row of C

In the same way, if you wish to access the entire third column of matrix **C**, enter the following command:

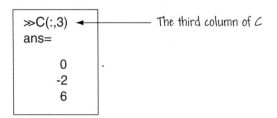

```
≫C(:,3)
ans=

         0
        -2
         6
```

The third column of C

Suppose you want to extract the 2×2 submatrix from **C** consisting of rows 2 and 3 and columns 1 and 2. You use brackets to specify groups of rows and columns to access a subset of data as follows:

```
>>C([2 3], [1 2])          A submatrix of C
ans=

    4   10
    1    0
```

5. **Calling functions in MathScript:** You can call MathScript functions from the Command Window. Consider the creation of a vector of a certain number of elements that are equally distributed in a given interval. To accomplish this in MathScript syntax, you can use the built-in function linspace.

Using the command help linspace you find that this function uses the syntax

$$\text{linspace}(a, b, n)$$

where a specifies the start of the interval, b specifies the end of the interval, and n identifies the number of elements. Thus, to create a vector of $n = 13$ numbers equally distributed between $a = 1$ and $b = 10$, use the following command:

```
>>G = linspace (1, 10, 13)
G=

    1   1.75   2.5   3.25   4   4.75   5.5   6.25   7
 7.75   8.5   9.25   10
```

If you do not specify a value for n, the linspace command will automatically return a vector of 100 elements.

To select a subset of **G** that consists of all elements after a specified index location, you can use the syntax described in guideline 4 and the length function to specify the end of the vector. For example, the following command will return all elements of **G** from the fifth element to the final element:

```
>>H=G(5:length(G))
H =

    4   4.75   5.5   6.25   7   7.75   8.5   9.25   10
```

*The function **linspace** is an example of a built-in MathScript function. Calling user-defined functions is covered in Section 3.5.2.*

6. **Assigning data types to variables:** MathScript variables adapt to data types. For example, if

$$a = \sin(3 * \text{pi}/2)\;,$$

then *a* is a double-precision floating-point number. If

$$a = \text{'temperature'}\;,$$

then *a* is a string.

7. **Using complex numbers:** You can use either *i* or *j* to represent the imaginary unit equal to the square root of -1.

If you assign values to either i or j in your scripts, then those variable names are no longer complex numbers. For example, if you let $y = 4 + j$, then y is a complex number with real part equal to 4 and imaginary part equal to $+j$. If however, you first assign $j = 3$, and then compute $y = 4 + j$, the result is $y = 7$, a real number.

8. **Matrix operations:** Many of the same mathematical functions used on scalars can also be applied to matrices and vectors. Consider adding two matrices **K** and **L**, where

$$\mathbf{K} = \begin{bmatrix} -1 & 2 & 0 \\ 4 & 10 & -2 \\ 1 & 0 & 6 \end{bmatrix} \text{ and } \mathbf{L} = \begin{bmatrix} 1 & 0 & 0 \\ 0 & 1 & 0 \\ 0 & 0 & 1 \end{bmatrix}.$$

To add the two matrices **K** and **L**, element by element, enter the following MathScript command:

```
≫K+L
ans =

     0    2    0
     4   11   -2
     1    0    7
```

In a similar fashion, you can also multiply two matrices **K** and **L**, as follows:

```
>K*L
ans =

      -1    2    0
       4   10   -2
       1    0    6
```

Consider the 3×1 matrix **M** (column vector) and the 1×3 matrix **N** (row vector)

$$\mathbf{M} = \begin{bmatrix} 1 \\ 2 \\ 3 \end{bmatrix} \text{ and } \mathbf{N} = [0 \quad 1 \quad 2].$$

Then, the product $\mathbf{M} * \mathbf{N}$ is the 3×3 matrix

```
>M*N
ans =

       0    1    2
       0    2    4
       0    3    6
```

and the product $\mathbf{N} * \mathbf{M}$ is a scalar

```
>N*M
ans =

       8
```

To multiply two matrices, they must be of compatible dimensions. For example, suppose a matrix **M** is of dimension $m \times n$, and a second matrix **N** is of dimension $n \times p$. Then you can multiply $\mathbf{M} \times \mathbf{N}$ resulting in an $m \times p$ matrix. In the example above, the 3×1 matrix **M** (column vector) was multiplied with the 1×3 matrix **N** (row vector) resulting in a 3×3 matrix. You cannot multiply $\mathbf{N} \times \mathbf{M}$ unless $m = p$. In the example above, the 1×3 matrix **N** (row vector) was multiplied with the 3×1 matrix **M** (column vector) resulting in a 1×1 matrix (a scalar), so in this case, $m = p = 1$.

When working with vectors and matrices in MathScript, it is often useful to perform mathematical operations *element-wise*. For example, consider the two vectors

$$\mathbf{M} = \begin{bmatrix} -1 \\ 4 \\ 0 \end{bmatrix} \text{ and } \mathbf{N} = \begin{bmatrix} 2 \\ -2 \\ 1 \end{bmatrix}.$$

In light of our previous discussion, it is not possible to compute $\mathbf{M} * \mathbf{N}$ since the dimensions are not compatible. However, you can multiply the vectors element-wise using the syntax '.*' for the multiplication operator, as follows:

$$\mathbf{M} .* \mathbf{N} = \begin{bmatrix} -1*2 \\ 4*(-2) \\ 0*1 \end{bmatrix} = \begin{bmatrix} -2 \\ -8 \\ 0 \end{bmatrix}.$$

By definition, matrix addition and subtraction occurs element-wise. However, it is also possible to perform division element-wise. For $\mathbf{M}$ and $\mathbf{N}$ as above, we find that element-wise division yields

$$\mathbf{M} ./ \mathbf{N} = \begin{bmatrix} -1/2 \\ 4/(-2) \\ 0/1 \end{bmatrix} = \begin{bmatrix} -0.5 \\ -2 \\ 0 \end{bmatrix}.$$

Element-wise operations are useful in plotting functions. For example, suppose that you wanted to plot $y = t \sin(t)$ for $t = [0:0.1:10]$. This would be achieved via the commands

```
t=[0:0.1:10];
y=t .* sin(t);        ⟵———— Element-wise multiplication using .*
plot(t,y)
```

9. **Logical expressions:** MathScript can evaluate logical expressions such as EQUAL, NOT EQUAL, AND, and OR. To perform an equality comparison, use the statement

```
a = = b
```
.

If a is equal to b, MathScript will return a 1 (indicating True); if a and b are not equal, MathScript will return a 0 (indicating False). To perform an inequality comparison, use the statement

| a ~ = b | .

If a is not equal to b, MathScript will return a 1 (indicating True); if a and b are equal, MathScript will return a 0 (indicating False).

In other scenarios, you may want to use MathScript to evaluate compound logical expressions, such as when at least one expression of many is True (OR), or when all of your expressions are True (AND). The compound logical expression AND is executed using the '&' command. The compound logical expression OR is executed using the '|' command.

10. **Control flow constructs:** Table 3.4 provides the MathScript syntax for commonly used programming constructs.

The MathScript syntax is similar to the widely used m-file script syntax used by The MathWorks, Inc. MATLAB®, COMSOL Script^{TM}, and others. In MathScript, you generally can execute scripts written in the m-file script syntax; however, some scripts or functions may not be compatible with the MathScript engine.

11. **Adding comments:** To add comments to your scripts, precede each line of documentation with a % character. For example, consider a script that has two inputs, x and y, and computes the addition of x and y as the output variable z,

```
% In this script, the inputs are x and y
% and the output is z.
z=x+y; % z is the addition of x and y
```
⎫
⎬ Comments start with %
⎭

The script shown above has three comments, all preceded by the % character. In the next section, we will discuss more details on how to use comments to provide help documentation.

Some considerations that have a bearing on your usage of LabVIEW 8 MathScript follow:

1. You cannot define variables that begin with an underscore, white space, or digit. For example, you can name a variable time, but you cannot name a variable 4time or _time.

2. MathScript variables are case sensitive. The variables X and x are not the same variables.

TABLE 3.4   MathScript Syntax for Commonly Used Constructs

Construct	Grammar	Example

Case-Switch Statement	switch expression
	case expression
	statement-list
	[**case** expression
	statement-list]
	...
	[otherwise
	statement-list]
	end
		switch mode
		case 'start'
		a = 0;
		case 'end'
		a = −1;
		otherwise
		a = a+1;
		end
		When a case in a case-switch statement executes, LabVIEW does not select the next case automatically. Therefore, you do not need to use break statements as in C.

For Loop	for expression statement-list end	for k = 1:10 a(k) = sin(2*pi*k/10) end
If-Else Statement	if expression statement-list [elseif expression statement-list] ... [else statement-list] end	if b = = 1 c = 3 else c = 4 end
While Loop	while expression statement-list end	while k < 10 a(k) = cos(2*pi*k/10) k = k + 1; end

3. LabVIEW 8 MathScript does not support n-dimensional arrays where $n > 2$, cell arrays, sparse matrices, or structures.

3.4.1   Key MathScript Functions

MathScript offers more than five hundred textual functions for math, signal processing, and analysis. These are in addition to the more than six hundred graphical functions for signal processing, analysis, and math that are available as VIs within LabVIEW. Table 3.5 lists many of the key areas with supporting MathScript functions. For a comprehensive function list, visit the National Instruments website at http://www.ni.com/mathscript or see the online help.

TABLE 3.5 MathScript Function Summary

Function Classes	Brief Description
Plots (2D and 3D)	Standard x-y plot; mesh plot; 3D plot; surface plot; subplots; stairstep plot; logarithmic plot; stem plot and more
Digital Signal Processing (DSP)	Signal synthesis; Butterworth, Chebyshev, Parks-McClellan, windowed FIR, elliptic (Cauer), lattice and other filter designs; FFT (1D/2D); inverse FFT (1D/2D); Hilbert transform; Hamming, Hanning, Kaiser-Bessel and other windows; pole/zero plotting and others
Approximation (Curve Fitting & Interpolation)	Cubic spline, cubic Hermite and linear interpolation; exponential, linear and power fit; rational approximation and others.
Ordinary Differential Equation (ODE) Solvers	Adams-Moulton, Runge-Kutta, Rosenbrock and other continuous ordinary differential equation (ODE) solvers
Polynomial Operations	Convolution; deconvolution; polynomial fit; piecewise polynomial; partial fraction expansion and others
Linear Algebra	LU, QR, QZ, Cholesky, Schur decomposition; SVD; determinant; inverse; transpose; orthogonalization; solutions to special matrices; Taylor series; real and complex eigenvalues and eigenvectors; polynomial eigenvalue and more
Matrix Operations	Hankel, Hilbert, Rosser, Vandermonde special matrices; inverse; multiplication; division; unary operations and others
Vector Operations	Cross product; curl and angular velocity; gradient; Kronecker tensor product and more
Probability and Statistics	Mean; median; Poisson, Rayleigh, chi-squared, Weibull, T, gamma distributions; covariance; variance; standard deviation; cross correlation; histogram; numerous types of white noise distributions and other functions
Optimization	Quasi-Newton, quadratic, Simplex methods and more
Advanced Functions	Bessel, spherical Bessel, Psi, Airy, Legendre, Jacobi functions; trapezoidal, elliptic exponential integral functions and more
Basic	Absolute value; Cartesian to polar and spherical and other coordinate conversions; least common multiple; modulo; exponentials; logarithmic functions; complex conjugates and more
Trigonometric	Standard cosine, sine and tangent; inverse hyperbolic cosine, cotangent, cosecant, secant, sine and tangent; hyperbolic cosine cotangent, cosecant, secant, sine and tangent; exponential; natural logarithm and more
Boolean and Bit Operations	AND, OR, NOT and other logic operations; bitwise shift, bitwise OR and other bitwise operations
Data Acquisition/ Generation	Perform analog and digital I/O using National Instruments devices
Other	Programming primitives such as If, For and While Loops; unsigned and signed datatype conversions; file I/O; benchmarking and other timing functions; various set and string operations and more

3.5 DEFINING FUNCTIONS AND CREATING SCRIPTS

You can define functions and create scripts to use in the MathScript Interactive Window or in the MathScript Node. In this section, we will focus on the use of the MathScript Interactive Window.

Functions and scripts can be created in the Script Editor Window on the MathScript Interactive Window (see Figure 3.2). You can also use your favorite text editor to create functions and scripts. Once your function or script is complete, you should save it for use later. The filename for a function must be the same as the name of the function and must have a lowercase .m extension. For example, the filename for a user-defined **starlight**[1] function must be **starlight.m**. Remember that MathScript is case sensitive.

Use unique names for all functions and scripts and save them in a directory that you specified in the Path section of the File≫MathScript Preferences dialog box. Entering the Path command in the Command Window returns the current search path list. The command path(a) changes the current path to a and path(path, b) adds b to the end of the search path list.

3.5.1   User-Defined Functions

As noted in Section 3.4, MathScript offers more than five hundred textual functions for math, signal processing, and analysis. But what if you have a special-purpose function that you want to add to your personal library? This function may be particular to your area of study or research, and is one that you need to call as part of a larger program. With MathScript it is simple to create a function once you understand the basic syntax.

A MathScript function definition must use the following syntax:

```
function outputs = function_name(inputs)
% documentation
script
```

An example of a user-defined function definition utilizing the proper syntax is

```
function ave = compute_average(x,y)
% compute_average determines the average of the two inputs x and y.
ave = (x + y)/2;
```

Begin each function definition with the term function. The *outputs* term lists the output variables of the function. If the function has more than one output variable, enclose the variables in square brackets and separate the variables with white space or commas. The *function_name* is the name of the function you want to define and is the name that you use when calling the function. The *inputs* term lists the input variables to the function. Use commas to separate the input variables. The *documentation* is the set of comments that you want MathScript to return for the function when you execute the help command. Comments are preceded with a % character. You can place comments anywhere in the function; however, LabVIEW returns only the first comment block in the Output Window to

---

provide help to the user. All other comment blocks are for internal documentation. The *script* defines the executable body of the function.

Checking the help on the function compute_average.m and then executing the function with $x = 2$ and $y = 4$ as inputs yields

```
≫help compute_average
compute_average determines the average of the two inputs x and y.

≫x=2;y=4;compute_average(x,y)
ans=

    3
```

Note that there is a MathScript function named mean that can also be used to compute the average of two inputs, as follows:

```
≫mean([2 4])
ans =

    3
```

Custom functions can be edited in the Script Editor window and saved for later use. In Figure 3.8, the buttons **Save & Compile**, **Save**, **Load**, **Clear**, and **Run** are shown. You can compile a function when you save it to decrease the run-time compilation time. Selecting **Load** will open a window to browse for the desired function (or script) to load into MathScript. Similarly, selecting **Save** will open a browser to navigate to the desired folder to save the function.

In Figure 3.8, the function compute_average is used to compute the average of two arrays. Notice that the function computes the average element-wise. If compute_average had inadvertently been named mean, then LabVIEW would execute the user-defined function instead of the built-in function. Generally it is not a good idea to redefine LabVIEW functions, and students should avoid doing so.

If you define a function with the same name as a built-in MathScript function, LabVIEW executes the function you defined instead of the original MathScript function. When you execute the help command, LabVIEW returns help content for the function you defined and not the help content for the original MathScript function.

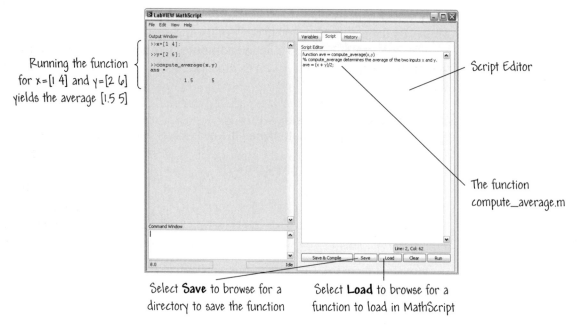

Running the function for x=[1 4] and y=[2 6] yields the average [1.5 5]

Script Editor

The function compute_average.m

Select **Save** to browse for a directory to save the function

Select **Load** to browse for a function to load in MathScript

FIGURE 3.8
Loading and saving functions

Other examples of valid function syntax for the **starlight** function include:

function starlight	% No inputs and no outputs
function a = starlight	% No inputs and one output
function [a b] = starlight	% No inputs and two outputs
function starlight(g)	% One input and no outputs
function a = starlight(g)	% One input and one output
function [a b] = starlight(g)	% One input and two outputs
function starlight(g, h)	% Two inputs and no outputs
function a = starlight(g, h)	% Two inputs and one output
function [a b] = starlight(g, h)	% Two inputs and two outputs

There are several restrictions on the use of functions. First, if you define multiple functions in one MathScript file, all functions following the first are **subfunctions** and are accessible only to the main function. A function can call only those functions that you define below it. Second, you cannot call functions recursively. For example, the function starlight cannot call starlight. And third, LabVIEW does not allow circular recursive function calls. For example, the function starlight cannot call the function bar if bar calls starlight.

3.5.2   Scripts

A script is a sequence of MathScript commands that you want to perform to accomplish a task. For convenience and reusability, once you have created a script, you can save it and load it into another session of LabVIEW at a later time. Also, often you can use a script designed for a different task as a starting point for the development of a new script. Since the scripts themselves are saved as common ascii text and editable with any text editor (including the one found in the MathScript Interactive Window), it is easy to do this. The MathScript functions as well as the user-defined functions can be employed in scripts.

Continuing the example above, suppose that we used a script to compute the average of two numbers. The compute_average function could be used within the script. Once saved, the script can subsequently be loaded into MathScript for use in another session. A script using the compute_average function is shown in Figure 3.9.

Saving and Loading Scripts

Being able to save scripts is an important feature giving you the capability to develop a library of scripts that you can readily access in future MathScript sessions. To save a script that you have created in the Script Editor window, select File≫Save Script, as illustrated in Figure 3.10(a). You can also save your script by clicking the **Save** button on the Script Editor window of the MathScript Interactive Window, as illustrated in Figure 3.10(b). In both cases, a file dialog

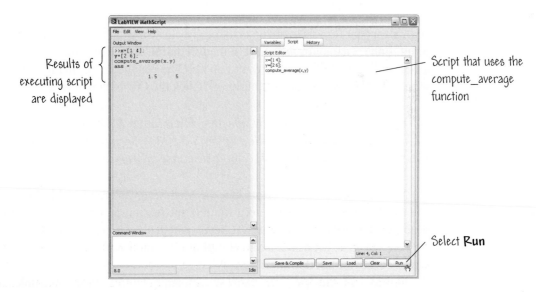

FIGURE 3.9
Editing, saving, and running a script to compute the average of two arrays element-wise

FIGURE 3.10
(a) Saving a script using the File≫Save Script pull-down menu (b) Saving a script using the **Save** button on the MathScript Interactive Window

box will appear for you to navigate to the directory in which you want to save the script. Enter a name for the script in the **File name** field. The name must have a lowercase .m extension if you want LabVIEW to run the script (in this example, we use the name average_example.m). Click the **OK** button to save the script.

 You can compile a script by selecting File≫Save & Compile Script or by clicking on the Save & Compile button in the MathScript Interactive Window. This will save and compile the script to decrease the run-time compilation time.

You can load existing scripts into the MathScript Interactive Window. This will be useful upon returning to a MathScript session or if you want to use a script in the current session that was developed in a previous session. To load an existing script, select File≫Load Script or click the **Load** button on the Script Editor of the MathScript Interactive Window.

Figure 3.11 illustrates the process of loading scripts. In the example, the script compute_average.m is loaded into a MathScript session, and then using the **Run** button, the script is executed.

FIGURE 3.11
(a) Loading a script using the File≫Load Script pull-down menu (b) Loading a script using the **Load** button on the MathScript Interactive Window

*You can find a working version of the script **average_example.m** and the function **compute_average.m** in the **Chapter 3** folder of the **Learning** directory.*

Practice with Scripts

In this example you will load an existing MathScript script and use it to smooth a noisy signal. To begin, open the MathScript Interactive Window (see Section 3.2 for more information on accessing MathScript). Load in the script **signal smoothing.m** found in the **Chapter 3** folder of the **Learning** directory. The script is shown in Figure 3.12. Selecting **Run** will produce a plot of a noisy signal and a smoothed version of the same signal.

The script has three inputs: smoothing factor, p, time span of interest, t, and noise strength, *sigma*. By default, the values are set at $p = 0.95$; $t=[0:0.05:10]$; and *sigma* $= 0.1$.

It is easy to investigate the effect of varying the smoothing parameter. In the Script Editor window, change the value of $p = 0.95$ to $p = 0.90$. Then click the **Run** button (see Figure 3.12). You should find that the smoothed signal is

FIGURE 3.12
Smoothing a noisy signal using a script

smoother for $p = 0.95$ than for $p = 0.90$, as illustrated in Figures 3.13(a) and 3.13(b). In fact, if you set the value of $p = 0.001$, the smoothed signal will be almost identical to the original noisy signal. Changing the value of *sigma* will produce signals with more or less noise, depending on the size of *sigma*. Raising the value of *sigma* to *sigma* $= 0.5$ results in a very noisy signal, as shown in Figure 3.13(c). ◆

3.6 SAVING, LOADING, AND EXPORTING DATA FILES

3.6.1 Saving and Loading Data Files

In MathScript you can save and load data files in the MathScript Interactive Window. A data file contains numerical values for variables. Being able to save and load data gives you the flexibility to save important data output from a MathScript session for use in external programs. There are two ways to save data files. The first method saves the data for *all* the variables in the workspace, and the second method allows you to select the variables to save to a file.

To save all the variables in the workspace, select File≫Save Data in the MathScript Interactive Window. You also can right-click the variable list on the Variables window and select **Save Data** from the shortcut menu. In the file dialog

FIGURE 3.13
(a) Smoothing with $p = 0.95$ and *sigma* $= 0.1$ (b) Smoothing with $p = 0.90$ and *sigma* $= 0.1$ (c) Smoothing with $p = 0.95$ and *sigma* $= 0.5$

box, navigate to the directory in which you want to save the data file. Enter a name for the data file in the **File name** field and click the **OK** button to save the data file.

The second method allows you to select the variables to save. In this case, in the Command Window, enter the command save filename var1, var2, ..., varn, where filename is the name of the file to store the data in and var1, var2, ..., varn are the variables that you want to save. In this case, the data will be saved in filename in the LabVIEW Data folder in the path specified in File≫MathScript Preferences. In Figure 3.14 the process of saving data is illustrated. In Figure 3.14(a), all the variables are saved in the file save_all.mlv after navigating to the folder LabVIEW Data. In Figure 3.14(b), the variable x is saved in the file save_x.mlv.

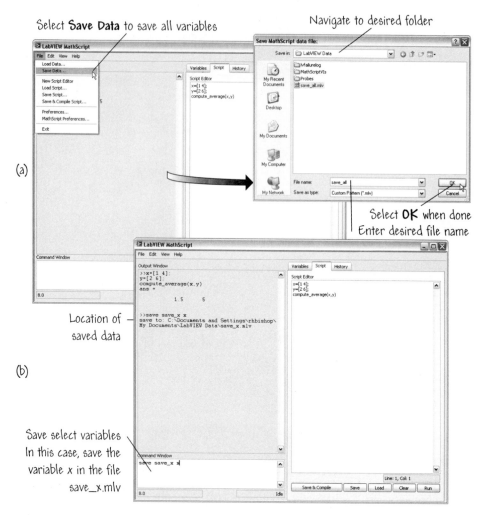

FIGURE 3.14
Saving data files: (a) Saving all the variables in the workspace (b) Saving select variables

You also can load existing data files into your MathScript session. In the MathScript Interactive Window, select File≫Load Data or right-click the variable list on the Variables window and select **Load Data** from the shortcut menu to load the data file you want, as illustrated in Figure 3.15.

You must save data files before you can load them into the MathScript Interactive Window.

MathScript can also load .mat and .mlv files created by other technical computing tools that support the m-file script syntax, as long as they only include data types that MathScript supports. Refer to Table 3.2 for MathScript data types.

FIGURE 3.15
Loading data from a previous MathScript session

3.6.2 Exporting to Microsoft Excel

In some instances, it may be useful to export your data to an external application, such as Microsoft Excel. In MathScript this is readily accomplished. On the MathScript Interactive Window, navigate to the Variables window. Select one of the variables that you desire to export to a spreadsheet. Exporting to MS Excel is accomplished from the Preview Pane and can be done if the display type is **Numeric**, **Graph**, **XY Graph**, or **Picture**. To do so, right-click the Preview Pane and select **Copy Data to Excel Clipboard** from the shortcut menu to copy the data for the desired variable, as illustrated in Figure 3.16. You then can paste the data in Excel. In Figure 3.16, the time vector t is being exported to Excel. The next step would be to highlight the variable y and export it to Excel. Then, both t and y would exist in the spreadsheet for further analysis.

Since data in Excel is generally in column format, make sure that the variables that you export to Excel are in array format of size $n \times 1$, where n is the length of the array. If the array is a $1 \times n$ array, then when exported to Excel, it will paste in a row rather than a column.

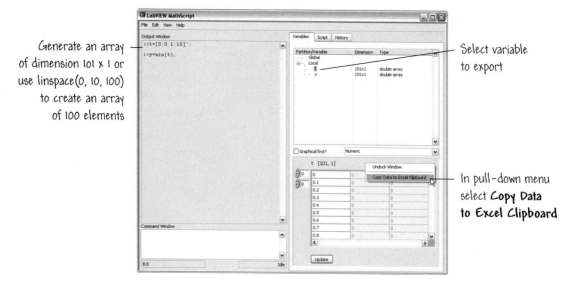

Generate an array of dimension 101 x 1 or use linspace(0, 10, 100) to create an array of 100 elements

Select variable to export

In pull-down menu select **Copy Data to Excel Clipboard**

FIGURE 3.16
Saving data to Microsoft Excel

3.7 MATHSCRIPT NODES

Working in the MathScript environment, you can choose a textual approach (with command-line prompt and scripts) as described in the previous sections, or you can choose a graphical approach by employing scripts within VIs in a special node called a **MathScript Node**. The MathScript Node provides a means of combining graphical and textual code within LabVIEW by entering script text directly or importing it from your .m or text files. An example of a MathScript Node is shown in Figure 3.17.

As seen in Figure 3.17, you define inputs and outputs on the border of the MathScript Node to specify the data that is to be transferred in and out of the node (between the graphical data flow programming and the textual MathScript code). Your scripts within the MathScript Node can access features from traditional LabVIEW graphical data flow programming. In Figure 3.17, the MathScript Node has a variable input on the left side of the node, named x. The right side of the MathScript Node contains the output variables y, y1, and d. By default, the MathScript Node also includes an error in input and an error out output on the bottom left and right of the node frame, respectively.

In many cases, your text-based algorithms that have been verified and used successfully in other situations may be utilized directly rather than re-coded using graphical programming constructs. The idea of the MathScript Node is to provide a mechanism for introducing text-based programming into the graphical programming arena. In the discussion of Structures in Chapter 6, we will further discuss the use of nodes, with emphasis on the Formula Node, which is another implementation of the same idea of using text-based sequential programming from within the VI.

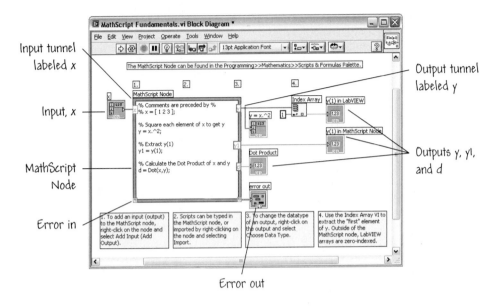

FIGURE 3.17
The MathScript Node

3.7.1   Accessing the MathScript Node

The MathScript Node is accessed on the **Programming≫Structures** palette or **Mathematics≫Scripts & Formulas** palette, as illustrated in Figure 3.18. First select the MathScript Node from the palette and place it on the block diagram.

FIGURE 3.18
The MathScript Node is on the (a) **Programming≫Structures** palette and (b) **Mathematics≫Scripts & Formulas** palette

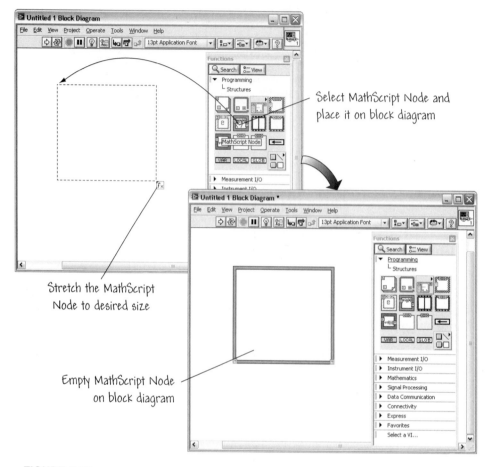

FIGURE 3.19
Placing the MathScript Node on the block diagram

Keeping the left mouse button depressed you can extend the node frame as illustrated in Figure 3.19. You can also use the **Positioning** tool to extend the MathScript Node to the desired size once it has been placed on the block diagram.

3.7.2   Entering Scripts into the MathScript Node

There are three ways to place a script in the MathScript Node. You can use the cursor tool to type the script directly in the MathScript Node. For short and simple scripts this is a convenient way to utilize the MathScript Node. The syntax is the same as described in Section 3.4.

You can also import an existing script from a file into the MathScript Node. To import a script, right-click the MathScript Node border and select **Import** from the shortcut menu to display the **Select a Script** dialog box. When the dialog box appears, select the file you want to import and click **Open**. The MathScript

Right-click on
MathScript Node
frame

Imported script

Select **Import** to browse
directories for desired script

FIGURE 3.20
Importing a script into the MathScript Node

text will then appear in the script node, as shown in Figure 3.20. You can also cut and paste a script into a MathScript Node using any text editor, including the one in the MathScript Interactive Window.

It is suggested that you write your script and run it within the MathScript Interactive Window environment for testing and debugging purposes before you import it into LabVIEW. Once the script is in the node, you will need to wire up the required inputs and outputs.

3.7.3   Input and Output Variables

The script within the MathScript Node most likely will require inputs to carry out the desired computations, and the results will need to be output.

To add an input variable, right-click the MathScript Node border and select **Add Input** from the shortcut menu, as shown in Figure 3.21. When the input variable appears on the node, you should add the name of the variable. For example, in Figure 3.21, the input p was added to provide a mechanism for changing the smoothing parameter (see the example Practice with Scripts in Section 3.5.2 for a description of the smoothing parameter). You can also use the **Labeling** tool to edit the variable names at any time by double-clicking on the variable name inside of the tunnel and editing, as desired.

FIGURE 3.21
Adding inputs to the MathScript Node

The MathScript Node will only execute the data types shown in Table 3.2. Right-click an output variable to change its data type. Input variables automatically adapt to the data type being used, even if that is an unsupported data type, and cannot be changed from the shortcut menu. Thus, input variables that are integer or other non-supported data types must be converted to an accepted data type prior to wiring to the MathScript Node.

By default the MathScript Node includes one input and one output for the error in and error out parameters. To take advantage of the error-checking parameters for debugging information, it is suggested that you create an indicator for the error out terminal on the MathScript Node before you run the VI. This allows you to view the error information generated at runtime.

To add an output variable, right-click the MathScript Node border and select **Add Output** from the shortcut menu, as illustrated in Figure 3.22. You can create controls and indicators for each input and output on the MathScript Node. For example, to create an indicator on an output terminal, right-click the output terminal and select Create≫Indicator from the shortcut menu. LabVIEW will create an indicator on the front panel and wire terminals to the output on the block diagram, as shown in Figure 3.21 and Figure 3.22.

Add "global stars" in the MathScript Node   **stars** is now a global variable

Select
stars

Select
XY Graph

Graph of signal produced
programmatically in the VI

Enter "global stars"
in the Command Window

XY Graph of **stars** within the
MathScript Interactive Window
appears via global variable feature

FIGURE 3.28
Sharing a global variable between the MathScript Interactive Window and the MathScript Node

As an example, consider the VI shown in Figure 3.28. The block diagram uses a Simulate Signal Express VI, a Convert from Dynamic Data Express VI, a waveform graph, and a MathScript Node. The Simulate Signal data type must be converted from a variant to a 1D array of DBL using the Convert from Dynamic Data Express VI (found on the **Functions≫Express≫Signal Manipulation** palette) with the default configuration.

 The MathScript Global.vi pictured in Figure 3.28 is located in **Chapter 3** *of the* **Learning** *directory.*

The global command should appear in the MathScript Node, as shown in Figure 3.28, to declare the desired variables as being global and therefore accessible in the MathScript Interactive Window. To illustrate the global variable feature, execute the MathScript Global VI containing the MathScript Node. When the VI execution ends, navigate to the Command Window in the MathScript Interactive Window environment, type global stars and press **Enter**. This will allow you to see the data passed to the variable stars on the MathScript Node.

In the Variables window of the MathScript Interactive Window, you will see the variable **stars**, as illustrated in Figure 3.28. Select **stars**, and then (following the discussion in Section 3.2.3) choose to view the **stars** data in graphical form. The XY Graph of the data, as shown in the Preview Pane of Figure 3.28, resembles the plot of the data obtained programmatically in the MathScript Global VI. The data is identical, but the plot parameters (such as color) may differ between the Preview Pane and the Waveform Graph.

Having ready access to the same variables in the graphical programming VI and in the MathScript Interactive Window environment provides a superb environment in which to design, test, and verify algorithms.

 BUILDING BLOCK

3.9 BUILDING BLOCKS: PULSE WIDTH MODULATION

In this Building Block, you will modify both the Rising Edge and Falling Edge VIs that you constructed in Chapter 2. Then you will utilize a MathScript Node to compute the rising and falling edge.

The basic function needed by the MathScript Node to achieve the desired task is the modulus function. The **mod(A,B)** function computes the remainder of A/B, when the quotient is rounded toward $-\infty$. The modulus function syntax in MathScript is

$$X = \text{mod(A,B)} \quad .$$

The **mod** function will be placed within the MathScript Node to replace the graphical programming functions (e.g., Round to Nearest) employed in the Building Blocks section in Chapter 2.

The block diagram for the Rising Edge VI using the MathScript Node is shown in Figure 3.29 and should be used as a guide in the development of this Building Block VI. To begin, open the Rising Edge VI that you constructed in Chapter 2. Save the VI as Rising Edge–MathScript VI in the **Users Stuff** folder in the **Learning** directory.

If you successfully completed Building Blocks in Chapter 2, you should be able to find working versions of **Rising Edge.vi** *and* **Falling Edge.vi** *in the* **Users Stuff** *folder of the* **Learning** *directory. If not, you can find working versions in the* **Building Blocks** *folder of the* **Learning** *directory.*

MathScript does not support integer arithmetic, therefore all inputs and outputs will have the double-precision numeric (DBL) data type. You need to convert

The To Double
Precision
Float functions

The mod
function

MathScript
Node

Inputs: *C* and *P*

Output: **redge**

FIGURE 3.29
The Rising Edge VI using the mod function within a MathScript Node

integer inputs to DBL and convert numeric outputs to Boolean to complete this Building Block. All changes should be made without modifying the front panel of either VI from Chapter 2.

Switch to the block diagram of Rising Edge–MathScript VI. At this point, it should be exactly the same as the block diagram for Rising Edge VI from the Chapter 2 Building Block. Delete all the functions and wires on the block diagram, but leave the inputs and outputs unchanged. The functions deleted include a Multiply, Subtract, Divide, Equal?, and Round to Nearest.

Using Figure 3.19 as a guide, place a MathScript Node on the block diagram and enlarge it sufficiently to allow the placement of three lines of code, as illustrated in Figure 3.29. As discussed in Section 3.7.2, add the script within the MathScript Node. The inputs to the script are Clock and Period, and the output is redge. Since Clock and Period are integers, you need to add the To Double Precision Float functions and wire the inputs Clock and Period to the MathScript Node border.

To add input and output variables to the MathScript Node, right-click on the border and select **Add Input**, as described in Section 3.7.3. Add two input variables denoted C and P, for Clock and Period, respectively. Follow the same procedure to add the output variable redge.

The block diagram for the Rising Edge VI is illustrated in Figure 3.29. Once the VI is wired, set the inputs to and select **Run**. Verify that the VI is operating as expected by comparing the output with the Rising Edge VI when considering the same inputs.

When done, save the VI as Rising Edge–MathScript.vi in the Users Stuff folder in the Learning directory.

Following the same procedure, edit the Falling Edge VI that you created in Chapter 2 Building Blocks. Verify that the VI is operating as expected by comparing the output of the Falling Edge–MathScript VI with the Falling Edge VI when considering the same inputs. Remember to save the VI as Falling Edge–MathScript.vi in the Users Stuff folder in the Learning directory.

Working versions of Rising Edge–MathScript.vi and Falling Edge–Math Script.vi are in the Building Blocks folder of the Learning directory.

3.10   RELAXED READING: DESIGNING A PRODUCTION TEST SYSTEM FOR A NEW GENERATION OF GAMING

In May 2005, Microsoft announced the Xbox 360, the latest innovation for digital entertainment and gaming. The Xbox 360 system includes wired and wireless controllers. The challenge to engineers at the Microsoft Corporation was to create a comprehensive, low-cost production test system for the Microsoft Xbox 360 wired and wireless controllers. The solution that was developed employed a flexible, automated test system based on Microsoft Windows XP, Microsoft SQL Server, LabVIEW, and NI PXI modular instruments.

Functional testing is a key component to any production line. The challenge in developing a production line functional tester is to package as many parallel test scenarios as possible within the given production cycle time. With the new

FIGURE 3.30
Microsoft uses PXI and LabVIEW to ensure a quality gaming experience with the Xbox 360

functional test system for the Xbox 360 controller, Microsoft implemented a test strategy that resulted in a 50 percent increase in test throughput per test station.

The controller functional test system demands high-performance signal capture to qualify the signal integrity of the new controller and ensure a high-quality user experience. With the latest NI modular instruments, including the NI PXI-5124 12-bit, 200 MS/s digitizer, the Microsoft engineers met the increased functional test requirements for the Xbox 360 controller. Using the LabVIEW graphical development environment, they created more than one hundred tests, implemented Ethernet communication, and incorporated a data storage interface to our Microsoft SQL Server database.

The end result was the successful and reliable production line testing and storage of all parameters to the Microsoft SQL Server. The PXI-5124 high-resolution digitizer was a key component permitting the engineers to verify the signal integrity of the USB communication between the controller and the Xbox 360 console with confidence. Using the high resolution input and high sampling rate of the PXI-5124 digitizer, the engineers were able to acquire test signals with 12 bits of resolution at data rates up to 200 MS/s, leading to a low-cost automated test system.

For more information, contact:

D.J. Mathias
Microsoft
One Microsoft Way
Redmond, WA 98052

3.11   SUMMARY

With the release of LabVIEW 8, a new text-based programming component called MathScript is introduced. The approach is a natural complement to the LabVIEW graphical data flow programming approach for algorithm development, signal processing, analysis, and mathematical computations. Through the use of MathScript, users have a single environment in which they can choose the most effective syntax, whether textual, graphical, or a combination of the two. Students can now perform calculations at the command-prompt level, in addition to accessing the vast library of existing graphical virtual instruments.

KEY TERMS

Class: A grouping of MathScript functions that perform mathematical or signal processing calculations and analysis using a text-based language.

Command History Window: Displays a history of the commands you executed.

Command Window: Specifies the MathScript command you want LabVIEW to execute.

Command Line: The field (or line) in the Command Window where commands are entered.

Function: MathScript includes more than five hundred built-in functions, and users can develop new functions by creating **m**-file script text files that contain text that specifies the desired functionality.

Global variables: A variable that can be accessed and passed among several VIs and between the MathScript Interactive Window and MathScript Nodes.

MathScript: A high-level, text-based programming language with an easily accessible syntax and ample functionality to address programming tasks related to signal processing, analysis, and mathematics.

MathScript Interactive Window: An interactive interface used to edit and execute mathematical commands, create mathematical scripts, and view numerical and graphical representations of variables.

MathScript Node: A LabVIEW structure that is used to programmatically execute scripts within a VI.

Output Window: Displays the commands you entered in the Command Window and the output that MathScript generates from those commands.

Preview Pane: Displays in numercial, string, graphical, or sound format the variables you select in the variable list (in the Variables window).

Script: A sequential program containing groups of commands (generally) saved in a file called an **m**-file available for use in LabVIEW sessions.

Script Editor window: Displays the scripts that you create.

Subfunction: All functions following the first function in a MathScript file are subfunctions and are accessible only to the main function.

Variables window: Displays a list of all variables you define and previews variables that you select.

E3.1 Open the MathScript Interactive Window. In the Command Window, use the help command to learn about the different usages of the plot function. In your own words, how will m and n be plotted with the syntax plot(m,n)?

E3.2 Write a script to generate a 3 × 2 matrix **M** of random numbers using the rand function. Use the help command for syntax help on the rand function. Verify that each time you run the script the matrix **M** changes. Save the script in the Users Stuff folder in the Learning directory.

E3.3 In the MathScript Interactive Window, create a script that generates a time vector over the interval 0 to 10 with a step size of 0.5 and creates a second vector, y, according to the equation

$$y = e^{-2t}\left(0.3\sin(t) + 0.7\cos(0.1t)\right)$$

Then, add the plot function to generate a graph of y versus t.

Type the script in the Script Editor window and, when done, use the **Save** button in the Script Editor to save the script in the Users Stuff folder in the Learning directory. Clear the script from the Script Editor window, and then **Load** the script back into the Script Editor and select **Run**.

E3.4 Using a MathScript Node, generate a stem plot of a cosine function, $y = \cos(t)$, where t varies from 0 to π, with an increment of $\pi/20$.

E3.5 Open the MathScript Interactive Window. In the Command Window, create the matrices **A** and **B**:

$$A = \begin{bmatrix} 1 & 5 \\ 2 & 9 \\ 5 & 1 \end{bmatrix} \text{ and } B = \begin{bmatrix} 7 & 4 & 3 \\ 4 & 3 & 1 \end{bmatrix}$$

Is it possible to perform the following math operations on the matrices? If so, what is the result?

(a) **A*****B**

(b) **B*****A**

(c) **A**+**B**

(d) **A**+**B**$'$ (where **B**$'$ is the transpose of **B**)

(e) **A**./**B**$'$

PROBLEMS

P3.1 Using MathScript, generate a plot of a sine wave of frequency $\omega = 10$ rad/sec. Use the linspace function to generate the time vector starting at $t = 0$ and ending at $t = 10$. Label the x-axis as **Time (sec)**. Label the y-axis as **sin(w*t)**. Add the following title to the plot: **Sine wave with frequency w = 10 rad/sec**.

P3.2 The rand function generates uniformly distributed random numbers between 0 and 1. This means that the average of all of the random values generated by the rand function should approach 0.5 as the number of random numbers increases. Using the rand function, generate random vectors of length 5, 100, 500, and 1000. Confirm that as the number of elements increases, the average of the random numbers approaches 0.5. Generate a plot of the average of the random numbers as a function of the number of random numbers. Use both rand and mean functions in your script.

P3.3 Using the MathScript Interactive Window, generate a random 2D matrix using the syntax z = peaks(100). View the variable z as a 3D surface using the Variables Window in the MathScript Interactive Window. Undock the 3D plot and rotate the plot in 3D space with the mouse so that you can estimate the maximum value of the plot on the z-axis. Use the max function to calculate the value you just estimated.
    (**Hint**: You may want to use the max function twice.)

P3.4 Open a new VI and place a MathScript Node on the block diagram. Create a numeric input on the MathScript Node frame and name it w. Generate the sine wave using Problem 3.1 as a reference and using MathScript syntax within the node. Replace the frequency with the variable w as an input controlled from a front panel knob. Create two outputs on the MathScript Node frame named t and y, respectively, where t is the time history and y is the sine function associated with t. Plot the sine wave using an XY Graph. Explain the behavior of the sine wave as you increase the frequency.

DESIGN PROBLEMS

D3.1 The MathScript script in Figure D3.1 simulates the trajectory of a projectile fired at a specified angle. Using Figure D3.1 as a guide, develop a VI using a MathScript Node to create an interactive simulator for calculating projectile trajectory. The two inputs to the simulator are the initial angle of the trajectory and the initial velocity. Utilize knob controls on the front panel and wire them to input variables deg and v0. The output vectors x and y should be combined and graphed on an XY Graph to show the trajectory.
    What angle would allow you to shoot over a 3 meter wall standing 1 meter from the wall while hitting a target on the ground about 6 meters away in 1.5 seconds? What is the initial velocity of the projectile to accomplish this targeting goal?

```
% D3.1 -- Projectile Trajectory Simulator
t_max = 1.5;              % final simulation time
dt = 0.1;                 % length of simulation vector
t = 0:dt:t_max;           % create the time vector for the simulation
deg = 79;                 % initial angle of the trajectory
g = 9.81;                 % acceleration due to gravity (m/s^2)
v0 = 15.0;                % initial velocity
theta =deg * pi/180;      % convert angle of trajectory to degrees
v0x = v0*cos(theta);      % calculate initial velocity in the x plane
v0y = v0*sin(theta);      % calculate initial velocity in the y plane
x = v0x*t;                % calculate x position with respect to time
y = v0y*t - g*t.^2;       % calculate y position with respect to time
plot(x,y);                % plot the result
grid on;                  % turn on the grid so that you can see the zero crossing
```

FIGURE D3.1
MathScript to simulate the motion of a projectile

D3.2 The script shown in Figure D3.2 simulates an AM modulated signal. Verify that the script functions properly using the MathScript Interactive Window. You will need to type the script into the Script Editor. Once you have verified that the script runs properly, open a new VI and using a MathScript Node, develop a VI simulation of the AM modulated signal.

```
% D3.2 -- Amplitude Modulation (AM) Simulator
t_max = 1;                        % final simulation time
dt = 0.001;                       % length of simulation vector
t = 0:dt:t_max;                   % create the time vector for the simulation
signal_freq = 20;                 % signal frequency
carrier_freq = 100;               % carrier frequency
signal = sin(signal_freq*2*pi*t); % create signal wave (typically music or speech)
carrier = sin(carrier_freq*2*pi*t); % create carrier wave
mod = carrier.*(1+signal);        % AM modulate the signal with the carrier
subplot(2,1,1)                    % define a single window with two plots
plot(t,mod);                      % plot the time domain modulated signal
real_freq = abs(fft(mod));        % calculate the frequency domain
subplot(2,1,2);                   % define a single window with two plots
plot(real_freq);                  % plot the frequency domain modulated signal
```

FIGURE D3.2
AM modulated signal simulation

Create front panel knob controls for the signal_freq and carrier_freq variables labeled **Signal Frequency** and **Carrier Frequency**, respectively. Use separate Waveform Graphs for both the mod and modified real_freq result. When the VI is ready, set signal_freq = 20 and carrier_freq = 100 and run the VI. What happens to the signals on either side of the carrier signal (the center frequency) as you move the signal_freq control on the plot of real_freq? What changes when you move the carrier_freq control?

CHAPTER 4

Editing and Debugging Virtual Instruments

Like text-based computer programs, virtual instruments are dynamic. VIs change as their applications evolve (usually increasing in complexity). For instance, a VI that initially only performs addition may at a later time be updated to add a multiplication capability. You need debugging and editing tools to verify and test VI coding changes. Since programming in G is graphical in nature, editing and debugging are also graphical, with options available in pull-down and shortcut menus and various palettes. The graphical nature of G programming is in contrast to the command-line programming with MathScript (discussed in Chapter 3) that is text-based and inherently sequential. In this chapter, we discuss how to create, select, delete, move, and arrange objects on the front panel and block diagrams. The important topic of selecting and deleting wires and locating and eliminating bad wires is also presented. Debugging subjects covered include execution highlighting (you can watch the code run!), single-stepping through code, and inserting probes to view data as the VI executes.

GOALS

1.   Learn to access and practice with VI editing tools.
2.   Learn to access and practice with VI debugging tools.

4.1   EDITING TECHNIQUES

4.1.1   Creating Controls and Indicators on the Block Diagram

As discussed in previous chapters, when building a VI you can create controls and indicators on the front panel and know that their terminals will automatically appear on the block diagram. Switching to the block diagram, you can begin wiring the terminals to functions (such as addition or multiplication functions), subVIs, or other objects. In this section we present an alternative method to create and wire controls and indicators in *one* step on the block diagram. For the following discussions, you should open a new VI and follow along by repeating the steps as presented.

1. Open a new VI and switch to the block diagram.
2. Place the Square Root function located in the **Programming≫Numeric** subpalette of the **Functions** palette on the block diagram, as shown in Figure 4.1.

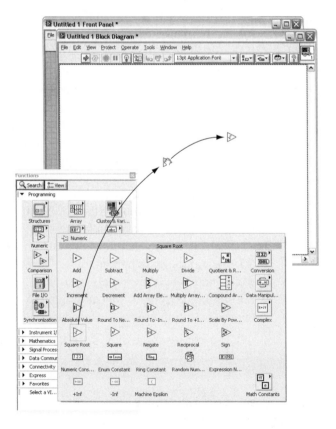

FIGURE 4.1
Add the Square Root function to the block diagram

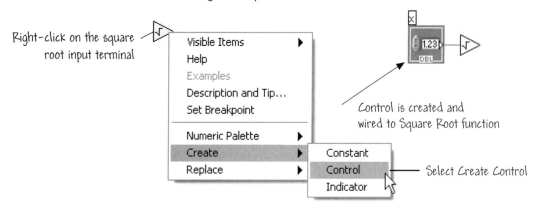

Right-click on the square
root input terminal

Control is created and
wired to Square Root function

Select Create Control

FIGURE 4.2
Front panel control created for the Square Root function

3. Now we want to add (and wire) a control terminal to the Square Root function. As shown in Figure 4.2, you right-click on the left side of the square root function and select **Control** from the **Create** menu. The result is that a control terminal is created and automatically wired to the Square Root function. Switch to the front panel and notice that a numeric control has appeared!

*To change a control to an indicator (or vice versa), right-click on the terminal (on the block diagram) or on the object (on the front panel) and select **Change to Indicator** (or **Change to Control**).*

4. In fact, you can wire indicators and constants by right-clicking on the node terminal and choosing the desired selection. For example, right-clicking on the right side of the Square Root function and selecting **Indicator** from the **Create** menu creates and automatically wires a front panel indicator, as illustrated in Figure 4.3. In many situations, you may wish to create and automatically wire a constant to a terminal, and you accomplish this by right-clicking and choosing **Constant** from the **Create** menu.

*To change a control or indicator (for example, from a knob to a dial), right-click on the object on the front panel and select **Replace**. The **Controls** palette will appear, and you can navigate to the desired new object.*

5. The resulting control and indicator for the Square Root function are shown in Figure 4.4.

4.1.2 Selecting Objects

The **Positioning** tool selects objects in the front panel and block diagram windows. In addition to selecting objects, you use the **Positioning** tool to move and resize

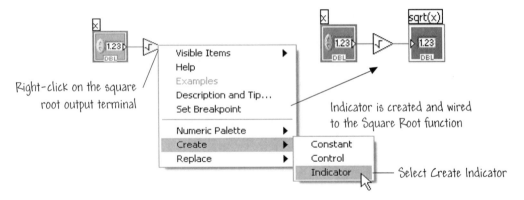

FIGURE 4.3
Front panel indicator created for the Square Root function

FIGURE 4.4
Front panel and block diagram for the Square Root function

objects (more on these topics in the next several sections). To select an object, click the left mouse button while the **Positioning** tool is over the object. When the object is selected, a surrounding dashed outline appears, as shown in Figure 4.5. To select more than one object, shift-click (that is, hold down <Shift> and simultaneously click) on each additional object you want to select. You also can select multiple objects by clicking in a nearby open area and dragging the cursor until all the desired objects lie within the selection rectangle that appears (see Figure 4.6).

Use the
Positioning
tool

A moving dashed
outline appears

Select the
numeric control

FIGURE 4.5
Selecting an object using the **Positioning** tool

Place cursor here
and click-and-drag
dashed box
to surround the
desired object

Objects to be selected

Cursor

Selected objects

FIGURE 4.6
Selecting a group of objects

Sometimes after selecting several objects, you may want to deselect just one of the objects (while leaving the others selected). This is accomplished by shift-clicking on the object you want to deselect—the other objects will remain selected, as desired.

4.1.3   Moving Objects

You can move an object by clicking on it with the **Positioning** tool and dragging it to a desired location. The objects of Figure 4.6 are selected and moved as shown in Figure 4.7. Selected objects can also be moved using the up/down and right/left arrow keys. Pressing the arrow key once moves the object one pixel; holding down the arrow key repeats the action. Holding down the <Shift> key while using the arrow keys moves the object several pixels at a time. In this manner, you can move and locate your objects very precisely.

FIGURE 4.7
Selecting and moving a group of objects

    The direction of movement of an object can be restricted to being either horizontal or vertical by holding down the <Shift> key when you move the object. The direction you initially move decides whether the object is limited to horizontal or vertical motion.

If you change your mind about moving an object while you are in the midst of dragging it to another location, continue to drag until the cursor is outside all open windows and the dashed line surrounding the selected object disappears—then release the mouse button. This will cancel the move operation and the object will not move. Alternatively, if the object is dragged and dropped to an undesirable location, you can select **Undo Move** from the **Edit** menu to undo the move operation.

4.1.4   Deleting and Duplicating Objects

You can delete objects by selecting the object(s) and choosing **Delete** from the **Edit** menu or pressing <Backspace> or <Delete> (Windows) or <delete> (Macintosh). Most objects can be deleted; however you cannot delete certain components of a control or indicator, such as the label or digital display. You must hide these components by right-clicking and deselecting **Visible Items≫ Label** or **Visible Items≫Digital Display** from the shortcut menu.

Most objects can be duplicated, and there are three ways to duplicate an object—by copying and pasting, by cloning, and by dragging and dropping. In all three cases, a complete new copy of the object is created, including, for example, the terminal belonging to a front panel control and the control itself.

You can copy text and pictures from other applications and paste them into LabVIEW.

To clone an object, click the **Positioning** tool over the object while pressing <Ctrl> for Windows (or <option> on the Mac) and drag the object to its new location. After you drag the selection to a new location and release the mouse button, a copy of the object appears in the new location, and the original object remains in the original location. When you clone or copy objects, the copies are labeled by the same name as the original with an incrementing number at the end of each label (copy2, copy3, and so forth).

You also can duplicate objects using Edit≫Copy and then Edit≫Paste from the **Edit** menu. First, select the desired object using the **Positioning** tool and choose Edit≫Copy. Then click at the location where you want the duplicate object to appear and choose Edit≫Paste from the **Edit** menu. You can use this process to copy and paste objects within a VI or even copy and paste objects between VIs.

The third way to duplicate objects is to use the drag-and-drop capability. In this way, you can copy objects, pictures, or text between VIs and from other applications. To drag and drop an object, select the object, picture, or text file with the **Positioning** tool and drag it to the front panel or block diagram of the target VI. You can also drag VIs from the file system (in Windows and on the Mac) to the active block diagram to create subVIs (we will discuss subVIs in Chapter 5).

4.1.5   Resizing Objects

You can easily resize most objects. **Resizing handles** appear when you move the **Positioning** tool over a resizable object, as illustrated in Figure 4.8. On rectangular objects the resizing handles appear at the corners of the object; resizing circles appear on circular objects. Passing the **Positioning** tool over a resizing handle transforms the tool into a resizing cursor. To enlarge or reduce the size of an object, place the **Positioning** tool over the resizing handle and click and drag the resizing cursor until the object is the desired size. When you release the mouse button, the object reappears at its new size. The resizing process is illustrated in Figure 4.8.

FIGURE 4.8
Resizing rectangular and circular objects

To cancel a resizing operation, continue dragging the frame corner outside the active window until the dotted frame disappears—then release the mouse button and the object will maintain its original size. Alternately, if the object has already been resized and you want to undo the resizing, you can use the **Undo Resize** command found in the **Edit** pull-down menu.

Some objects only change size horizontally or vertically, or keep the same proportions when you resize them (e.g., a knob). In these cases, the resizing cursor appears the same, but the dotted resize outline moves in only one direction. To restrict the resizing of any object in the vertical or horizontal direction (or to maintain the current proportions of the object) hold down the <shift> key as you click and drag the object.

You may want to resize multiple objects to have similar dimensions, either horizontally, vertically, or both. To do this, use the **Resize Objects** menu on the LabVIEW toolbar as will be described in Section 4.1.10.

4.1.6   Labeling Objects

Labels are blocks of text that annotate components of front panels and block diagrams. There are two kinds of labels—free labels and owned labels. Owned labels belong to and move with a particular object and describe that object only. You can hide these labels but you cannot copy or delete them independently of their owners. Free labels are not attached to any object, and you can create, move, or dispose of them independently. Use them to annotate your front panels and block diagrams. Free labels are one way to provide accessible documentation for your VIs.

You use the **Labeling** tool to create free labels or to edit either type of label. To create a free label, double-click anywhere in an open area and type the desired text in the bordered box that appears. An example of creating a free label is shown in Figure 4.9.

When finished entering the text, click on the **Enter Text** button on the toolbar, which appears on the toolbar to remind you to end your text entry. You can also end your text entry by pressing the <Enter> key on the numeric keypad (if you have a numeric keypad). Remember that if you do not type any text in the label, the label disappears as soon as you click somewhere else.

When you add a control or an indicator to the front panel, an owned label automatically appears. The label box is ready for you to enter the desired text.

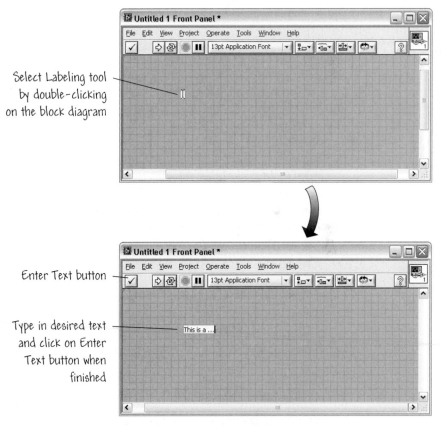

FIGURE 4.9
Creating a free label

If you do not enter text immediately, the default label remains. To create an owned label for an existing object, right-click on the object and select **Visible Items**≫**Label** from the shortcut menu (see Figure 1.13). You can then enter your text in the bordered box that appears. If you do not enter the text immediately, the label disappears.

If you place a label or any other object over (or partially covering) a control or indicator, it slows down screen updates and could make the control or indicator flicker. To avoid this problem, do not overlap a front panel object with a label or other objects.

You can copy the text of a label by double-clicking on the text with the **Labeling** tool or dragging the **Labeling** tool across the text to highlight the desired text. When the desired text is selected, choose **Edit**≫**Copy** to copy the text onto the clipboard. You can then highlight the text of a second label and use **Edit**≫**Paste** to replace the highlighted text in the second label with the text from the clipboard. To create a new label with the text from the clipboard, double-click on the screen where you want the new label positioned and then select **Edit**≫**Paste**.

The resizing technique described in the previous section also works for labels. You can resize labels as you do other objects by using the resizing cursor. Labels normally autosize; that is, the label box automatically resizes to contain the text you enter. If for some reason you do not want the labels to automatically resize to the entered text, right-click on the label and select **Size to Text** to toggle autosizing off.

The text in a label remains on one line unless you enter a carriage return to resize the label box. By default, the <Enter> (or <return>) key is set to add a new line. This can be changed in Tools≫Options≫Environment so that the text input is terminated with the <Enter> (or <return>) key.

4.1.7   Changing Font, Style, and Size of Text

Using the **Text Settings** in the toolbar, you can change the font, style, size, and alignment of any text displayed in a label or on the display of controls and indicators. Certain controls and indicators display text in multiple locations—for example, on graphs (a type of indicator), the graph axes scale markers are made up of many numbers, one for each axes tick. You have the flexibility to modify each text display independently.

Text settings were discussed in Chapter 1 and are shown in Figure 1.7. Notice the word **Application** showing in the **Text Settings** pull-down menu, which also contains the **System**, **Dialog**, and **Current** options. The last option in the list—

the **Current Font**—refers to the last font style selected. The predefined fonts are used for specific portions of the various interfaces:

- The **Application** font is the default font. It is used for the **Controls** and **Functions** palettes.

- The **System** font is the font used for menus.

- The **Dialog** font is the font used for text in dialog boxes.

These fonts are predefined so that they map "best" when porting your VIs to other platforms.

Text Settings has size, style, justify, and color options. Selections made from any of these submenus (that is, size, style, etc.) apply to all selected objects. For example, if you select a new font while you have a knob selected, the labels, scales, and digital displays all change to the new font. Figure 4.10 illustrates the situation of changing the style of all the text associated with a knob from plain text to bold text.

If you select any objects or text and make a selection from the **Text Settings** pull-down menu, the changes apply to everything selected. The process of selecting just the text of an owned label of a knob is illustrated in Figure 4.11. Once the desired text is selected (in this case, click on the label once to select it), you can make any changes you wish by selecting the proper pull-down submenu from **Text Settings**. If no text is selected, the font changes apply to the default font, so that labels created from that point on will reflect the new default font, while not affecting the font of existing labels.

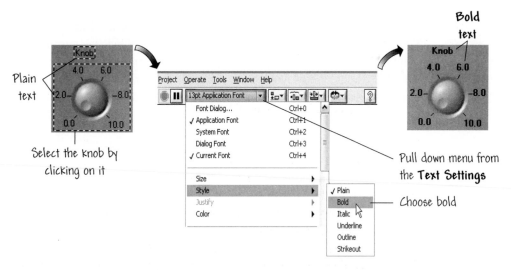

FIGURE 4.10
Changing the font style on a knob control

Highlight the text display by clicking on it and use the options from **Text Settings**

FIGURE 4.11
Selecting text for modification of style, font, size, and color

When working with objects that have multiple pieces of text (e.g. slides and knobs), remember that text settings selections affect the objects or text currently selected. For example, if you select the entire knob while selecting bold text, the scale, digital display, and label all change to a bold font, as shown in Figure 4.10. As shown in Figure 4.12(a), when you select the knob label, followed by selecting bold text from the **Style** submenu of the **Text Settings** pull-down menu, only the knob label changes to bold. Similarly, when you select text from a scale marker while choosing bold text, all the markers change to bold, as shown in Figure 4.12(b).

If you select **Font Dialog** in the **Text Settings** while a front panel is active, the dialog box shown in Figure 4.13 appears. If a block diagram is active instead, the **Diagram default** option at the bottom of the dialog box is checked. With either the **Panel default** or **Diagram default** checkboxes selected, the other selections made in this dialog box will be used with new labels on the front panel or block diagram. In other words, if you click the **Panel default** and/or **Diagram default** checkboxes, the selected font becomes the current font for the front panel, the block diagram, or both. The current font is used on new labels. The checkboxes allow you to set different fonts for the front panel and block diagram. For example, you could have a small font on the block diagram and a large one on the front panel.

4.1.8   Selecting and Deleting Wires

A single horizontal or vertical piece of wire is known as a **wire segment**. The point where three or four wire segments join is called a **junction**. A **wire branch** contains all the wire segments from one junction to another, from a terminal to the next junction, or from one terminal to another if there are no junctions in between. You select a wire segment by clicking on it with the **Positioning** tool. Clicking twice selects a branch, and clicking three times selects the entire wire. See Figure 4.14 for an example of selecting a branch, segment, or an entire wire.

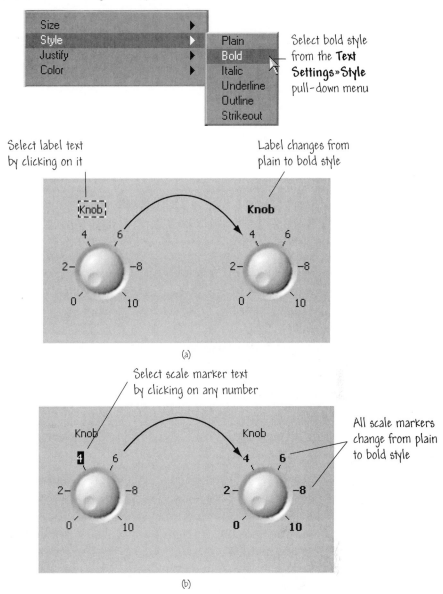

FIGURE 4.12
Changing text attributes on a knob

4.1.9   Wire Stretching and Broken Wires

Wired objects can be moved individually or in groups by dragging the selected objects to a new location with the **Positioning** tool. The wires connecting the objects will stretch automatically. If you want to move objects from one diagram to another, the connecting wires will not move with the selected objects, unless you select the wires as well. After selecting the desired objects and the connecting wires, you can cut and paste the objects as a unit to another VI.

FIGURE 4.13
Using the **Font Dialog** box to change font, size, alignment, color, and style

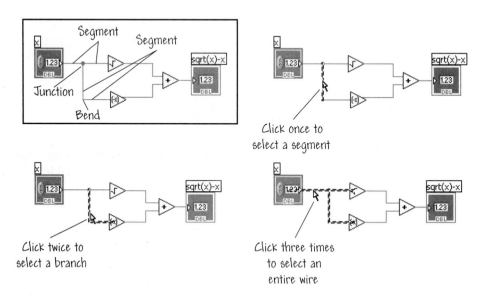

FIGURE 4.14
Selecting a segment, branch, or an entire wire

Select wired object and drag it
to the new desired location

FIGURE 4.15
Moving wired objects

An example of stretching a wire is shown in Figure 4.15. In the illustration, the object (in this case, a numeric indicator) is selected with the **Positioning** tool and stretched to the desired new location.

Wire stretching occasionally leaves behind loose ends. Loose ends are wire branches that do not connect to a terminal. Your VI will not execute until you remove all loose ends. An example of a loose end is shown in Figure 4.16. Loose ends can be easily removed using the Edit≫Remove Broken Wires command.

Loose end

FIGURE 4.16
Loose ends

When you make a wiring mistake, a broken wire (indicated by a dashed line with a red X) appears. Figure 4.17 shows a dashed line that represents a broken wire. It is inevitable that broken wires will occur in the course of programming

Broken wire (black dashed line
with red x)   Good wire (solid orange line)

(a) Broken wiring   (b) Good wiring

FIGURE 4.17
Locating broken wires

in G. One common mistake is to attempt to connect two control terminals together or to connect a control terminal to an indicator terminal when the data types do not match (for example, connecting a Boolean to a numeric). If you have a broken wire, you can remove it by selecting it with the **Positioning** tool and eliminate it by pressing the <Delete> key. If you want to remove all broken wires on a block diagram at one time, choose **Remove Broken Wires** from the **Edit** menu.

There are many different conditions leading to the occurrence of broken wires. Some examples are:

- **Wire type, dimension, unit, or element conflicts**—A wire type conflict occurs when you wire two objects of different data types together, such as a numeric and a Boolean, as shown in Figure 4.18(a).

- **Multiple wire sources**—You can wire a control to multiple output destinations (or indicators), but you cannot wire multiple data sources to a single destination. In the example shown in Figure 4.18(b), we have attempted to wire two data sources (that is, the random number and the constant ln 2) to one indicator. This produces a broken wire and must be fixed by disconnecting the random number (represented by the dice) or disconnecting the ln 2 constant. Another example of a common multiple sources error occurs during front panel construction when you inadvertently place a numeric control on the front panel when you meant to place a numeric indicator. If you do this and then try to wire an output value to the terminal of the front panel control, you will get a multiple sources error. To fix this error, just right-click on the terminal and select **Change To Indicator**.

- **No wire source**—An example of a wire with no source is shown in Figure 4.18(c). This problem is addressed by providing a control. Another example of a situation leading to a no-source error is if you attempt to wire two

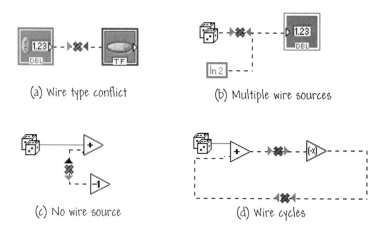

(a) Wire type conflict

(b) Multiple wire sources

(c) No wire source

(d) Wire cycles

FIGURE 4.18
Typical wiring errors leading to broken wire indications

front panel indicators together when one should have been a control. To fix this error, just right-click on a terminal and select **Change to Control**.

- **Wire cycles**—In most cases, wires must not form closed loops of icons or structures, as shown in Figure 4.18(d). These closed loops are known as cycles. LabVIEW will not execute cycles because each node waits on the other to supply it data before it executes (remember data flow!). Shift register feedback nodes (discussed in Chapter 6) provide the proper mechanism to feed back data in a repetitive calculation or loop. LabVIEW has a Simulation Module (available as an add-on feature) that permits users to integrate continuous-time and discrete-time simulations including general feedback loops within their LabVIEW block diagrams.

Sometimes you have a faulty wiring connection that is not visible because the broken wire segment is very small or is hidden behind an object. If the **Broken Run** button appears in the toolbar, but you cannot see any problems in the block diagram, select Edit≫Remove Broken Wires—this will remove all broken wires in case there are hidden, broken wire segments. If the **Run** button returns, you have corrected the problem. If the wiring errors have not all been corrected after selecting Edit≫Remove Broken Wires, click on the **Broken Run** button to see a list of errors. Click on one of the errors listed in the **Error List** dialog box, and you will automatically be taken to the location of the erroneous wire in the block diagram. You can then inspect the wire, detect the error, and fix the wiring problem.

LabVIEW automatically finds a wire route around existing objects on the block diagram when you are wiring components. The natural inclination of the automatic routing is to decrease the number of bends in the wire. When possible, automatically routed wires from control terminals exit the right side of the terminal, and automatically routed wires to indicator terminals enter the left side of the terminal.

If you find that you have a messy wiring situation on your block diagram you can easily fix it. You can right-click any wire and select **Clean Up Wire** from the shortcut menu to automatically route an existing wire. This may help in debugging your VI since it will be easier to see the wire routes on the block diagram.

If for some reason you want to temporarily disable automatic wire routing and route a wire manually, first use the **Wiring** tool to click a terminal and release the mouse. Then press the <A> key to temporarily disable automatic wire routing for the current wire. Click another terminal to complete the wiring. Once the wiring is complete, the automatic wire routing is resumed. You also can temporarily disable automatic routing after you click to start wiring by holding down the mouse button while you wire to another terminal and then releasing the mouse button. After you release the mouse button, automatic wire routing resumes.

4.1.10   Aligning, Distributing, and Resizing Objects

To align a group of objects, first select the desired objects. Then choose the axis along which you want to align them from the **Align Objects** pull-down menu in the toolbar. You can align objects along the vertical axis using left, center, or right edge. You also can align objects along a horizontal axis using top, center, or bottom edge. Open a new VI and place three objects (such as, three numeric controls) on the front panel and experiment with different aligning options. Figure 4.19 illustrates the process of aligning three objects by their left edges.

In a similar fashion, you can distribute a group of objects by selecting the objects and then choosing the axis along which you want to distribute the selected objects from the **Distribute Objects** pull-down menu in the toolbar (see Figure 1.9). In addition to distributing selected objects by making their edges or centers equidistant, four menu items at the right side of the ring let you distribute

FIGURE 4.19
Aligning objects

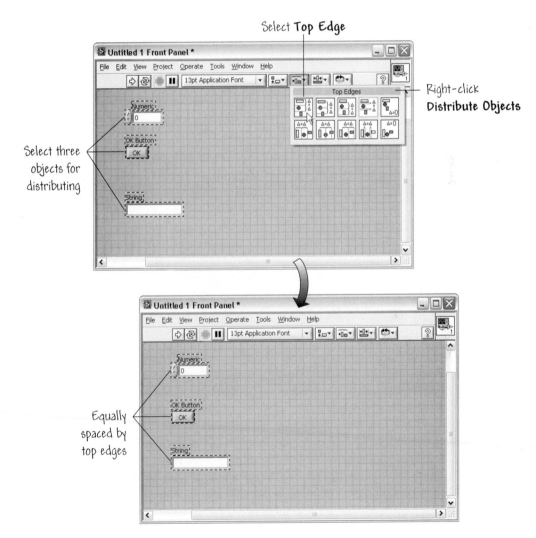

the gaps between the objects, horizontally or vertically. For example, the three objects shown in Figure 4.20 are distributed at different distances from each other. If we want them to be equally spaced by their top edges, we can use the **Distribute Objects** pull-down menu, as shown in Figure 4.20, to rearrange the elements at equal spacing.

You can use the **Resize Objects** pull-down menu to resize multiple front panel objects to the same size. For example, in Figure 4.21, three objects of different widths are selected. Then using the **Resize Objects** pull-down menu, the **Maximum Width** icon is selected resulting in the three objects taking the width of the String, which originally had the largest width.

FIGURE 4.20
Distributing objects

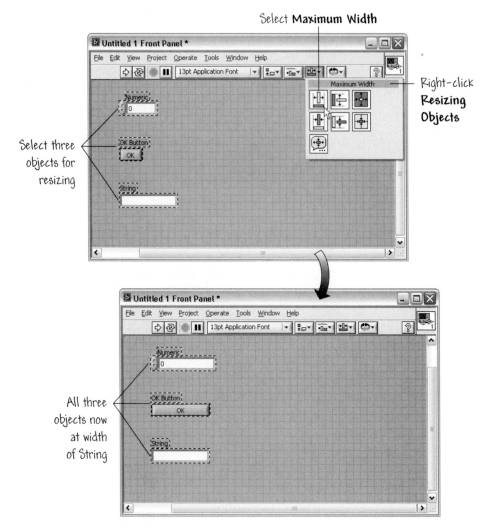

FIGURE 4.21
Resizing objects

4.1.11 Coloring Objects

You can customize the color of many LabVIEW objects. However, the coloring of objects that convey information via their color is unalterable. For example, block diagram terminals of front panel objects and wires use color codes for the type and representation of data they carry, so you cannot change their color.

To change the color of an object (or the background of a window), right-click on the object of interest with the **Coloring** tool from the **Tools** palette, as seen in Figure 4.22. Choose the desired color from the selection palette that appears. If you keep the mouse button pressed as you move through the color selection

FIGURE 4.22
Customizing the color of objects

palette, the object or background you are coloring redraws with the color the cursor currently is touching. This gives you a "real-time" preview of the object in the new color. If you release the mouse button on a color, the selected object retains the chosen color. To cancel the coloring operation, move the cursor out of the color selection palette before releasing the mouse button, or use the Edit≫Undo Color Change command after the undesired color change has been made.

If you select the box with a T in it, the object is rendered transparent. One use of the T (transparent) option is to hide the box around labels. Another interesting use is to create numeric controls without the standard three-dimensional border. Transparency affects the appearance but not the function of the object!

Some objects have both a foreground and a background color that you can set separately. The foreground color of a classic knob, for example, is the main dial area, and the background color is the base color of the raised edge. On the **Coloring** tool, you select the foreground or background, as illustrated in Figure 4.22. Clicking the upper left box on the **Coloring** tool and then clicking on an object on the front panel will color the object's foreground. Similarly, clicking the lower right box on the **Coloring** tool and then clicking on an object on the front panel will color the object's background.

Click on the color palette to select desired color

Range of 0-255

FIGURE 4.23
The **Color** dialog box

Selecting the button on the lower right-hand side in the palette (see Figure 4.23) accesses a dialog box with which you can customize the colors. Each of the three color components, red, green, and blue, describes eight bits of a 24-bit color (in the lower right-hand side of the **Color** dialog box). Therefore, each component has a range of 0 to 255. The last color you select from the palette becomes the current color. Clicking on an object with the **Color** tool sets that object to the current color.

Using the **Color Copy** tool, you can copy the color of one object and transfer it to a second object without using the **Color** palette. To accomplish this, click with the **Color Copy** tool on the object whose color you want to transfer to another object. Then, select the **Color** tool and click on another object to change its color.

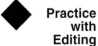

Practice with Editing

In this exercise you will edit and modify an existing VI to look like the panel shown in Figure 4.24. After editing the VI, you will wire the objects in the block diagram and run the program.

1. Open the Editing VI by choosing **Open** from the **File** menu (or using the **Open** button on the Getting Started screen) and searching in the Chapter 4 folder in the Learning directory. The front panel of the Editing VI contains a number of objects depicted in Figure 4.25. The objective of this exercise is to make the front panel of the Editing VI look like the one shown in Figure 4.24. If you have a color monitor, you can see the final VI in color by opening the Editing Done VI located in the Chapter 4 folder in the Learning directory.

2. Add an owned label to the numeric control using the **Positioning** tool by right-clicking on the numeric control and selecting Visible Items≫Label from the menu. Type the text Temperature offset inside the bordered box and click the mouse outside the label or click the **Enter Text** button on the left-hand side of the toolbar.

FIGURE 4.24
The front panel for the Editing VI

FIGURE 4.25
The Editing VI with various objects: knob, waveform chart, thermometer, and numeric control

3. Reposition the waveform graph and numeric control. Click on the waveform graph and drag it to the lower center of the front panel (see Figure 4.24 for the approximate location). Then click on the numeric control and drag it to the upper right of the front panel (see Figure 4.24 for approximate location).

   Notice that as you move the numeric control, the owned label moves with the control. If the control is currently selected, click on a blank space on the front panel to deselect the control and then click on the label and drag it to another location. Notice that in this case the control does not follow the move. You can position an owned label anywhere relative to the control. If you move the owned label to an undesirable location, you can move it back by selecting Edit≫Undo Move to place the label back above the numeric control.

4. Reposition the stop button, the knob, and the thermometer to the approximate locations shown in Figure 4.24.

5. Add labels to the knob and to the thermometer. To accomplish this task, right-click on the object and choose Visible Items≫Label. When the label box appears, type in the desired text. In this case, we want to label the knob Temperature variation and the thermometer Average Temperature.

6. Move the numeric indicator associated with the thermometer to a location near the bottom of the thermometer. Then reposition the thermometer label so that it is better centered above the thermometer (see Figure 4.24).

7. In this next step, we will reformat the waveform graph.

   (a) To remove the waveform graph legend, right-click on the waveform graph and select **Properties**. On the **Appearance** menu, deselect the check mark next to **Show Plot Legend**.

   (b) To remove the waveform graph palette, deselect **Show Graph Palette** from the **Appearance** menu.

   (c) Remove the waveform x-axis scale by selecting the **Scales** menu. On the **Scales** menu, be sure to select the x-axis from the drop-down box. Deselect the **Show Scale** option.

   (d) Add a label to the wavefrom graph—label the object Temperature data history. Select the owned label and move it to the bottom of the waveform graph, as illustrated in Figure 4.24. Note that this can also be done on the **Properties** screen.

8. Align the stop button and the thermometer.

   (a) Select both the stop button and the thermometer using the **Positioning** tool. Pick a point somewhere to the upper left of the stop button and drag the dashed box down until it encloses both the stop button and the thermometer. Upon release of the mouse button, both objects will be surrounded by moving dashed lines.

(b) Click on **Align Objects** and choose **Left Edges**. The stop button and the thermometer will then align to the left edges. If the objects appear not to move at all, this indicates that they were essentially aligned to the left edges already.

9. Align the stop button, the knob, and the numeric control horizontally by choosing the **Vertical Centers** axis from the **Align Objects** pull-down menu in the toolbar. Remember to select all three objects beforehand!

10. Space the stop button, the knob, and the numeric control evenly by choosing the **Horizontal Centers** axis from the **Distribute Objects** pull-down menu in the toolbar. Again, remember to select all three objects beforehand!

11. Change the color of the stop button.

    (a) Using the **Coloring** tool, right-click on the stop button to display the color palette.

    (b) Choose a color from the palette. The object will assume the last color you selected. In the Editing Done VI you will see that a dark blue color was selected—you can choose a color that you like.

12. Change the color of the knob. Using the **Coloring** tool, right-click on the knob to display the color palette and then choose the desired color from the color palette. In the Editing Done VI you will find that a red/pink color was selected—you can choose a color that suits you.

13. Change the color of the waveform graph. In the Editing Done VI you will see that a gold color was selected.

14. Change the color of the thermometer bulb. Using the **Coloring** tool, right-click on the thermometer bulb (at the bottom of the thermometer) to display the color palette and then choose the desired color. In the Editing Done VI you will find that a green color was selected.

15. Change the font style of the owned labels. Use the cursor to select each of the labels and then select Style≫Bold from **Text Settings**. See if you can do this for all three of the owned labels at once: thermometer, knob, and numeric control.

At this point, the front panel of the Editing VI should look very similar to the front panel shown in Figure 4.24.

The block diagram for the Editing VI is shown in Figure 4.26 after the editing of the front panel is finished—but before wiring the block diagram.

We can now wire together the various objects to obtain a working VI. Notice that several additional objects are on the block diagram: a uniform white noise subVI, a mean subVI that computes the mean (or average) of a signal, a While Loop, and an Add function. In later chapters, you will learn to use While Loops

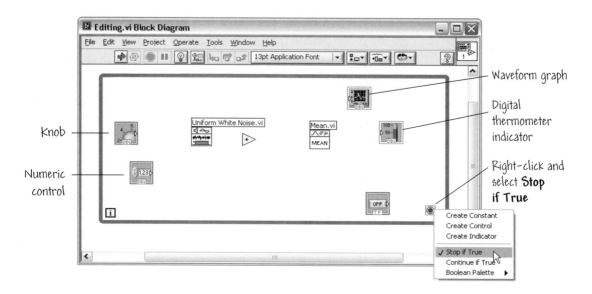

FIGURE 4.26
The block diagram for the Editing VI—before wiring

and learn how to use preexisting subVIs packaged with LabVIEW. For instance, the Uniform White Noise.vi is located on the **Functions** palette under the **Signal Processing»Signal Generation** subpalette.

Go ahead and wire the Editing VI block diagram so that it looks like the block diagram shown in Figure 4.27. If you run into difficulties, you can open and examine the block diagram for the Editing Done VI (which can be found in Chapter 4 of the Learning directory).

FIGURE 4.27
The block diagram for the Editing Done VI—after wiring

A few wiring tips:

1. To wire the objects together with the **Wiring** tool, click and release on the source terminal and drag the **Wiring** tool to the destination terminal. When the destination terminal is blinking, click and release the left mouse button.

2. To identify terminals on the Add function, right-click on the icon and select Visible Items≫Terminal to see the connector. When the wiring is finished, right-click again on the function and choose Visible Items≫Terminal to show the icon once again.

3. To bend the wires as you connect two objects, click with the left mouse button on the bend location with the **Wiring** tool.

After you have finished wiring the objects together, switch to the front panel by selecting **Show Front Panel** from the **Window** menu. Use the **Operating** tool to change the value of the front panel controls. Run the VI by clicking on the **Run** button on the toolbar.

The Average Temperature indicator should be approximately the value that you select as the Temperature offset. The amount of variation in the temperature history, as shown on the Temperature data history waveform graph, should be about the same as the setting on the Temperature variation knob.

When you are finished editing and experimenting with your VI, save it by selecting **Save As** from the **File** menu. Remember to save all your work in the Users Stuff folder in the Learning directory. Close the VI by selecting **Close** from the **File** menu. ◆

4.2 DEBUGGING TECHNIQUES

In this section we discuss LabVIEW's basic debugging elements, which provide an effective programming debugging environment. Most features commonly associated with good interactive debugging environments are provided—and in keeping with the spirit of graphical programming, the debugging features are accessible graphically. Execution highlighting, single-stepping, breakpoints, and probes help debug your VIs easily by tracing the flow of data through the VI. You can actually watch your program code as it executes!

4.2.1 Finding Errors

When your VI cannot compile or run due to a programming error, a **Broken Run** button appears on the toolbar. Programming errors typically appear during VI development and editing stages and remain until you properly wire all the objects in the block diagram. You can list all your program errors by clicking on the **Broken Run** button. An information box called **Error List** appears listing all the errors. This box is shown in Figure 4.28 for the Editing Done VI with a broken wire.

FIGURE 4.28
Locating program errors

Warnings make you aware of potential problems when you run a VI, but they do not inhibit program execution. If you want to be notified of any warnings, click the **Show Warnings** checkbox in the **Error List** dialog box. A warning button then appears on the toolbar whenever a warning condition occurs.

If your program has any errors that prevent proper execution, you can search for the source of a specific error by selecting the error in the **Error List** (by clicking on it) and then clicking on **Show Error** (lower right-hand corner of the **Error List** dialog box). This process will highlight the object on the block diagram that reported the error, as illustrated in Figure 4.28. Double-clicking on an error in the error list will also highlight the object reporting the error.

Some of the most common reasons for a VI being broken during editing are:

1. A function terminal requiring an input is unwired. For example, an error will be reported if you do not wire all inputs to arithmetic functions.

2. The block diagram contains a broken wire because of a mismatch of data types or a loose, unconnected end.

3. A subVI is broken.

4.2.2 Highlight Execution

You can animate the VI block diagram execution by clicking on the **Highlight Execution** button located in the block diagram toolbar, shown in Figure 4.29.

FIGURE 4.29
The **Highlight Execution** and step buttons located on the toolbar

For debugging purposes, it is helpful to see an animation of the VI execution in the block diagram, as illustrated in Figure 4.30. When you click on the **Highlight Execution** button, it changes to a bright light to indicate that the data flow will be animated for your visual observation when the program executes. Click on the **Highlight Execution** button at any time to return to normal running mode.

Execution highlighting is commonly used with single-step mode (more on single stepping in the next section) to trace the data flow in a block diagram in an effort to gain an understanding of how data flows through the block diagram. Keep in mind that when you utilize the highlight execution debugging feature, it greatly reduces the performance of your VI—the execution time increases significantly! The data flow animation shows the movement of data from one node to another using "bubbles" to indicate data motion along the wires. This process is illustrated in Figure 4.30. Additionally, in single-step mode, the next node to be executed blinks until you click on the next step button.

4.2.3 Single-Stepping Through a VI and Its SubVIs

For debugging purposes, you may want to execute a block diagram node by node. This is known as single-stepping. To run a VI in single-step mode, press any of the debugging step buttons on the toolbar to proceed to the next step. The step buttons are shown on the toolbar in Figure 4.29. The step button you press determines where the next step executes. You click on either the **Step Into** or **Step Over** button to execute the current node and proceed to the next node. If

FIGURE 4.30
Using the highlight execution mode to watch the data flow through a VI

the node is a structure (such as a While Loop) or a subVI, you can select the **Step Over** button to execute the node, but not single-step through the node. For example, if the node is a subVI and you click on the **Step Over** button, you execute the subVI and proceed to the next node, but cannot see how the subVI node executed internally. To single-step through the subVI you would select the **Step Into** button.

Click on the **Step Out** button to finish execution of the block diagram nodes or finish up the single-step debugging session. When you press any of the step buttons, the **Pause** button is pressed as well. You can return to normal execution at any time by releasing the **Pause** button.

If you place your cursor over any of the step buttons, a tip strip will appear with a description of what the next step will be if you press that button.

You might want to use highlight execution as you single-step through a VI, so that you can follow data as it flows through the nodes. In single-step mode and highlight execution mode, when a subVI executes, the subVI appears on the

FIGURE 4.31
Single stepping into a subVI with execution highlighting selected

main VI diagram with either a green or red arrow in its icon, as illustrated in Figure 4.31. The diagram window of the subVI is displayed on top of the main VI diagram. You then single-step through the subVI or let it complete executing.

You can save a VI without single-stepping or highlight execution capabilities. This compiling method typically reduces memory requirements and increases performance by 1–2%. To do this, right-click in the icon pane (upper right corner of the front panel window) and select **VI Properties**. As in Figure 4.32, from the **Execution** menu, deselect the **Allow Debugging** option to hide the **Highlight Execution** and **Single Step** buttons.

4.2.4   Breakpoints and Probes

You may want to halt execution (set **breakpoints**) at certain locations of your VI (for example, subVIs, nodes, or tool, wires). Using the **Breakpoint** click on any item in the block diagram where you want to set or clear a breakpoint. Breakpoints are depicted as red frames for nodes and red dots for wires.

Deselect **Allow debugging** to increase
program speed and reduce memory usage

FIGURE 4.32
Turning off the debugging options using the **VI Setup** menu

You use the **Probe** tool to view data as it flows through a block diagram wire. To place a **probe** in the block diagram, click on any wire in the block diagram where you want to place the probe. Probes are depicted as numbered yellow boxes with associated interactive windows in which the values passing through the wire at runtime are displayed. You can place probes all around the block diagram to assist in the debugging process.

Conditional probes allow you to set conditional breakpoints for common cases of each data type. For example, creating a conditional probe on a numeric allows you to set a pause condition using Equal to, Greater than, or Less than. This gives you flexibility in your debugging.

A new feature in LabVIEW 8 is the ability to retain wire values. This feature can be enabled and disabled with the **Retain Wire Values** button on the toolbar of the block diagram, as shown in Figure 4.29. Once enabled, if you create a probe on the block diagram *after* executing your VI, the probe displays the data that flowed through the wire during the last VI execution. The VI must run sucessfully at least once before you can collect data from any wires using probes. Retaining wire values is a handy feature when you need to debug a VI with a complex block diagram.

Practicing with Debugging

A nonexecutable VI—named **Debug.vi**—has been developed for you to debug and to fix. You will get to practice using the single-step and execution highlighting modes to step through the VI and inserting breakpoints and probes to regulate the program execution and to view the data values.

1. Open the Debug VI in **Chapter 4** of the **Learning** directory by choosing **Open** from the **File** menu. Notice the **Broken Run** button in the toolbar indicating the VI is not executable.

2. Switch to the block diagram by choosing **Show Block Diagram** from the **Window** pull-down menu. You should see the block diagram shown in Figure 4.33.

   (a) The Random Number (0-1) function (represented by the two die) can be found in the **Functions≫Programming Numeric** subpalette and returns a random number between zero and one.

   (b) The Add function (which can also be found in the **Functions≫ Programming Numeric** subpalette) adds the random number to a bias number represented by the variable **Temperature bias**, which accepts input by the user on the front panel.

   (c) The subVI Temp (**Digital Thermometer.vi** found in the **Activity** folder in the **LabVIEW 8** directory.) simulates acquiring temperature data one value at a time from a data acquisition board. It uses the Demo Voltage Read subVI to acquire the simulated data. As you single step through the code, you will see the program execution inside both subVIs.

   (d) The second Add function adds the simulated temperature data point to the random number plus the bias number to create a variable named **Actual Temperature**.

FIGURE 4.33
Block diagram showing the debugging exercise

3. Investigate the source of the programming error by clicking on the **Broken Run** button to obtain a list of the programming errors. You should find one error—**Add: contains unwired or bad terminal**.

4. Highlight the error in the **Error List** and then click on **Show Error** to locate the source of the error within the block diagram. You should find that the second Add function (the one on the right side) is shown to be the source of the programming error.

5. Fix the error by properly wiring the two Add functions together. Once this step is successfully completed, the **Broken Run** button should reappear as the **Run** button.

6. Select **Highlight Execution** and then run the VI by pressing the **Run Continuously** button. You can watch the data flow through the code! Returning to the front panel, you will see the waveform chart update as the new temperature is calculated and plotted. Notice that the simulation runs very slowly in highlight execution mode. Click the **Highlight Execution** button off to see everything move much faster!

7. Terminate the program execution by clicking on the **Abort Execution** button.

8. Enable single-stepping by clicking on one of the step buttons. You can enable **Highlight Execution** if you want to see the data values as they are computed at each node.

9. Use the step buttons to single-step through the program as it executes. Remember that you can let the cursor idle over the step buttons and a tip strip will appear with a description of what the next step will be if you press that button. Press the **Pause** button at any time to return to normal execution mode.

10. Enable the probe by right-clicking on the wire connecting the two Add functions and selecting **Probe**. Continue to single-step through the VI and watch how the probe displays the data carried by the wire. Alternatively, you could have used the **Probe** tool to place the probe on the wire. Try placing a probe on the block diagram (say between the Temp subVI and the Add function) using the **Probe** tool.

11. Place a few more probes around the VI and repeat the single-stepping process to see how the probes display the data.

12. When you are finished experimenting with the probes, close all open probe windows.

13. Set a breakpoint by selecting the **Breakpoint** tool from the **Tools** palette and clicking on the wire between the two Add functions. You will notice that a red ball appears on the wire indicating that a breakpoint has been set at that location.

14. Run the VI by clicking on the **Run Continuously** button. The VI will pause each time at the breakpoint. To continue VI execution, click on the **Pause** button. It helps to use highlight execution when experimenting with breakpoints; otherwise the program executes too fast to easily observe the data flow.

15. Terminate the program execution by pressing the **Abort Execution** button. Remove the breakpoint by clicking on the breakpoint (that is, on the red ball) with the **Breakpoint** tool.

16. Save the working VI by selecting **Save** from the **File** menu. Remember to save the working VI in the folder Users Stuff within the Learning directory. Close the VI and all open windows by selecting **Close** from the **File** menu.
◆

4.2.5   Navigation Window

When working with complicated VIs, the block diagrams and front panels may be too unwieldy to easily navigate. LabVIEW provides the feature of the **Navigation** window as a tool to display an overview of the active front panel in edit mode or the active block diagram. To activate the **Navigation** window, select View≫Navigation Window. Once the **Navigation** window is open, you click an area of the image in the **Navigation** window to display that area in the front panel or block diagram window.

Consider the Train Wheel PtByPt VI depicted in Figure 4.34 with the **Navigation** window in view. The portion of the block diagram in the center of the **Navigation** window has been selected and the associated block diagram code appears in the block diagram window. You can also click and drag the image in the **Navigation** window to scroll through the front panel or block diagram. As illustrated in Figure 4.34, portions of the front panel or block diagram that are not visible appear dimmed in the **Navigation** window.

4.3   A FEW SHORTCUTS

Frequently used menu options have equivalent command key shortcuts. For example, to save a VI you can choose **Save** from the **File** menu, or press the control key equivalent <Ctrl-S> (Windows) or <command-S> (Macintosh). Some of the main key equivalents are shown in Table 4.1.

Shortcut access to the **Tools** palette on the front panel and block diagram is given by:

■ **Windows**—Press <shift> and the right mouse button

■ **Macintosh**—Press <command-shift> and the mouse button

FIGURE 4.34
Using the Navigation window

TABLE 4.1   Frequently Used Command Key Shortcuts

Windows	Macintosh	Function
<Ctrl-S>	<command-S>	Save a VI
<Ctrl-R>	<command-R>	Run a VI
<Ctrl-E>	<command-E>	Toggle between the front panel and the block diagram
<Ctrl-H>	<command-shift-H>	Toggle the **Context Help** window on and off
<Ctrl-B>	<command-B>	Remove all bad wires
<Ctrl-W>	<command-W>	Close the active window
<Ctrl-F>	<command-F>	Find objects and VIs

BUILDING BLOCK

4.4   BUILDING BLOCKS: PULSE WIDTH MODULATION

The Rising Edge and Falling Edge VIs that you constructed in Chapter 2 will be improved with the editing techniques learned in this chapter. Navigate to the Users Stuff in the Learning directory and open both VIs.

Working versions of Falling Edge.vi and Rising Edge.vi can be found in the Building Blocks folder of the Learning directory. These working VIs are provided in case you did not save them in the Users Stuff folder after working in Chapter 2.

It is important that you create documentation of VIs you make so that information is available which may not be readily apparent to users interacting with your VI for the first time. You may wish to use free text labels to clarify features of either the front panel or the block diagram. Write a VI description within the documentation under **VI Properties** for the Falling Edge VI. Use the following description:

This VI is used in Pulse Width Modulation to determine when a high output signal should switch to low, based on the current input values of Duty Cycle, Period, and Clock.

Controls such as knobs and slides are useful in situations where there is a preferred or ideal input range for any given variable, since each control limits the user to its specified input range. Replace the numeric control for Duty Cycle with a vertical pointer slide in the Falling Edge VI. Open the **Properties** window for the slide pointer control to edit the data range and scale range. We want to set the limits so that the user can only enter values between 0 and 1 on the control. Click the **Data Range** tab and set the increment value to 0.01. Then select **Coerce to nearest** so that any input value specified past the hundredths place will be rounded to the nearest hundredth. On the **Appearance** tab, make the Digital Display visible. This gives the user the flexibility of adjusting the duty cycle by sliding the pointer or by entering a new value in the digital display.

Recall that in Chapter 2 Building Blocks we assumed the user will always provide valid inputs for Duty Cycle and Period. Now we can guarantee that the user does not input invalid values for Period and Clock by enforcing the limits on the inputs. To adjust the data range for Period and Clock so that no negative value can be input to Clock and the minimum value that can be input to Period is 100, right-click on the respective input to open the **Properties** window and click the **Data Range** tab. In the case of the Clock, deselect **Use Default Range** and set the minimum value to zero. In the case of the Period, follow the same steps, except set the minimum value to 100.

After modifying the front panel, you may find that you need to rearrange some objects for ease of use of the inputs. Use the **Align Objects** pull-down menu to organize objects on both the front panel and the block diagram to improve the appearance of your VIs. A smartly organized VI makes debugging more straightforward and enhances the ease with which others can understand and utilize the program.

Your VI should now resemble the one shown in Figure 4.35. Run the VI with the new Duty Cycle control. Note that as you slide the control, the values increment by hundredths, and if you enter an invalid input on the numeric display, it is rounded to a valid value.

When you are done experimenting with your new VI, save it as Falling Edge Edited.vi in the Users Stuff folder in the Learning directory. You will use this VI as a building block in later chapters—so make sure to save your work!

A working version of Falling Edge Edited.vi can be found in the Building Blocks folder of the Learning directory.

FIGURE 4.35
Editing the Falling Edge VI

4.5   RELAXED READING: LABVIEW HELPS FABRICATION PROCESS OF NEXT GENERATION MICROPROCESSORS

Moore's Law speculates that the numbers of transistors on a microchip will double every 18 months. Extreme Ultraviolet Lithography (EUVL) is leading the race to push Moore's Law to its current physical limitations. By using smaller wavelengths of light in the chip lithography process, smaller traces can be etched into a semiconductor wafer. This will allow chips to be produced with nanoscale circuitry, meaning microprocessors with a billion transistors. Each transistor on the chip can be as small as 40 atoms in width. The EUVL process uses a mirror coated with a non-reflective negative image, or mask, of a microprocessor circuit. An EUV light source, which produces light peaking at 134 Angstroms, is used to illuminate the mask. A magnified image of the circuit on the mask is projected onto the semiconductor wafer, burning away conductive material and leaving a microprocessor circuit.

The mask mirror is a critical component in the EUVL process. Lawrence Livermore National Labs (LLNL) is one of the only producers of EUVL mask mirrors. To test these mirrors, LLNL called upon EUV Technology (www.euvl.com), a small Silicon Valley-based company, to build the first reflectometer that was small enough to be placed in a clean room and able to meet LLNL's testing

FIGURE 4.36
The EUVL laser plasma reflectometer (LPR) measures the reflectivity of the mask mirrors vs. wavelength

requirements. EUV Technologies partnered with Cal-Bay Systems to develop the reflectometer's control and data acquisition system. This small reflectometer, known as a laser plasma reflectometer (LPR), uses a laser-induced gold plasma as its EUV light source. It is one of the keys to bringing EUVL out of the lab and into the semiconductor fab for large-scale chip production. Previously, EUVL mask mirrors were tested for reflectivity using a particle accelerator ring known as a synchrotron as an EUV light source. A synchrotron costs hundreds of millions of dollars and is the size of a football field. Test time had to be scheduled months in advance, took a setup time of one hour, and had a turnaround time of eight hours. EUV Technology's LPR, shown in Figure 4.36, occupies only 1.5 square meters, easily fits into a clean room, has a test setup time of ten seconds, and can test the mirrors in five minutes. The EUV Technologies LPR received the prestigious 2005 R&D 100 award.

The research and development of any complicated first-generation system requires a flexible set of tools. The LPR's laser must be fired in synchrony with data acquisition and motion control for precise measurements to be made. Two stepper motors and five servomotors are used to position various filters, the EUV light monochromator, a gold target, and the mask mirror. Digital timing signals must be generated to trigger cascading events such as flashing the laser lamp, firing the laser, and acquiring analog data. Motorized micrometers and New Focus, Inc. Picomotors are used to align optical components with nanometric precision to focus the laser and EUV light. Analog voltages from optical power sensors must be read to acquire the reflectivity data. System interlocks had to be checked to ensure that the system was in a safe operating state. Finally, the system had to communicate with other production tools over an Ethernet network.

National Instruments data acquisition and motion control hardware was selected because of their advanced triggering features and the ability to synchronize events via the Real Time Systems Integration (RTSI) bus. The PCI-6052E multifunction data acquisition card was used to create several precisely spaced timing signals that cause the laser to fire and analog signal acquisition to occur. Seven axes of coordinated motion with the two PCI-7344 FlexMotion control cards were achieved. The entire control system runs on a rack-mounted industrial PC and is housed in a 19 in industrial-rack enclosure.

LabVIEW was chosen as the software development environment because of its ease of use, flexibility, network-ready functionalities, and tight integration of motion control and data acquisition. The LabVIEW data acquisition and motion control example programs were used to quickly prototype a solution to the complex hardware timing issues. After the prototyping software was written, we developed an operator interface, which allowed scanning optical power levels vs. any motion axis (see Figure 4.37). This provided characterization information that allowed the LPR to be fine tuned. This was an invaluable feature. Having all of the motion control functions available in LabVIEW via FlexMotion allowed

FIGURE 4.37
The EUV Technology Laser Plasma Reflectometer LabVIEW User Interface

the code to be modular and flexible. Finally a network interface was developed to allow remote operation of the device by other tools in the fabrication system.

As the LPR's physical and functional design evolved, the LabVIEW graphical code was quickly modified to meet the new needs of the system; motion control axes were added and removed, scanning methodology changed, subsystems and interlocks evolved, and calibration and characterization methods were refined. Having all of the hardware routines developed in LabVIEW allowed all the flexibility we required. LabVIEW and National Instruments hardware proved to be an excellent platform for developing a successful system in a highly experimental R&D environment.

For further information contact:

Jim Kring
jim@jimkring.com

4.6   SUMMARY

The subjects of editing and debugging VIs were the main subjects of this chapter. Just as you would edit a C program or debug a Fortran subroutine, you must know how to edit and debug VIs. We discussed how to create, select, delete, move, and arrange objects on the front panel and block diagram. The important topic of selecting and deleting wires and locating and eliminating bad wires was also discussed. The program debugging topics covered included execution highlighting, how to step into, through, and out of the code, and how to use probes to view data as it flows through the code.

KEY TERMS

Breakpoint: A pause in execution used for debugging. You set a breakpoint by clicking a VI, node, or wire with the **Breakpoint** tool.

Breakpoint tool: Tool used to set a breakpoint on a node or wire.

Broken VI: A VI that cannot compile and run.

Coloring tool: Tool used to set foreground and background colors.

Execution highlighting: Debugging feature that animates the VI execution to illustrate the data flow within the VI.

Label: Text object used to name or describe other objects or regions on the front panel or block diagram.

Labeling tool: Tool used to create labels and enter text.

Navigation window: Displays an overview of the active front panel in edit mode or the active block diagram.

Operating tool: Tool used to enter data into controls and operate them—resembles a pointing finger.

Positioning tool: Tool used to move, select, and resize objects.

Probe: Debugging feature for checking intermediate values in a VI during execution.

Probe tool: Tool used to create probes on wires.

Resizing handles: Angled handles on the corners of objects that indicate resizing points.

EXERCISES

E4.1   Construct a VI to accept five numeric inputs, add them up and display the result on a gauge, and light up a round light if the sum of the input numbers is less than 8.0. The light should light up in green, and the gauge dial should be yellow. The VI in Figure E4.1 can be used as a guide.

FIGURE E4.1
A VI to add five numeric inputs and light up a round LED if the sum is less than 8.0

E4.2 Open Change to Indicator.vi found in the Chapter 4 subfolder in the Exercises&Problems folder within the Learning directory. Click on the **Broken Run** button. Select the error, and click on the **Show Error** button. This wire connects more than one data source. The wire between the output of the Add function and the control result is highlighted and has an "x" over it. Both the output and the control are "sources of data." You can only connect a data source to a data display. Right-click on the control result and select **Change to Indicator**. The error disappears and you can run the VI. Input $x = 2$ and $y = 3$ and run the VI. Verify that the output result is 5.

E4.3   Open the virtual instrument **Multiple Controls-1 Terminal.vi**. You can find this VI in the **Chapter 4** subfolder in the **Exercises&Problems** folder within the **Learning** directory. Click on the **Broken Run** button, select the error **This Wire Connects to More than One Data Source**, and click on the **Show Error** button. The wires from the two controls are highlighted. If you look closely, you will see that both wires are going to the same terminal. That means two controls are defining the value of one input value. Delete the wire from the y control. Wire the y control to the other input terminal on the Add function. Both errors disappear and you can run the VI. Input $x = 1$ and $y = 4$ and run the VI. Verify that the output $x + y = 5$.

E4.4   Open the virtual instrument **Crazy Wires.vi**. You can find this VI in the **Chapter 4** subfolder in the **Exercises&Problems** folder within the **Learning** directory. Go to the block diagram and see the wires running everywhere. There is an easy way to fix this! Right-click on the wire running from **Knob** to the top terminal of the Comparison function. Select **Clean Up Wire** and watch the wire get straightened instantly. Do the same for all the wires on the block diagram.

E4.5   Open the Oil Change VI in the **Chapter 4** subfolder in the **Exercises&Problems** folder within the **Learning** directory and shown in Figure E4.5. The VI assists in determining whether or not a car requires an oil change based on the mileage and elapsed time since the last oil change. The VI assumes that a car needs an oil change after either 6 months or 5,000 miles, whichever comes first. Can you see the problem with the VI by inspection of the block diagram? To confirm and fix the problem, perform the following steps:

FIGURE E4.5
The Oil Change VI is broken

(a) Click the broken arrow to see what errors exist in the VI.

(b) Fix the error and run the VI for various values of months and miles.

E4.6 Open Broken VI found in the **Chapter 4** subfolder in the **Exercises&Problems** folder within the **Learning** directory. Click the broken run arrow to see a list of errors in the program. Determine why this VI fails to compile and run. Modify the VI as necessary so that the errors are repaired.

PROBLEMS

P4.1 Construct a VI that generates two random numbers (between 0 and 1) and displays both random numbers on meters. Label the meters **Random number 1** and **Random number 2**, respectively. Make the face of one meter blue and the face of the other meter red. When the value of the random number on the red meter is greater than the random number on the meter with the blue face, have a square LED show green; otherwise have the LED show black. Run the VI several times and observe the results. On the block diagram select **Highlight Execution** and watch the data flow through the code.

P4.2 In this problem you will construct a stop light display. Create a dial control that goes from 0 to 2, with three LED displays: one green, one yellow, and one red. Have the VI turn on the green LED when the dial is on 0, the yellow LED when the dial is on 1, and the red LED when the dial is on 2.

P4.3 Open the block diagram of the Waveform Min Max example VI located in the library **Operations.llb.** in the **Waveform** subfolder within the **examples** folder. See Figure P4.3 for reference.

FIGURE P4.3
The block diagram of the Waveform Min Max example.vi showing the Simple Sine Waveform subVI terminals.

(a) On the block diagram, delete the numeric control labeled frequency by clicking on it and then selecting **Delete** on the **Edit** pull-down menu. Clean up the broken wire left behind by right-clicking it and selecting **Delete Wire Branch** from the shortcut menu or by selecting **Remove Broken Wires** from the **Edit** menu.

(b) The control can be replaced two different ways. On the block diagram, right-click the node labeled n waves on the Simple Sine Waveform subVI and select Create≫Control from the shortcut menu. Alternatively, in the **Controls** palette on the front panel, go to the **Numeric** subpalette. Select a numeric control and place it on the front panel. Then wire the control to the n waves input to the Simple Sine Waveform VI on the block diagram. Using one of these methods, replace the numeric control and label it frequency.

(c) Move to the front panel and right-click the frequency control. Select Replace≫ Modern≫Knob to change the control to a knob. Try moving different components around on the front panel to make space for it, using the **Align Objects** tools if necessary.

(d) Run the VI and experiment with different input values.

If you want to save your changes, be sure not to replace the original VI with your edited one. Save the VI you create with a new name in the Users Stuff *folder in the* Learning *directory.*

P4.4 In this problem, we want to open and run an existing VI. In LabVIEW, go to Help≫Find Examples and click the **Search** tab and type "probes." Select "probes" to display the example VIs that include "probes" in the title. Find the VI entitled Using Supplied Probes.vi and open it.

(a) Change to the block diagram. Follow the instructions written on the diagram for each probe.

(b) What probes are available on the block diagram?

(c) On the Conditional Double Array probe, what happens if you set the number of elements equal to 250?

(d) To continue after a break, press the **Pause** button on the menu bar.

(e) Continue probing each red box. What information is supplied to you in the Dynamic Data Type probe?

(f) What happens if you implement a conditional breakpoint of zero on the Error Probe?

(g) With the VI running, stop the VI from the front panel. What is the state of the Boolean probe?

P4.5 Open the Temperature Converter VI in the Chapter 4 subfolder in the Exercises&Problems folder within the Learning directory. The block diagram is

FIGURE P4.5
The block diagram for the Temperature Converter VI

shown in Figure P4.5. This VI takes an input in degrees Celsius and converts the input to Kelvins and degrees Fahrenheit, displaying the converted temperatures on numeric indicators. Explain how this VI could have been created differently to make its purpose more readily evident to other programmers modifying the code and also more user friendly. Then fix the VI to reflect these changes.

Save the VI as Temperature Converter.vi in the Users Stuff folder in the Learning directory.

P4.6 Develop a VI that converts an input value in degrees to radians with four digits of precision. Refer to Figure P4.6 for help on changing the output display to four digits.

FIGURE P4.6
Converting degrees to radians with four digits of output precision

DESIGN PROBLEMS

D4.1   In this problem you will use the editing techniques from this chapter to improve the appearance of both the front panel and block diagram of **Average.vi**. This will aid in making the purpose and functionality of the Average VI more readily apparent to users who are unfamiliar with the VI.

Open the Average VI in the **Chapter 4** subfolder in the **Exercises&Problems** folder within the **Learning** directory. The block diagram is shown in Figure D4.1.

Perform the following editing steps:

(a) Resize the knobs on the front panel so that they are all the same size. You may wish to use some of the tools on the **Resize Objects** menu.

(b) Evenly align and distribute the objects on both the front panel and the block diagram. It is conventional to place inputs (controls) on the left side and outputs (indicators) on the right side. As you rearrange objects on the block diagram, you will most likely need to clean up the wires.

FIGURE D4.1
The unedited Average VI

(c) Hide the labels on the knob controls and change the label on the indicator to read **average**.

(d) Place a free label on the front panel which briefly explains the purpose of the VI.

D4.2   Create a front panel that has 8 LED indicators and a vertical slider control that is an 8 bit unsigned integer. Display a digital indicator for the slider, and make sure that the LEDs are evenly spaced and aligned at the bottom. The problem is to turn the 8 LEDs into a binary (base 2) representation for the number in the slider. For example, if the slider is set to the number 10 (which in base 2 is $00001010 = 1 * (2^3) + 1 * (2^1)$), the LEDs 1 and 3 should be on. To test your solution, check the number 131. LEDs 0, 1 and 7 should be on since 131 is 10000011 in base 2.

D4.3   Construct a VI that generates a random number between -10 and 10. The VI should display the random number on an indicator on the front panel and then light an LED corresponding to the range in which the number falls. Use one LED for negative values of the random number, one LED for values in the range of 0 to 4, and one LED for values greater or equal to 5. Change the properties of the LEDs so that they are bright yellow when turned on and dark yellow when off. Use descriptive labels for the indicators and use the tools on the **Align Objects** and **Distribute Objects** menus to space the objects evenly on the block diagram and front panel.

C H A P T E R 5

SubVIs

In this chapter we learn to build subVIs. SubVIs are VIs used by other VIs—like subroutines. One of the keys to constructing successful VIs is understanding how to build and use subVIs. The hierarchical design of LabVIEW applications (that is, of virtual instruments) depends on the use of subVIs. We will discuss two basic ways to create and use a subVI: creating subVIs from VIs and creating subVIs from selections. The **Icon Editor** is presented as a way to personalize subVI icons so that information about the function of the subVI is apparent from visual inspection. The editor has a tool set similar to that of most common paint programs. The **VI Hierarchy** window will be introduced as a helpful tool for managing the hierarchical nature of your programs.

GOALS

1. Learn to build and use subVIs.
2. Understand the hierarchical nature of VIs.
3. Practice with the **Icon Editor** and with assigning terminals.

5.1   WHAT IS A SUBVI?

It is important to understand and appreciate the hierarchical nature of virtual instruments. **SubVIs** are critical components of a hierarchical and modular VI that is easy to debug and maintain. A subVI is a stand-alone VI that is called by other VIs—that is, a subVI is used in the block diagram of a **top-level VI**. SubVIs are analogous to subroutines in text-based programming languages like C or Fortran, and the subVI node is analogous to a subroutine call statement. The **pseudocode** and block diagram shown in Figure 5.1 demonstrate the analogy between subVIs and subroutines. There is no limit to the number of subVIs you can use in a calling VI. Using subVIs is an efficient programming technique in that it allows you to reuse the same code in different situations. The hierarchical nature of programming in G follows from the fact that you can call a subVI from within a subVI.

You can create subVIs from VIs, or create them from selections (selecting existing components of a VI and placing them in a subVI). When creating a new subVI from an existing VI, you begin by defining the inputs and outputs of the

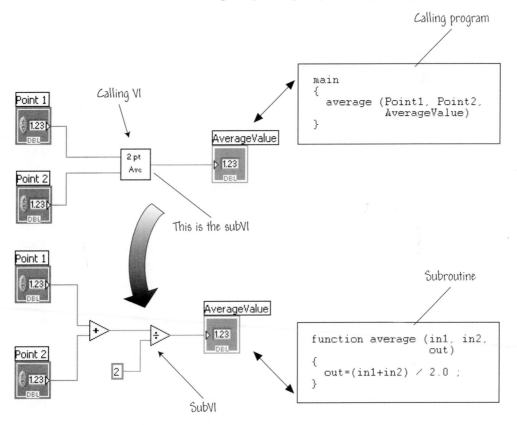

FIGURE 5.1
Analogy between subVIs and subroutines

subVI, and then you "wire" the subVI connector properly. This allows calling VIs to send data to the subVI and to receive data from the subVI. On the other hand, if an existing complex block diagram has a large number of icons, you may choose to group related functions and icons into a lower-level VI (that is, into a subVI) to maintain the overall simplicity of the block diagram. This is what is meant by using a modular approach in your program development.

5.2   REVIEW OF THE BASICS

Before moving on to the subject of subVIs, we present a brief review of some of the basics presented in the first three chapters as part of an exercise in constructing a VI. Your VI will ultimately be used as a subVI later in the chapter.

Three parameters associated with a circle of given radius are circumference, diameter, and area. For a radius of length r we have the following relationships:

$$C = 2\pi r$$
$$A = \pi r^2$$
$$D = 2r$$

where C denotes circumference, A denotes area, D denotes diameter, and π is a constant. In this chapter we will build a subVI that computes and outputs C, A, and D, given the circle radius r as input. To begin the design process, we will construct a VI that performs the desired calculations. Then, later in the chapter we will modify the VI so that it can be used as a subVI in other applications.

The task is to construct a VI to utilize the formulas given above to compute C, A, and D, given r. The following list serves as a step-by-step guide. Feel free to deviate from the provided list according to your own style. As you proceed through this book you will find that the step-by-step lists will slowly disappear, and you will have to construct your VIs on your own (with help, of course!).

1. Open a new VI. Save the untitled VI as **Circle Measurement.vi**. In this case you are saving the VI before any programming has actually occurred. This is a matter of personal choice, but saving a VI frequently during development may save you a great deal of rework if your system freezes up or crashes.

2. Add a numeric control to the front panel and label it **r**. This will be where the radius is input.

3. Place a numeric indicator on the front panel and label it **C**. This will be where the circumference is displayed.

4. Make two duplicates (or clones) of the numeric indicator. Using the **Positioning** tool, select the numeric control while pressing the <Ctrl> (Windows) or <option> (Macintosh) key and drag the numeric control to a new location. When you release the mouse button, a copy labeled **C 2** appears in the new

location. Repeat this process so that three numeric indicators are visible on the front panel.

5. Label the new numeric indicators **A** and **D**. These indicators will display area and diameter.

6. Place the numeric indicators in a column on the right side of the front panel. Using the **Align Objects** tool, align the three numeric indicators by their left edges.

7. Switch to the block diagram. Arrange the control and indicator terminals so that the three indicator terminals appear aligned in a vertical column on the right side and the control appears on the left side.

8. Use functions and constants from the **Programming≫Numeric** palette to program the formulas for circumference, area, and diameter. You can use the VI shown in Figure 5.2 as a guide.

9. Once the block diagram has been wired, switch back to the front panel and input a value into the radius control.

10. Run the program with **Run Continuously** and select **Highlight Execution** to watch the data flow through the program. Change the value of the radius

FIGURE 5.2
The Circle Measurements VI

FIGURE 5.3
The Circle Measurements icon terminal

and see how the outputs change. When you are finished experimenting, you can stop the program execution.

11. Practice single-stepping through the program using the step buttons on the block diagram.

12. Once the VI execution is complete, switch back to the front panel and right-click on the icon in the upper right corner. Select **Show Connector** from the menu. Observe the connector that appears. When we are done, it will indicate one input terminal and three output terminals, as illustrated in Figure 5.3. We will be learning how to edit the icon and connector in the next section.

13. Save the VI by selecting **Save** from the **File** menu.

This exercise provided the opportunity to practice placing objects on the front panel, wiring them on the block diagram, using editing techniques to arrange the objects for easy visual inspection of the code, and using debugging tools to verify proper operation of the VI. The Circle Measurements VI can be used as a subVI (requiring some effort to wire the VI connector properly), and that is the subject of the following sections.

    A working version of the Circle Measurements VI can be found in the **Chapter 5** *folder of the* **Learning** *directory. It is called* **Circle Measurements.vi.**

5.3   EDITING THE ICON AND CONNECTOR

A subVI is represented by an icon in the block diagram of a calling VI. The subVI also must have a connector with terminals properly wired to pass data to and from the top-level VI. In this section we discuss icons and connectors.

FIGURE 5.4
Activating the **Icon Editor**

5.3.1 Icons

Every VI has a default icon displayed in the upper right corner of the front panel
and block diagram windows. The default icon is a picture of the LabVIEW logo
and a number indicating how many new VIs you have opened since launching
LabVIEW. You use the **Icon Editor** to customize the icon. To activate the editor,
right-click on the default icon in the front panel and select **Edit Icon** as illustrated
in Figure 5.4. You can also display the **Icon Editor** by double-clicking on the
default icon.

 *You can only open the **Icon Editor** when the VI is not in **Run** mode. If
necessary, you can change the mode of the VI by selecting **Change to Edit Mode**
from the **Operate** menu.*

The **Icon Editor** dialog box is shown in Figure 5.5. You use the tools on the
palette at the left of the dialog box to create the icon design in the pixel editing
area. An image of the icon (in actual size) appears in a vertical column just right
of center in the dialog box and to the right of the pixel editing area. Depending
on the type of monitor you are using, you can design and save a separate icon
for B&W, 16-color, and 256-color modes. The editor defaults to **256 Colors**, but
you can click on one of the other color options to switch modes. To copy the
icon from 256 colors to 16 colors, select the 16 Colors box and then click on the
Copy from: 256 Colors button located at the upper right side of the **Icon Editor**
dialog box. Similarly, you can use the appropriate **Copy from** menu items to
copy icons from 16 Colors to 256 Colors, 16 Colors to B&W, and so forth.

The tools in the **Icon Editor** palette to the left of the editing area perform
many functions. If you have used any other paint program, you should be familiar
with the tools available in the editor. Table 5.1 describes the various tools. The
options at the right of the editing screen perform the following functions:

Icon Editor
tool palette

Use these buttons to copy icon from B&W
to 16 Colors, 16 Colors to 256 Colors, and so forth

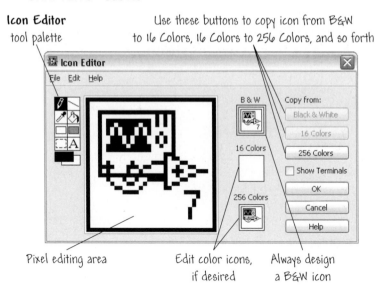

Pixel editing area

Edit color icons,
if desired

Always design
a B&W icon

FIGURE 5.5
The **Icon Editor**

TABLE 5.1   Icon Editor Tools

	Draws and erases pixel by pixel.
	Draws straight lines. Use <Shift> to restrict drawing to horizontal, vertical, and diagonal lines.
	Selects the foreground color from an element in the icon. Use the <Shift> key to select the background color.
	Fills an outlined area with the foreground color.
	Draws a rectangle bordered with the foreground color. Double-click on this tool to frame the icon in the foreground color.
	Draws a rectangle bordered with the foreground color and filled with the background color. Double-click on this tool to frame the icon in the foreground color and fill it with the background color.
	Selects an area of the icon for moving, cloning, deleting, or performing other changes. Double-click on this tool and press <Delete> on the keyboard to delete the entire icon at once.
	Enters text into the icon. Double-click on this tool to select a different font. Small Fonts is commonly used as the font in icons.
	Displays the current foreground and background colors. Click on each to get a palette from which you can choose new colors.

- **Show Terminals**—Click on this option to display the terminal pattern of the connector overlaid on the icon.

- **OK**—Click on this button to save your drawing as the VI icon and return to the front panel window.

- **Cancel**—Click on this button to return to the front panel window without saving any changes.

You can use the **Icon Editor** menu to cut, copy, and paste images from and to the icon. When you paste an image and a portion of the icon is selected, the image is resized to fit into the selection.

As an exercise, open Circle Measurements.vi and edit the icon. Remember that you developed the Circle Measurements.vi at the beginning of this chapter—it should be saved in Users Stuff. In case you did not save it before or cannot find it now, you can open the VI located in the Chapter 5 folder in the Learning directory. An example of an edited icon is shown in Figure 5.6. Can you replicate the new icon? Check out the Circle Measurements Edited.vi to see the edited icon shown in Figure 5.6.

5.3.2   Connectors

The connector is a set of terminals that correspond to the VI controls and indicators. This is where the inputs and outputs of the VI are established so that the VI can be used as a subVI. A connector receives data at its input terminals and passes the data to its output terminals when the VI is finished executing. Each terminal corresponds to a particular control or indicator on the front panel. The connector terminals act like parameters in a subroutine parameter list of a function call.

If you use the front panel controls and indicators to pass data to and from subVIs, these controls or indicators need terminals on the connector pane. In this section, we will discuss how to define connections by choosing the number of

FIGURE 5.6
Editing the Circle Measurements VI icon

terminals you want for the VI and assigning a front panel control or indicator to each of those terminals.

You can view and edit the connector pane from the front panel only.

To define a connector, you select **Show Connector** from the icon pane menu on the front panel window as shown in Figure 5.7. The connector replaces the icon in the upper right corner of the front panel window. By default, LabVIEW displays a terminal pattern based on the number of controls and indicators on your front panel, as shown in Figure 5.7. Control terminals are on the left side of the connector pane, and indicator terminals are on the right. If desired, you can select a different terminal pattern for your VI.

5.3.3   Selecting and Modifying Terminal Patterns

To select a different terminal pattern for your VI, right-click on the connector and choose **Patterns**. This process is illustrated in Figure 5.8. To change the pattern, click on the desired pattern on the palette. If you choose a new pattern, you will lose any assignment of controls and indicators to the terminals on the old connector pane.

FIGURE 5.7
Defining a connector

FIGURE 5.8
Changing your terminal pattern

The maximum number of terminals available for a subVI is 28.

If you want to change the spatial arrangement of the connector terminal patterns, choose one of the following commands from the connector pane menu: **Flip Horizontal**, **Flip Vertical**, or **Rotate 90 Degrees**. If you want to add a terminal to the connector pattern, place the cursor where the terminal is to be added, right-click on the connector pane window, and select **Add Terminal**. If you want to remove an existing terminal from the pattern, right-click on the terminal and select **Remove Terminal**.

Think ahead and plan your connector patterns well. For example, select a connector pane pattern with extra terminals if you think that you will add additional inputs or outputs at a later time. With these extra terminals, you do not have to change the connector pane for your VI if you find you want to add another input or output. This flexibility enables you to make subVI changes with minimal effect on your hierarchical structure. Another useful hint is that if you create a group of subVIs that are often used together, consider assigning the subVIs a consistent connector pane with common inputs. You then can easily remember each input location without using the **Context Help** window—this will save you time. If you create a subVI that produces an output that is used as the input to another subVI, try to align the input and output connections. This technique simplifies your wiring patterns and makes debugging and program maintenance easier.

Place inputs on the left and outputs on the right of the connector when linking controls and indicators—this prevents complex wiring patterns.

5.3.4   Assigning Terminals to Controls and Indicators

Front panel controls and indicators are assigned to the connector terminals using the **Wiring** tool. The following steps are used to associate the connector pane with the front panel controls and indicators.

1. Click on the connector terminal with the **Wiring** tool. The terminal turns black, as illustrated in Figure 5.9(a).

2. Click on the front panel control or indicator that you want to assign to the selected terminal. As shown in Figure 5.9(b), a dotted-line **marquee** frames the selected control.

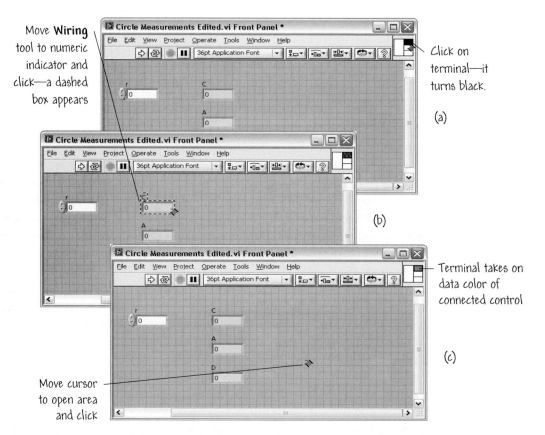

FIGURE 5.9
Assigning connector terminals to controls and indicators

3. Position the cursor in an open area of the front panel and click. The marquee disappears, and the selected terminal takes on the data color of the connected object, indicating that the terminal is assigned. This process is illustrated in Figure 5.9(c).

The connector terminal turns white to indicate that a connection was not made. If this occurs, you need to repeat steps 1 through 3 until the connector terminal takes on the proper data color.

4. Repeat steps 1–3 for each control and indicator you want to connect.

The connector terminal assignment process also works if you select the control or indicator first with the **Wiring** tool and then select the connector terminal. As already discussed, you can choose a pattern with more terminals than you need since unassigned terminals do not affect the operation of the VI. It is also true that you can have more front panel controls or indicators than terminals. Once the terminals have been connected to controls and indicators, you can disconnect them all at one time by selecting **Disconnect All Terminals** in the connector pane shortcut menu. Note that although the **Wiring** tool is used to assign terminals on the connector to front panel controls and indicators, no wires are drawn.

5.4   THE HELP WINDOW

You enable the **Context Help** window by selecting Help≫Show Context Help. When you do this, you find that whenever you move an editing tool across a subVI node, the **Context Help** window displays the subVI icon with wires attached to each terminal. An example is shown in Figure 5.10 in which the cursor was moved over the subVI Temp & Vol.vi and the **Context Help** window appeared and showed that the subVI has two outputs: Temp and Volume.

LabVIEW has a help feature that can keep you from forgetting to wire subVI connections—indications of required, recommended, and optional connections in the connector pane and the same indications in the **Context Help** window. For example, by classifying an input as **Required**, you can automatically detect whether you have wired the input correctly and prevent your VI from running if you have not wired correctly. To view or set connections as **Required, Recommended**, or **Optional**, right-click a terminal in the connector pane on the VI front panel and select **This Connection Is**. A check mark indicates its status, as shown in Figure 5.11. By default, inputs and outputs of VIs you create are set to **Recommended**—if a change is desired, you must change the default to either **Required** or **Optional**.

Move cursor over subVI and the **Context Help** window will appear

Context Help window for Temp & Vol.vi

FIGURE 5.10
The **Context Help** window

Right-click on connector terminal

Select **This Connection Is**

This connector terminal is **Recommended**

FIGURE 5.11
Showing the status of terminal connectors

When you make a connection **Required**, then you cannot run the VI as a subVI unless that connection is wired correctly. In the **Context Help** window, **Required** connections appear in bold text. When you make a connection **Recommended**, then the VI can run even if the connection is not wired, but the error list window will list a warning. In the **Context Help** window, **Recommended** connections appear in plain text. A VI with **Optional** connections can run without the optional terminals being wired. In the **Context Help** window, **Optional** connections are grayed text for detailed diagram help and hidden for simple diagram help.

**Building
a SubVI**

A VI designed to compute the radius, circumference, and area of a circle was presented in Section 5.2. In this exercise you will assign the connector terminals of that VI to the numeric controls and numeric indicator so that the VI can be used as a subVI by other programs. Begin by opening **Circle Measurements.vi** in the **Chapter 5** folder of the **Learning** directory. Recall that the difference between the two VIs Circle Measurements Edited and Circle Measurements is that the edited version has a custom icon rather than the default icon. A subVI with an edited icon (rather than the default icon) enables easier visual determination of the purpose of the subVI because you can inspect the icon as it exists in the code.

When you have completed assigning the connector terminals, the front panel connector pane should resemble the one shown in Figure 5.12. To display the terminal connectors, right-click on the VI icon in the front panel and select **Show Connector**. You should notice that a terminal pattern appropriate for the VI has been automatically selected with one input on the left side of the connector pane and three outputs on the right. The number of terminals selected by default depends on the number of controls and indicators on the front panel.

Using the **Wiring** tool, assign the input terminal to the numeric control r. Similarly, assign the three output terminals to the numeric indicators C, A, and D.

There are two levels of code documentation that you can pursue. It is important to document your code to make it accessible and understandable to other users. You can document a VI by choosing File≫VI Properties≫Docu-

FIGURE 5.12
Assigning terminals for the Circle Measurements VI

Add description in this block area

Click here when satisfied with the description

FIGURE 5.13
Typing the description for the subVI

mentation. A **VI description** box will appear as illustrated in Figure 5.13. You type a description of the VI in the dialog box, and then whenever **Show Context Help** is selected and the cursor is placed over the VI icon, your description will appear in the **Context Help** window.

You can also document the objects on the front panel by right-clicking on an object and choosing **Description and Tip** from the object shortcut menu. Type the object description in the dialog box that appears as in Figure 5.14. You cannot edit the object description in Run Mode—change to Edit Mode.

Continuing with the exercise, you should add a description for each object on the front panel. Some suggested descriptions follow:

- r: Radius of circle to be analyzed.

- C: The circumference of a circle with radius equal to r.

- A: The area of a circle with radius equal to r.

- D: The diameter of a circle with radius equal to r.

When you have finished assigning the connector terminals to the controls and indicators, and you have documented the subVI and each of the controls and indicators, save the subVI as Circle Measurements subVI.vi. Make sure to save your work in the Users Stuff folder.

FIGURE 5.14
Documenting objects on a subVI

*A working version of Circle Measurements subVI can be found in the **Chapter 5** folder in the **Learning** directory. You may want to open the VI, read the documentation, and take a look at the icon and the connector just to verify that your construction of the same subVI was done correctly.* ◆

5.5   USING A VI AS A SUBVI

There are two basic ways to create and use a subVI: creating subVIs from VIs and creating subVIs from selections. In this section we concentrate on the first method, that is, using a VI as a subVI. Any VI that has an icon and a connector can be used as a subVI. In the block diagram, you can select VIs to use as subVIs from **Functions≫Select a VI** palette. Choosing this option produces a file dialog box from which you can select any available VI in the system, as shown in Figure 5.15.

The *LabVIEW Student Edition* comes with many ready-to-use VIs. In Chapter 1 you searched around the LabVIEW examples and found many of the example and demonstration VIs. In the following example you will use one of the preexisting VIs called **Temp & Vol.vi** as a subVI.

FIGURE 5.15
The **Select a VI** palette

Using a VI as a SubVI

1. Open a new front panel.

2. Select a vertical slide switch control from the **Controls≫Modern≫Boolean** palette and label it Temperature Scale. Place free labels on the vertical slide switch to indicate Fahrenheit and Celsius using the **Labeling** tool, as shown in Figure 5.16, and arrange the labels as shown in the figure.

FIGURE 5.16
Calling the subVI Temp & Vol.vi

3. Select a thermometer from the **Controls≫Modern≫Numeric** palette and place it on the front panel. Label the thermometer Temperature.

4. Change the range of the thermometer to accommodate values ranging between 0.0 and 100.0. With the **Operating** tool, double-click on the high limit and change it to 100.0 if necessary.

5. Switch to the block diagram by selecting Window≫Show Block Diagram.

6. Right-click in a free area of the block diagram and choose **Functions≫Select a VI** to access the dialog box. Select Temp & Vol.vi in the Chapter 5 folder of the Learning directory. Click **Open** in the dialog box to place Temp & Vol.vi on the block diagram.

7. Add the other objects to the block diagram using Figure 5.16 as a guide.

   - Place a Subtract function and a Divide function on the block diagram. These are located in the **Functions≫Programming≫Numeric** palette. On the Subtract function, add a numeric constant equal to 32. You can add the constant by right-clicking on an input of the Subtract function and selecting **Create Constant**. Then using the **Labeling** tool change the constant from the default 0.0 to 32.0. Similarly, add a numeric constant of 1.8 on the Divide function. These constants are used to convert from degrees Fahrenheit to Celsius according to the relationship

$$°C = \frac{°F - 32}{1.8}.$$

   - Add the Select function (located on the **Programmiing≫Comparisons** subpalette of the **Functions** palette). The Select function returns the value wired to the True or False input, depending on the Boolean input value. Use **Show Context Help** for more information on how this function works.

8. Wire the diagram objects as shown in Figure 5.16.

9. Switch to the front panel and click the **Run Continuously** button in the toolbar. The thermometer shows the value in degrees Fahrenheit or degrees Celsius, depending on your selection.

10. Switch the scale back and forth to select either Fahrenheit or Celsius.

11. When you are finished experimenting with your VI, save it as Thermometer Scale.vi in the Users Stuff folder.   ◆

A working version of Thermometer Scale.vi exists in the Chapter 5 folder located in the Learning directory. When you open this VI, notice the use of color on the vertical switch (orange denotes Fahrenheit and blue represents Celsius).

In the previous exercise, the stand-alone Temp & Vol.vi was used in the role of a subVI. Suppose that you have a subVI on your block diagram and you want to examine its contents—to view the code. You can easily open a subVI front panel window by double-clicking on the subVI icon. Once the front panel opens, you can then open the subVI block diagram by selecting **Show Block Diagram** from the **Window** menu. At that point, any changes you make to the subVI code alter only the version in memory—until you save the subVI. Also note that, even before saving the subVI, the changes affect all calls to the subVI and not just the node you used to open the VI.

5.6   CREATING A SUBVI FROM A SELECTION

The second way to create a subVI is to select components of the main VI and group them into a subVI. You capture and group related parts of VIs by selecting the desired section of the VI with the **Positioning** tool and then choosing **Create SubVI** from the **Edit** pull-down menu. The selection is automatically converted into a subVI, and a default icon replaces the entire section of code. The controls and indicators for the new subVI are automatically created and wired to the existing wires. Using this method of creating subVIs allows you to modularize your block diagram, thereby creating a hierarchical structure.

Building a SubVI Using the Selection Technique

In this exercise you will modify the Thermometer Scale VI developed in Section 5.5 to create a subVI that converts Fahrenheit temperature to Celsius temperature. The subVI selection process is illustrated in Figure 5.17.

Open the Thermometer Scale VI by selecting **Open** from the **File** menu. The VI is located in the Chapter 5 folder of the Learning directory. For reference, the front panel and block diagram are shown in Figure 5.16.

To create a subVI that converts Fahrenheit to Celsius, begin by switching to the block diagram window by choosing **Show Block Diagram** from the **Window** menu. The goal is to modify the existing block diagram to call a subVI created using the **Create SubVI** option.

Select the block diagram elements that comprise the conversion from Fahrenheit to Celsius, as shown in Figure 5.17. A moving dashed line will frame the chosen portion of the block diagram. Now select **Create SubVI** in the **Edit** menu. A default subVI icon will appear in place of the selected group of objects. You can use this selection method of creating subVIs to modularize your VI.

The next step is to modify the icon of the new subVI. Open the new subVI by double-clicking on the default icon called Untitled 1 (SubVI). Two front panel objects should be visible: one numeric control labeled Temp and one unlabeled numeric indicator. Right-click on the numeric indicator and select Visible Items≫Label from the shortcut menu. Type Celsius Temp in the text box and then align the two objects on the front panel by their bottom edges.

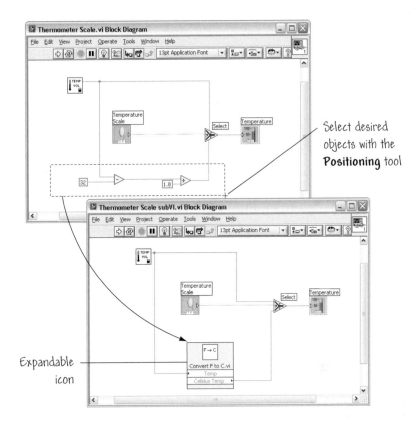

Select desired
objects with the
Positioning tool

Expandable
icon

FIGURE 5.17
Selecting the code that converts from Fahrenheit to Celsius

Use the **Icon Editor** to create an icon similar to the one shown in Figure 5.18. Invoke the editor by right-clicking in the **Icon Pane** in the front panel of the subVI and selecting **Edit Icon** from the menu. Erase the default icon by double-clicking on the **Select** tool and pressing <Delete>. Redraw the icon frame by double-clicking on the rectangle tool which draws a rectangle around the icon. The easiest way to create the text is using the **Text** tool and the font Small Fonts. The arrow can be created with the **Pencil** tool.

The connector is automatically wired when a subVI is created using the ***Create SubVI*** *option.*

When you are finished editing the icon, close the **Icon Editor** by clicking on **OK**. The new icon appears in the upper right corner of the front panel, as shown in Figure 5.18. Right-click on the Convert F to C.vi and click on **View as Icon** to remove the checkmark. Then use the **Positioning** tool to expand the icon to show the input Temp and the output Celsius Temp, as shown in Figure 5.17. Save the subVI by choosing **Save** from the **File** menu. Name the VI Convert F to C and save it in the Users Stuff folder.

FIGURE 5.18
The new subVI to convert degrees Fahrenheit to degrees Celsius

 You cannot build a subVI from a section of code with more than 28 inputs and outputs because 28 is the maximum number of inputs and outputs allowed on a connector pane. ◆

5.7   SAVING YOUR SUBVI

It is highly recommended that you save your subVIs to a file in a directory rather than in a library. While it is possible to save multiple VIs in a single file called a **VI library**, this is not desirable. Saving VIs as individual files is the most effective storage path because you can copy, rename, and delete files more easily than when using a VI library.

VI libraries have the same load, save, and open capabilities as other directories, but they are not hierarchical. That is, you cannot create a VI library inside of another VI library, nor can you create a new directory inside a VI library. After you create a VI library, it appears in the **File** dialog box as a file with an icon that is somewhat different from a VI icon.

5.8   THE VI HIERARCHY WINDOW

When you create an application, you generally start at the top-level VI and define the inputs and outputs for the application. Then you construct the subVIs that you will need to perform the necessary operations on the data as it flows through the block diagram. As discussed in previous sections, if you have a very complex block diagram, you should organize related functions and nodes into subVIs for desired block diagram simplicity. Taking a modular approach to program development creates code that is easier to understand, debug, and maintain.

The **VI Hierarchy** window displays a graphical representation of the hierarchical structure of all VIs in memory and shows the dependencies of top-level VIs and subVIs. There are several ways to access the **VI Hierarchy** window:

- You can select View≫VI Hierarchy to open the **Hierarchy Window** with the VI icon of the current **active window** surrounded by a thick red border.

- You can right-click on a subVI and select **Show VI Hierarchy** to open the **VI Hierarchy** window with the selected subVI surrounded by a thick red border.

- If the **VI Hierarchy** window is already open, you can bring it to the front by selecting it from the list of open windows under the **Window** menu.

The **VI Hierarchy** window for the Thermometer Scale subVI.vi is shown in Figure 5.19. The window displays the dependencies of VIs by providing information on VI callers and subVIs. As you move the **Operating** tool over objects in the window, the name of the VI is shown below the VI icon. This window also contains a toolbar, as shown in Figure 5.19, that you can use to configure several types of settings for displayed items.

You can switch the **VI Hierarchy** window display mode between horizontal and vertical display by pressing either the **Horizontal Layout** or **Vertical Layout** button in the hierarchy toolbar. In a horizontal display, subVIs are shown to the right of their calling VIs; in a vertical display, they are shown below their calling VIs. In either case, the subVIs are always connected with lines to their calling VIs. The window shown in Figure 5.19 is displayed vertically.

Arrow buttons and arrows beside nodes indicate what is displayed and what is hidden according to the following rules:

- A red arrow button pointing towards the node indicates some or all subVIs are hidden, and clicking the button will display the hidden subVIs.

- A black arrow button pointing towards the subVIs of the node indicates all immediate subVIs are shown.

- A blue arrow pointing towards the callers of the node indicates the node has additional callers in this VI hierarchy but they are not shown at the present time. If you show all subVIs, the blue arrow will disappear. If a node has no subVIs, no red or black arrow buttons are shown.

Double-clicking on a VI or subVI opens the front panel of that node. You also can right-click on a VI or subVI node to access a menu with options, such as showing or hiding all subVIs, opening the VI or subVI front panel, editing the VI icon, and so on.

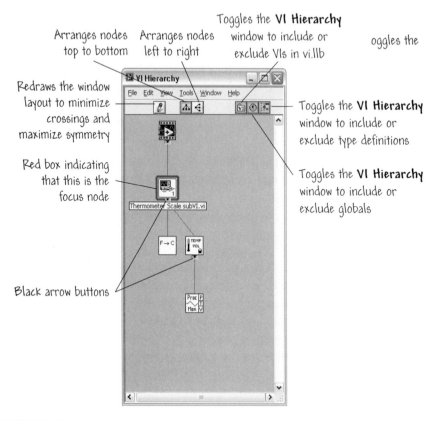

FIGURE 5.19
The **VI Hierarchy** window for Thermometer Scale subVI.vi

You can initiate a search for a given VI by simply typing the name of the node directly onto the **VI Hierarchy** window. When you begin to type in text from the keyboard a small search window will automatically appear displaying the text that has been typed and allowing you to continue adding text. The search for the desired VI commences immediately. The search window is illustrated in Figure 5.20. If the characters currently displayed in the search window do not match any node names in the search path, the system beeps, and no more characters can be typed. You can then use the <Backspace> or <Delete> key to delete one or more characters to resume typing. The search window disappears automatically if no keys are pressed for a certain amount of time, or you can press the <Esc> key to remove the search window immediately. When a match is found you can use the right or down arrow key, the <Enter> key on Windows, the <return> key on Macintosh to find the next node that matches the search string. To find the previous matching node, press the left or up arrow key, <Shift-Enter> on Windows, or <shift-return> on Macintosh.

In the summer of 2002, a significant amount of interesting data was collected during a closed experiment in growing a crop of dwarf soybeans. The ability of the NI-based system to integrate with the internet was key for rapid publication and dissemination of real-time data. Eventually, it will be possible for students of all levels to track and participate in real-time discussions of closed-system experiments as they develop from start to finish.

New features currently being added to the system include integrating cameras into the test module that will use vision software functionality to provide feedback to better understand and control the conditions of the sealed Laboratory Biosphere.

For more information, contact:

Gerard Houghton or Mark Van Thicco
Biosphere Foundation
9 Silver Hills Road
Santa Fe, NM 87505

5.11   SUMMARY

Constructing subVIs was the main topic of this chapter. One of the keys to constructing successful LabVIEW programs is understanding how to build and use subVIs. SubVIs are the primary building blocks of modular programs, that are easy to debug, understand, and maintain. They are analogous to subroutines in text-based programming languages like C or Fortran. Two methods of creating subVIs were discussed—creating subVIs from VIs and creating subVIs from VI selections. A subVI is represented by an icon in the block diagram of a calling VI and must have a connector with terminals properly wired to pass data to and from the calling VI. The **Icon Editor** was discussed as the way to personalize the subVI icon so that information about the function of the subVI is readily apparent from visual inspection of the icon. The editor has a tool set similar to that of most common paint programs. The **VI Hierarchy** window was introduced as a helpful tool for managing the hierarchical nature of your programs.

KEY TERMS

Active window: Window that is currently set to accept user input, usually the frontmost window. You make a window active by clicking on it or by selecting it from the **Window** menu.

Control: Front panel object for entering data to a VI interactively or to a subVI programmatically.

Description box: Online documentation for VIs.

Dialog box: An interactive screen with prompts in which you describe additional information needed to complete a command.

Function: A built-in execution element, comparable to an operator or statement in a conventional programming language.

Icon Editor: Interface similar to that of a paint program for creating VI icons.

Indicator: Front panel object that displays output.

Marquee: Moving, dashed border surrounding selected objects.

Pseudocode: Simplified, language-independent representation of programming code.

SubVI: A VI used in the block diagram of another VI—comparable to a subroutine.

Top-level VI: The VI at the top of the VI hierarchy. This term distinguishes a VI from its subVIs.

VI Hierarchy window: Window that graphically displays the hierarchy of VIs and subVIs.

VI library: Special file that contains a collection of related VIs.

EXERCISES

E5.1   Open a new VI and switch to the block diagram window. Select Write to Text File.vi from the **Functions≫Programming≫File I/O** palette and place it on the block diagram. Notice that the **Run** button breaks when this subVI is placed on the block diagram. Click the **Broken Run** button to access the **Error List**. The one listed error states that the VI has a required input. As you know from the chapter, each input terminal can be categorized as required, recommended, or optional. You must wire values to required inputs before you can run a VI. Recommended and optional inputs do not prevent the VI from running and use the default value for that input. Default values are shown when you open (in this case, double-click on the VI).

Select the error from the **Error List** dialog box by clicking on it and then choose **Show Error**. Notice that the Write to Text File.vi icon is highlighted by a marquee around the icon. Open **Context Help** (Help≫Show Context Help or <Ctrl-H>) to read about the VI, paying special attention to the inputs listed as **Required** (shown in bold). You should see that text is a required input. Right-click on the center terminal on the right side of the icon and select **Create Control**. Notice that once the string control is attached to the icon, the error disappears and you can run the VI.

Navigate to the **Functions≫Mathematics≫Integ & Diff** palette and select the Numeric Integration VI and place it on the block diagram. Notice that the **Run** button does not break. Use **Context Help** again and notice that none of the inputs to this VI are **Required**. You will be able to run the VI without any inputs wired to it.

E5.2 Open a new VI and switch to the block diagram window. Navigate to the **Functions≫Programming≫Timing** palette, select the Wait Until Next ms Multiple.vi and place it on the block diagram. Determine the errors that cause the **Broken Run** to appear. Determine a fix to the problem so that the VI can run properly.

E5.3 Open Slope.vi, which calculates the slope between two points, (X1, X2) and (Y1, Y2). The Slope.vi can be found in the Exercises&Problems folder in Chapter 5 of the Learning directory. The **Run** button is broken; hence, the VI is not executable. Click the **Run** button to access the **Error List** and you will find that the error is found within the subVI. You must open the subVI by double-clicking on it and correct the errors within it before the main VI will run. From within the SlopeSub.vi, go to the block diagram and wire the output of the Subtract terminal to the input of the Divide terminal. Save the subVI, return to the main VI, and notice that the **Run** button is not broken. Enter values into the four numeric controls in Slope.vi and run the VI to vary the slope between two points. Stop the VI.

Leave the SlopeSub.vi open and go to File≫New VI to open a new VI. Go to the block diagram of the new VI to practice the various ways to insert subVIs into a main VI. First, click on the icon of the SlopeSub.vi and drag it onto the block diagram of the new VI. Place the **Wiring** tool over the icon and see that all the input and output terminals are present. Second, from the block diagram, go to **Functions≫Select a VI** and browse to the location on your computer where you saved SlopeSub.vi. Click the **Open** button and place the icon of the Slope-Sub.vi on the block diagram. Remember that both methods of placing subVIs on block diagrams generate the same result.

E5.4 Open a new VI and go to Help≫Find Examples.... Click on the **Browse** tab and press the **Directory Structure** button. In the center panel, double-click on the express folder. This will locate shipping examples that predominantly use Express VIs to accomplish various tasks. Open the Lissajous2.vi by double-clicking on it and run the VI changing the controls on the front panel to alter the curve. Now go to the block diagram and see that there are not many Express VIs needed to perform this program.

From the block diagram, go to View≫VI Hierarchy and you will see that there are more additional subVIs used than you have seen from the block diagram. Use the tools you learned in this chapter to become familiar with the **VI Hierarchy Window** and explore the various subVIs used in Lissajous2.vi.

E5.5   Create a VI called Sign.vi which accepts a floating point input. It should have a Boolean output that returns True if the input is positive, and one that returns True if the input is negative. Make an icon and configure the connector pane so that the Sign VI can be used as a subVI with another VI. Use the icon in Figure E5.5 as a guide. The connector terminals should be assigned one floating point input and two Boolean outputs. Once the VI terminals have been assigned, add documentation so that the **Context Help** will display the following statement: "Tests if an input is positive or negative." Go to File≫VI Properties and then go to **Documentation** to enter the information.

FIGURE E5.5
The front panel for the Sign VI illustrating the edited icon and the connector terminals

E5.6   In this exercise you will modify a VI so that it can be employed as a subVI. Open the Find Greatest VI located in the Exercises&Problems folder in Chapter 5 of the Learning directory. Right-click the icon and notice that the connector pane does not have the right number of inputs and outputs and the terminals have not been correctly linked to the connector. To fix this, take the following steps:

(a) Right-click on the icon pane and select **Show Connector**.

(b) Right-click again on the icon pane and select **Disconnect All Terminals**.

(c) Right-click on the icon pane once again. On the **Patterns** menu, select the connector configuration with three terminals on the left and one terminal on the right.

(d) Assign the three controls to the terminals on the left. Assign the indicator to the terminal on the right.

(e) Open the **Icon Editor** by double-clicking on the icon and create an icon for the VI.

(f) Verify that you have set up the connector pane correctly by opening a new VI and placing Find Greatest within it as a subVI. Create controls and an indicator, and run the program several times with different inputs. The output should always be equal to the greatest value input to the subVI.

E5.7 An effective icon can convey information about the functionality of a subVI at a glance. The NI website has an icon art glossary containing a wide array of graphics which you may find useful to quickly design professional looking icons. Visit the Icon Art Glossary at http://www.ni.com/devzone/idnet/library/icon_art_glossary.htm and familiarize yourself with the available graphics. Follow the link titled Guidelines for Creating Icons to learn more about basic rules of thumb to keep in mind while creating your own icons.

The Calculator VI is a simple VI which performs four different operations on two numeric inputs. Open Calculator.vi found in the Exercises&Problems folder in Chapter 5 of the Learning directory. Right-click on the icon in the upper right hand corner of the front panel and select **Edit Icon** to create a new icon for this VI that is more descriptive. You can use graphics from the Icon Art Gallery or create an icon by hand. Now right-click on the icon and select **Show Connector** to assign terminals to the controls and indicators.

Verify that you have set up the connector pane correctly by opening a new VI and placing Calculator within it as a subVI. Create controls and an indicator, and run the program several times with different inputs.

PROBLEMS

P5.1 Construct a VI that computes the average of three numbers input by the user using native LabVIEW functions. One computation in your program should be summing the three input numbers followed by a division by 3. The resulting average should be displayed on the front panel. Also, add a piece of code that multiplies the computed average by a random number in the range $[0, \ldots, 1]$. Create a subVI by grouping the parts of the code that compute the average.

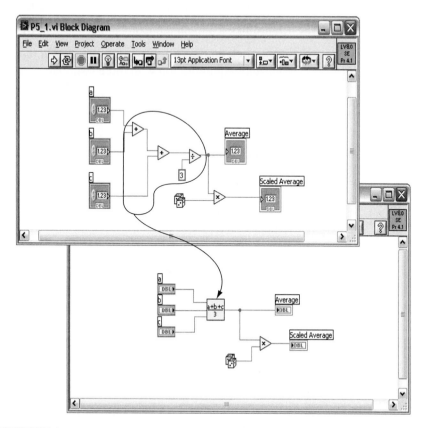

FIGURE P5.1
Computing the average of three numbers using a subVI to compute the average

Remember to edit the icon so that it represents the function of the subVI, namely, the average of three numbers. Figure P5.1 can be used as a guide.

P5.2   Open Check Limit.vi, which generates a sine wave and plots it on a waveform graph. Check Limit.vi can be found in the Exercises&Problems folder in Chapter 5 of the Learning directory. The values of the sine wave are compared with a numeric control, therefore evaluating whether the sine wave exceeds the set limit. It also uses a While Loop that we will learn about it in the next chapter.

Edit the connector icon so that it only has the necessary number of terminals and connect them to the appropriate controls and indicators so that all of the front panel objects can be accessed if this VI were used in the future as a subVI. Also, change the VI so the toolbar, menu bar, and scroll bars are not visible when the VI executes. (**Hint:** Navigate to File≫VI Properties and choose **Window Appearance**, then choose **Customize**.)

P5.3 Create a subVI that multiplexes four inputs to a single output. The subVI should have four floating-point numeric controls (denoted In1 thru In4), one floating-point numeric indicator (denoted by Out), and one unsigned 8-bit integer control (denoted by Select). If Select = 1, then Out = In1; if Select = 2, then Out = In2; if Select = 3, then Out = In3; and if Select = 4, then Out = In4. (**Hint:** The Select function VI from the **Programming≫Comparison** palette may be useful.)

P5.4 Create a VI that executes the Quit LabVIEW VI from the **Functions≫Programming ≫Application Control** palette. Open a new VI and place the Quit LabVIEW VI on the block diagram. To edit the VI properties, select File≫VI Properties and choose the **Execution** category. Check the box next to **Run when opened**. Save the VI in Users Stuff. Close the VI and then open it again. What happens? Try to figure out how you can edit the VI. (**Hint:** A subVI may be useful.)

P5.5 Construct a VI that solves the quadratic formula to find the roots of the equation

$$ax^2 + bx + c = 0$$

for the variable x, where the constants a, b, and c are real numbers. In general, the roots of the quadratic equation can be either two real numbers or two complex numbers. For this problem, your VI only needs to find the real roots. The solution of the quadratic equation is given by the relations:

$$x_1 = \left(-b + \sqrt{b^2 - 4ac}\right)/2a \text{ and } x_2 = \left(-b - \sqrt{b^2 - 4ac}\right)/2a$$

Construct a VI that computes the roots according to the above relationships for x_1 and x_2. Use Figure P5.5 as a guide for the front panel. There should be two numeric outputs for the two roots and three inputs for the constants a, b, and c. Once you are finished building the block diagram, create an icon and configure the terminal connections so that your VI can be used as a subVI. Verify that you have done this correctly by placing your quadratic formula subVI in a new VI and connecting the terminals to controls and indicators.

FIGURE P5.5
Front panel for the Quadratic Equation VI with edited icon

DESIGN PROBLEMS

D5.1 Construct a subVI to calculate an age in years given a birth date and the current date as inputs. Use Figure D5.1 as a guide for the front panel. You should use three separate integer controls to enter each date: one for the day of the month, one for the month, and one for the year. Enter months as numbers. For example, use the number 1 for January, the number 2 for February, and so forth. Be sure to create an icon for the VI and connect the output and inputs to terminals on the icon. When you are finished, create documentation for your VI under **VI Properties**. Add the following information to the documentation: "Finds age in years of a person given the current date and the person's birth date as input."

FIGURE D5.1
The front panel for the Age VI

D5.2 Create a subVI which will calculate the Body Mass Index (BMI) of an individual. The BMI is an internationally used measure of obesity. The subVI will have two numeric inputs: a weight in pounds and a height in inches. Calculate the BMI using the formula

$$BMI = (703 * W)/H^2$$

where W is the weight in pounds and H is the height in inches. Display the calculated BMI on a numeric indicator.

Classify the body weight according to Table D5.2 and output a string containing this information. There should also be a "warning" Boolean output that is True for any unhealthy input (i.e. underweight, overweight, or obese).

TABLE D5.2   BMI Classification

<18.5	Underweight
18.5–24.9	Healthy
25–29.9	Overweight
≥ 30	Obese

CHAPTER 6

Structures

Structures govern the execution flow in a VI. This chapter introduces you to four structures in LabVIEW: the For Loop, the While Loop, the Case structure, and the Flat Sequence structure. Formula Nodes will be introduced as an effective way to implement mathematical equations. We will also discuss the subject of controlling the execution timing of VIs.

GOALS

1. Study For Loops, While Loops, Case structures, and Flat Sequence structures.
2. Understand how to use timing functions in LabVIEW.
3. Understand the use of shift registers and feedback nodes.
4. Become familiar with Formula Nodes.
5. Appreciate common wiring errors with structures.

6.1   THE FOR LOOP

For Loops and **While Loops** control repetitive operations in a VI, either until a specified number of iterations completes (i.e., the For Loop) or while a specified condition is True (i.e., the While Loop). A difference between the For Loop and the While Loop is that the For Loop executes a predetermined number of times, and a While Loop executes until a certain conditional input becomes True or False. For Loops and While Loops are found on the **Structures** palette of the **Functions≫Programming** subpalette—see Figure 6.1. In this section, we concentrate on For Loops. While Loops are discussed in the next section.

A For Loop executes the code inside its borders (known as the **subdiagram**) a total of N times, where the N equals the value in the count terminal, as illustrated in Figure 6.2.

The For Loop has two terminals: the **count terminal** (an input terminal) and the **iteration terminal** (an output terminal). You can set the count explicitly by wiring a value from outside the loop to the count terminal, or you can set the count implicitly with auto-indexing (a topic discussed in Chapter 7). The value of the

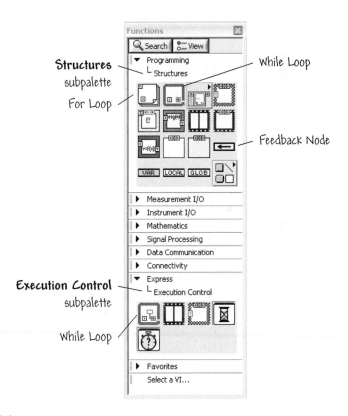

FIGURE 6.1
The For Loop and While Loop are found on the **Programming≫Structures** palette, the While Loop is also found on the **Express≫Execution Control** palette

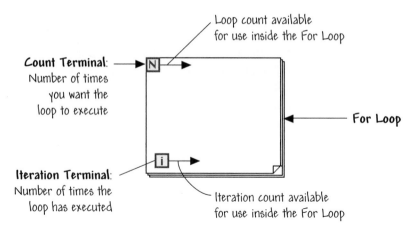

Loop count available for use inside the For Loop

Count Terminal:
Number of times you want the loop to execute

For Loop

Iteration Terminal:
Number of times the loop has executed

Iteration count available for use inside the For Loop

FIGURE 6.2
The For Loop

count terminal is also exposed to the inside of the loop, allowing you access to the value of the count. The iteration terminal (see Figure 6.2) contains the current number of completed loop iterations, where 0 represents the first iteration, 1 represents the second iteration, and continuing up to $N - 1$ to represent the Nth iteration.

Both the count and iteration terminals are long integers with a range of 0 through $2^{32} - 1$. You can wire a floating-point number to the count terminal, but it will be rounded off and **coerced** to lie within the range 0 through $2^{32} - 1$. The For Loop does not execute if you wire 0 to the count terminal!

The For Loop is located on the **Functions≫Programming≫Structures** palette. Unlike many other **objects**, the For Loop is not simply dropped on the block diagram. Instead, a small icon representing the For Loop appears in the block diagram, giving you the opportunity to size and position the loop. To do so, first click in an area above and to the left of all the objects that you want to execute within the For Loop, as illustrated in Figure 6.3. While holding down the mouse button, drag out a rectangle that encompasses the objects you want to place inside the For Loop. A For Loop is created upon release of the mouse button. The For Loop is a resizable box—use the **Positioning** tool for resizing by grabbing a corner of the For Loop and stretching to the desired dimensions. You can add additional block diagram elements to the For Loop by dragging and dropping them inside the loop boundary. The For Loop border will highlight as objects move inside the boundaries of the loop, and the block diagram border will highlight as you drag objects out of the loop.

If you move an existing structure (e.g., a For Loop) so that it overlaps another object on the block diagram, the partially covered object will be visible above one edge of the structure. If you drag an existing structure completely over another object, the covered object will display a thick shadow to warn you that the object is underneath.

Place For Loop icon
on block diagram

Drag box around
desired objects and
release mouse button

Pseudocode equivalent   Subdiagram

For *i* = 0 to N−1
number = rand( )

FIGURE 6.3
Placing a For Loop on the block diagram

6.1.1   Numeric Conversion

Most of the numeric controls and indicators you have used so far have been double-precision, floating-point numbers. Numbers can be represented as integers (byte [I8], word [I16], or long [I32]) or floating-point numbers (single, double, or extended precision). If you wire together two terminals that are of different data types, LabVIEW converts one of the terminals to the same representation as the other terminal. As a reminder, a **coercion dot** is placed on the terminal where the conversion takes place.

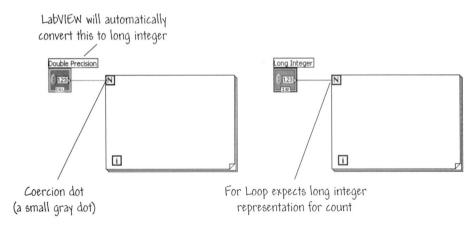

LabVIEW will automatically
convert this to long integer

Coercion dot
(a small gray dot)

For Loop expects long integer
representation for count

FIGURE 6.4
Converting double-precision, floating-point numbers at the count terminal

For example, consider the For Loop count terminal shown in Figure 6.4. The terminal representation is long integer. If you wire a double-precision, floating-point number to the count terminal, the number is converted to a long integer. Notice the small gray dot in the count terminal of the first For Loop—that is the coercion dot. To change the representation of the count terminal input, right-click on the terminal and select **Representation**, as illustrated in Figure 6.5. A palette will appear from which you can select the desired representation. In the case of the For Loop, you can change the count terminal input from double-precision, floating-point to long integer.

Right-click on
numeric control

Select long integer (I32)

FIGURE 6.5
Changing the representation of a front panel numeric object

Continue if True   Stop if True

Icon changes
to stop sign

(a)   (b)

FIGURE 6.9
(a) Continue if True and (b) Stop if True

The first time through a For Loop or a While Loop, the iteration count is zero. If you want to register how many times the loop has actually executed, you must add 1 to the count.

A While Loop Example

In this exercise we will place a random number object inside a While Loop and display the random numbers and While Loop counter on the front panel. This is very similar to the For Loop exercise in the previous section. The VI shown in Figure 6.10 can be constructed by following these steps:

1. Place a Random Number function on the block diagram (found in the **Functions≫Programming≫Numeric** palette). Create an indicator on the output of the Random Number function and label it number: 0 to 1.

2. Place the While Loop on the block diagram so that the Random Number function is enclosed within the loop, as shown in Figure 6.10. Use the **Functions≫ Express≫Execution Control** palette to locate the While Loop.

3. When you place the While Loop from the **Express≫Execution Control** palette, a stop button will automatically be created. This will be used to stop the While Loop iterations while in Run mode.

4. Create an indicator on the output of the iteration terminal and label it Loop number.

5. Click on the **Run** button to start the program execution, and run the program with **Highlight Execution** to observe the program data flow.

6. On the front panel you will see the loop counter continue to increment until you push the Stop button. This causes the conditional terminal to change to True, and the While Loop iterations cease.

Click Run to start VI

VI will run until STOP button is pushed

While Loop will execute until value is TRUE, which occurs when STOP button is pushed

FIGURE 6.10
Displaying a series of random numbers using the While Loop

*If you want to run a VI multiple times, place all the code in a loop rather than using the **Run Continuously** button. It is better to stop a VI with a Boolean control than with the **Abort Execution** button.*

7.   Save the VI as While Loop Demo.vi in the Users Stuff folder in the Learning directory.

*A working version of the VI called **While Loop Demo.vi** can be found in the **Chapter 6** folder in the **Learning** directory.*

If you place the terminal of the Boolean control outside the While Loop, as shown in Figure 6.11, you create either an infinite loop or a loop that executes only once, depending on the initial value. Why? Because the Boolean input data value is read before it enters the loop—remember data flow programming?—and not within the loop or after completion of the loop.

To experiment with this, modify your VI by placing the Boolean terminal outside of the While Loop. Connect the Boolean terminal to the While Loop border, as illustrated in Figure 6.11. A green rectangle will appear on the loop border. Then wire the green rectangle to the conditional terminal.

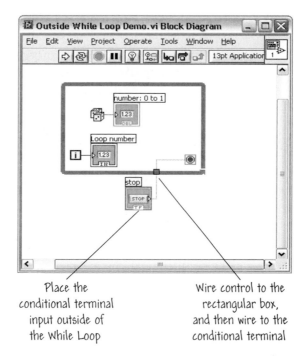

Place the
conditional terminal
input outside of
the While Loop

Wire control to the
rectangular box,
and then wire to the
conditional terminal

FIGURE 6.11
Placing the conditional terminal input outside of the While Loop

Conduct the following two numerical experiments: First, set the button state on the front panel to "on" and then run the VI (use **Highlight Execution** to watch the data flow). After a reasonable period of time, press the Stop button. What happens? The VI does not stop running—it is in an infinite loop. Can you explain this behavior using the notion of data flow programming? Since LabVIEW operates under data flow principles, inputs to the While Loop must pass their data before the loop executes. A While Loop passes data out only after the loop completes all iterations. The infinite loop experiment can be stopped by clicking on the **Abort Execution** button on the front panel toolbar.

For a second experiment, push the Stop button on the front panel and then run the VI (use **Highlight Execution** to watch the data flow). What happens in this case? You should see that the VI will execute once through the While Loop.

Save the VI with the conditional terminal outside the While Loop as Outside While Loop Demo.vi in the Users Stuff folder in the Learning directory.

 A working version of Outside While Loop Demo.vi can be found in the Chapter 6 folder in the Learning directory.   ◆

6.3   SHIFT REGISTERS AND FEEDBACK NODES

When programming with loops, you often need to access data from previous iterations of the loop. Two ways of accessing the data from previous iterations of the loop are by utilizing the shift register and the Feedback Node. Both methods are discussed in this section.

6.3.1   Shift Registers

Shift registers transfer values from one iteration of a For Loop or While Loop to the next. The shift register comprises of a pair of terminals directly opposite each other on the vertical sides of the loop border, as shown in Figure 6.12. You create a shift register by right-clicking on the left or right loop border and selecting **Add Shift Register** from the shortcut menu.

The right terminal (the rectangle with the up arrow) stores the data as each iteration finishes. The stored data from the previous iteration is shifted and appears at the left terminal (the rectangle with the down arrow) at the beginning of the next iteration, as shown in Figure 6.13. A shift register can hold any data type, including numeric, Boolean, strings (see Chapter 10), and arrays (see Chapter 7), but the data wired to the terminals of each register must be of the same type. The shift register conforms to the data type of the first object that is wired to one of its terminals.

Consider a simple illustrative example. Suppose we have the two situations depicted in Figure 6.14. Both cases look very similar—but the code in (b) contains a shift register. The code in (b) computes a running sum of the iteration count within the For Loop. Each time through the loop, the new sum is saved in the shift register. At the end of the loop, the total sum of 10 is passed out to the numeric indicator. Why 10? Because the sum of the iteration count is 10 ($= 0 + 1 + 2 + 3 + 4$). On the other hand, the code in (a) that does not contain the

FIGURE 6.12
Shift register containing a pair of terminals

**Computing
a Running
Average**

In this example, we will use shift registers to compute the running average of a sequence of random numbers. Since the Random Number function in LabVIEW provides random numbers from 0 to 1, we expect that the average will be 0.5. How many random numbers will it take to obtain an average near 0.5?

Figure 6.18 shows a VI that computes the running average of several random numbers. The length of the random number sequence is input via the slide control, as depicted on the front panel in Figure 6.18. Using the block diagram in the figure

FIGURE 6.18
Computing a running average of a sequence of random numbers

as a model, develop your own VI to compute the running average of a sequence of random numbers. The formula that is coded to compute the average is

$$Ave_i = \frac{i}{i+1} Ave_{i-1} + \frac{1}{i+1} RN_i,$$

where $i = 0, 1, \cdots, N - 1$, Ave_i is the computed average at the ith iteration, and RN_i is the current random number from the random number function. If you aren't familiar with computing a running average, just concentrate on the programming aspects of the VI shown in Figure 6.18 and do not worry about the formula. The main point is to understand how to use the shift registers in conjunction with For Loops and While Loops!

In this example, the shift register is used to pass the value of the variable Ave_{i-1} from one iteration to the next. You should notice as you run the code that for small values of N (e.g., $N = 3$) the average is generally not close to the expected value of $Ave_N = 0.5$; however, as you increase N, the average gets closer and closer to the expected value. Try it out!

The VI depicted in Figure 6.18 can be found in **Chapter 6** *of the* **Learning** *directory and is called* **Random Number Average.vi**. *Check it out if you can't get yours to work or if you want to compare your results.* ◆

6.3.4 Feedback Nodes

The Feedback Node can be utilized in a For Loop or While Loop if you wire the output of a subVI, function, or group of subVIs and functions to the input of that same VI, function, or group—that is, if you create a *feedback* path. Like a shift register, the Feedback Node stores data when the loop completes an iteration, sends that value to the next iteration of the loop, and transfers any data type. The Feedback Node is illustrated in Figure 6.19 inside of a While Loop.

Once the feedback is wired inside the loop, both the Feedback Node arrow and the Initializer Terminal will automatically appear. The Feedback Node arrow indicates the direction of data flow along the wire. Initializing a Feedback Node resets the initial value the Feedback Node passes the first time the loop executes when the VI runs. If you leave the input of the Initializer Terminal unwired, each time the VI runs, the initial input of the Feedback Node is the last value from the previous execution or the default value for the data type if the loop has never executed.

You also can select the Feedback Node on the **Programming≫Structures** palette and place it inside a For Loop or While Loop (see Figure 6.1 for location on the palette). If you place the Feedback Node on the wire before you branch the wire that connects the data to the tunnel, the Feedback Node passes each value

FIGURE 6.19
The Feedback Node

to the tunnel. If you place the Feedback Node on the wire after you branch the wire that connects data to the tunnel, the Feedback Node passes each value back to the input of the VI or function and then passes the last value to the tunnel.

Use the Feedback Node to avoid unnecessarily long wires in loops.

Practice with Feedback Nodes

In this example we compare the use of the shift register with the Feedback Node and illustrate the differences in placing the Feedback Node in the feedback path and the feedforward path.

Open the Feedback Node Demo VI located in the **Chapter 6** folder of the **Learning** directory. The block diagram is shown in Figure 6.20. The block diagram has three parts:

1. In the top section of the code, the Feedback Node is placed after the branch.

2. In the middle section of the code, the Feedback Node is placed before the branch.

3. In the bottom section of the code, shift registers are used instead of Feedback Nodes.

Wiring a 0 to the Initialization Terminal or to the shift register initializes each diagram. The Time Delay function slows the operation of the code for better viewing. Run the VI and observe the outputs on the front panel.

FIGURE 6.20
Different implementations of the Feedback Node

The top section of the code reads the initialized Feedback Node and passes this value to the Add function.

The middle section of the code reads the initialized Feedback Node and passes this value to the indicator. This Add function does not execute until the next iteration of the loop. This section of code will output a number always one less than the first section of code.

The Feedback Node implementation in the top section of the code and the shift register implementation in the bottom section of the code have the same functionality.

 If you want to replace a Feedback Node with a shift register, right-click the Feedback Node and select **Replace with Shift Register**. ◆

6.4 CASE STRUCTURES

A **Case structure** is a method of executing conditional text. This is analogous to the common If...Then...Else statements in conventional, text-based programming languages. You place the Case structure on the block diagram by selecting it from the **Structures** subpalette of the **Functions≫Programming** palette or from the **Execution Control** subpalette of the **Functions≫Express** palette, as shown in Figure 6.21. As with the For Loop and While Loop structures, you can either drag the Case structure icon to the block diagram and enclose the desired objects within its borders, or you can place the Case structure on the block diagram, resize it as necessary, and drag objects inside the structure.

The Case structure can have multiple subdiagrams. The subdiagrams are configured like a deck of cards of which only one card is visible at a time. At the top of the Case structure border is the selector label. The diagram identifier can be numeric, Boolean, string, or enumerated type control. More information on enumerated type controls can be found in the LabVIEW help. An enumerated type control is unsigned byte, unsigned word, or unsigned long and is selectable from the decrement and increment buttons, as depicted in Figure 6.22(a) and (b).

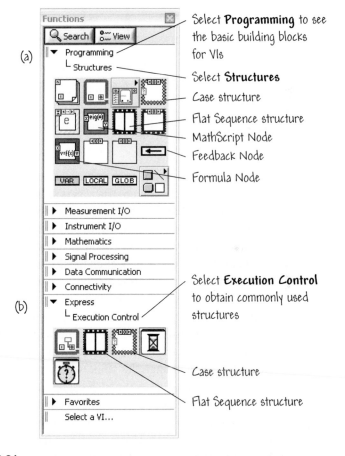

FIGURE 6.21
(a) Select Case structure from the **Programming≫Structures** palette (b) Selecting the Case structure from the **Express≫Execution Control** palette

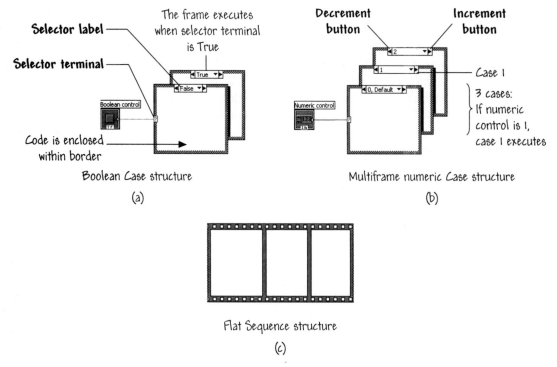

FIGURE 6.22
Overview of Case and Flat Sequence structures

The selector label displays the values that cause the corresponding subdiagrams to execute. The Flat Sequence structure is similar in form to the Case structure and is depicted in Figure 6.22(c). The selector label value is followed by a selector label range, which shows the minimum and maximum values for which the structure contains a subdiagram. To view other subdiagrams (that is, to see the subdiagrams within the "stack of cards"), click the decrement (left) or increment (right) button to display the previous or next subdiagram, respectively. Decrementing from the first subdiagram displays the last, and incrementing from the last subdiagram displays the first subdiagram—it wraps around!

You can position the selector terminal anywhere on the Case structure along the left border. The selector label automatically adjusts to the input data type. For example, if you change the value wired to the selector terminal from a numeric to a Boolean, then cases 0 and 1 change to False and True, respectively (in the selector label). Here is a point to consider in changing the data type: if the Case structure selector originally received numeric input, then n cases $0, 1, 2, \cdots, n$ may exist in the code. Upon changing the selector input data from numeric to Boolean, cases 0 and 1 change automatically to False and True. However, cases $2, 3, \cdots, n$ are not discarded! You must explicitly delete these extra cases before the Case structure can execute.

You can also type and edit values directly into the selector label using the **Labeling** tool. The selector values are specified as a single value, as a list, or as a range of values. A list of values is separated by commas, such as −1, 0, 5, 10. A range is typed in as 10..20, which indicates all numbers from 10 to 20, inclusively. You also can use open-ended ranges. For example, all numbers less than or equal to 0 are represented by ..0. Similarly, 100.. represents all numbers greater than or equal to 100.

The case selector can also use string values that display in quotes, such as "red," and "green." You don't need to type in the quotes when entering the values unless the string contains a comma or the symbol "..". In a case selector using strings, you can use special backslash codes (such as \r, \n, and \t) for nonalphanumeric characters (carriage return, line feed, tab, respectively).

If you type in a selector value that is not the same type as the object wired to the selector terminal, then the selector value displays in red, and your VI is broken. Also, because of the possible round-off error inherent in floating-point arithmetic, you cannot use floating-point numbers in case selector labels. If you wire a floating-point type to the case terminal, the type is rounded to the nearest integer, and a coercion dot appears. If you try to type in a selector value that is a floating-point number, then the selector value displays in red, and your VI is broken.

6.4.1   Adding and Deleting Cases

If you right-click on the Case structure border, the resulting menu gives you the many options shown in Figure 6.23. You choose **Add Case After** to add a case after the case that is currently visible, or **Add Case Before** to add a case before the currently visible case. You can also choose to copy the currently shown case by selecting **Duplicate Case** and to delete the current visible case by selecting **Delete This Case**. When you add or remove cases (i.e., subdiagrams) in a Case structure, the diagram identifiers are automatically updated to reflect the inserted or deleted subdiagrams.

Sometimes you may want to rearrange the listed order of the cases in the structure. For example, rather than have the cases listed (0, 1, 2, 3, ···), you might want them listed in the order (0, 2, 1, 3, ···). Resorting the order in which the cases appear in the case structure on the block diagram does not affect the run-time behavior of the Case structure! It is merely a matter of programming preference. You can change the order in which the cases are listed in the structure by selecting **Rearrange Cases** from the shortcut menu. When you do so, the dialog box shown in Figure 6.23 appears. The **Sort** button sorts the case selector values based on the first selector value. To change the location of a selector, click the selector value you want to move (it will highlight when selected) and drag it to the desired location in the stack. In the **Rearrange Cases** dialog box, the section entitled "Complete selector string" shows the selected case selector name

FIGURE 6.23
Adding and deleting cases in Case structures

in its entirety in case it is too long to fit in the "Case List" box located in the top portion of the dialog box. Online assistance can be found in the context-sensitive help in the **Help** menu.

The **Make This The Default Case** item in the shortcut menu specifies the case to execute if the value that appears at the selector terminal is not listed as one of the possible choices in the selector label. Case statements in other programming languages generally do not execute any case if a case is out of range, but in LabVIEW you must either include a default case that handles out-of-range values or explicitly list every possible input value. For the default case that you define, the word "Default" will be listed in the selector label at the top of the Case structure.

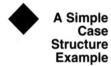

A Simple Case Structure Example

In this exercise you will build a VI that uses a Boolean Case structure, shown in Figure 6.24. The input numbers from the front panel will pass through **tunnels** to the Case structure, and there they are either added or subtracted, depending on the Boolean value wired to the selector terminal.

What is a tunnel? A tunnel is a data entry point or exit point on a structure. You can wire a terminal from outside the Case structure to a terminal within the structure. When you do this a rectangular box will appear on the structure border—this represents the tunnel. You can also wire an external terminal to the

FIGURE 6.24
A Boolean Case structure example to add or subtract two input numbers

structure border to create the tunnel, and then wire the terminal to an internal terminal in a second step. Tunnels can be found on other structures, such as Sequence structures, While Loops, and For Loops. The data at all input tunnels is available to all cases.

In this exercise you will need to create two tunnels to get data into the Case structure and one tunnel to pass data out of the structure. Using Figure 6.24 as a guide, construct a VI that utilizes the Case structure to perform the following task: if the Boolean wired to the selector terminal is True, the VI will add the numbers; otherwise, the VI will subtract the numbers. The Boolean value of the selector terminal is toggled using a horizontal toggle switch found on the front panel.

The VI depicted in Figure 6.24 can be found in **Chapter 6** *of the* **Learning** *directory and is called* **Simple Case Exercise.vi.** ◆

6.4.2   Wiring Inputs and Outputs

As previously mentioned, the data at all input terminals (tunnels and selection terminal) is available to all cases. Cases are not required to use input data or to supply output data, but if any one case supplies output data, all must do so. Forgetting this fact can lead to coding errors. Correct and incorrect wire situations are depicted in Figure 6.25, where for the FALSE case two input numbers are added and sent (through the tunnel) out of the structure, and for the TRUE case the computer system beeps.

When you create an output tunnel in one case, tunnels appear at the same position on the structure border in the other cases. You must define the output tunnel for each case. Unwired tunnels look like white squares, and when they occur, the **Broken Run** button will also appear, so be sure to wire to the output tunnel for each unwired case. You can wire constants or controls to unwired cases by right-clicking on the white square and selecting **Constant** or **Control** from the **Create** menu. When all cases supply data to the tunnel, it takes on a solid color consistent with the data type of the supplied data, and the **Run** button appears.

FIGURE 6.25
Wiring Case structure inputs and outputs

FIGURE 6.26
A VI that computes the ratio of two numbers and uses a Case structure to handle division by zero

**Using
Case
Structures**

In this exercise, you will build a VI that computes the ratio of two numbers. If the denominator is zero, the VI outputs ∞ to the front panel and causes the system to make a beep sound. If the denominator is not zero, the ratio is computed and displayed on the front panel.

Begin by opening a new VI. Build the simple front panel shown in Figure 6.26. The numeric control x **Numerator** supplies the numerator x, and the numeric control y **Denominator** supplies the denominator y. The **x/y** numeric indicator displays the ratio of the two input numbers. Switch to the block diagram and place the Case structure on the diagram.

By default, the Case structure selector terminal is Boolean—this is the format we desire in this situation. You can display only one case at a time, and we will start by considering the True case. Remember that to change cases, you need to click on the increment and/or decrement arrows in the top border of the Case structure.

Place the other diagram objects and wire them as shown in Figure 6.26. The main objects are:

1. The Not Equal to 0? function (found in the **Functions≫Programming≫ Comparison** menu) checks whether the denominator is zero. The function returns a TRUE if the denominator is not equal to 0.

2. The Divide function (found in the **Functions≫Programming≫Numeric** menu) returns the ratio of the numerator and denominator numbers.

3. The Positive Infinity constant indicates that a divide by zero has been attempted (see the **Functions≫Programming≫Numeric** menu).

4. The Beep.vi function (found in the library **Function≫Programming≫ Graphics & Sound** menu) causes the system to issue an audible tone when division by zero has been attempted. On Windows platforms, all input parameters to the beep function are ignored, while on the Macintosh, you can specify the tone frequency in Hertz, the duration in milliseconds, and the intensity as a value from 0 to 255, with 255 being the loudest. Double click on the icon to see the VI front panel.

You must define the output tunnel for each case. When you create an output tunnel in the TRUE case, an output tunnel will appear at the same position in the FALSE case. Unwired tunnels look like white squares. In this and all other VI development, be sure to wire to the output tunnel for each unwired case. In Figure 6.26, the constant ∞ is wired to the output tunnel in the FALSE case.

The VI will execute either the TRUE case or the FALSE case. If the denominator number is not equal to zero, the VI will execute the TRUE case and return the ratio of the two input numbers. If the denominator number is equal to zero, the VI will execute the FALSE case and output ∞ to the numeric indicator and make the system beep.

Once all objects are in place and wired properly, return to the front panel and experiment with running the VI. Change the input numbers and compute the ratio. Make the denominator input equal to zero and listen for the system beep. When you are finished, save the VI as Case Structure Demo.vi in the Users Stuff folder in the Learning directory. Close the VI.

A working version of this VI called Case Structure Demo.vi *can be found in* Chapter 6 *of the* Learning *directory.* ◆

To conclude Section 6.4, we present an example that illustrates how Case structures can be used to construct a VI that computes factorials.

Using Case Structures to Compute Factorials

The factorial $n!$ is defined for a positive integer n as

$$n! \equiv n * (n - 1) \ldots * 2 * 1.$$

The factorial of 0 is defined as 1. In this example, we will create a VI which performs this function, using a For Loop with a shift register.

1. Open a new blank VI and place a Case structure on the block diagram. The case structure can be found on the **Programming≫Structures** palette. We will use the Case structure to determine whether or not the user has entered a nonnegative integer.

2. Place a numeric control to the left of the Case structure and change the representation to I32. Label this control **n**. Use Figure 6.27 as a guide in the development of your VI.

3. Navigate to the **Programming≫Comparison** palette and find the Greater Than or Equal to 0? function. Connect the input of this function to **n** and

FIGURE 6.27
The block diagram for the Factorial VI

connect the output to the selector terminal of the Case structure. The True frame of the case structure will execute when the input is a value greater than or equal to zero. The False frame will execute in all other cases.

4. Click either the increment or decrement button on the selector terminal to view the False case. Go to the **Programming≫Dialog & User Interface** palette and place a One Button Dialog function within the Case structure. Create a constant on the message input to this node with a message prompting the user to enter a nonnegative input.

5. Now switch to the True case by clicking either the increment or decrement button on the selector terminal. Place a For Loop within the Case structure. The For Loop is used because we know that we need to perform the multiplication operation exactly n times.

6. Wire the input n to the count terminal of the For loop so that the loop will execute n times. When you do this, a tunnel will automatically appear on the border of the Case structure.

7. Place a Multiply function and an Increment function within the For Loop. These can be found on the **Programming≫Numeric** palette. The Multiply function will be used to implement the operation $n*(n-1)$. Wire the output of the iteration terminal to the increment function and then connect the output of the Increment function to one of the inputs to the Multiply function. The output of the iteration terminal must be incremented since we want to use the values $1, 2, 3, \ldots n$ in our operations, rather than the values $0, 1, 2, \ldots n-1$.

8. Right-click on the border of the For Loop and select **Add Shift Register** from the menu. You will need to use a shift register so that the VI stores each product and saves it to be used in the next loop iteration. This can be accomplished by connecting the output of the Multiply function to the input of the shift register terminal on the right side of the loop. Wire the other input of the Multiply function to the output of the shift register terminal on the left side of the For Loop.

The shift registers turn blue to indicate that they are handling an integer data type.

9. Since the factorial of 0 is 1, we want to initialize the shift register with the value 1. Initialize the input of the shift registers by right-clicking on the input to the shift register terminal on the left side of the For Loop and selecting **Create Constant**. Change the value from 0 to 1.

10. Create an indicator labeled n! on the output of the shift register terminal on the right side of the For Loop.

You have now created a VI which calculates factorials. Experiment with different input values. Try running the VI with highlight execution turned on so that you can see what values appear on the output of the iteration terminal and shift register with each loop iteration. Compare your VI to the factorial VI found on the **Mathematics≫Elementary & Special Functions≫Discrete Math** palette.

A working version of this VI called **Factorial.vi** *can be found in* **Chapter 6** *of the* **Learning** *directory.*   ◆

6.5   FLAT SEQUENCE STRUCTURES

The **Sequence structure** executes subdiagrams sequentially. The subdiagrams look like frames of film; hence they are known as **frames**. There are two classes of sequence structures: the Flat Sequence and the Stacked Sequence. Although the two classes of Sequence structures are similar, there are key differences between the two. For most purposes, the Flat Sequence structure is the better choice, for that reason we will only cover the Flat Sequence structure in this text. The Flat Sequence structure is depicted in Figure 6.22(c).

Determining the execution order of a program by arranging its elements in a certain sequence is called **control flow**. Most text-based programming languages (such as Fortran and C) have inherent control flow because statements execute in the order in which they appear in the program. In data flow programming, a node executes when data is available at all of the node inputs (this is known as **data dependency**), but sometimes you cannot connect one node to another. When data dependencies are not sufficient to control the data flow, the Sequence structure is a way of controlling the order in which nodes execute. You use the Sequence structure to control the order of execution of nodes that are not connected with wires. Within each frame, as in the rest of the block diagram, data dependency determines the execution order of nodes.

Data flow for the Flat Sequence structure differs from data flow for other structures. Frames in a Flat Sequence structure execute in order and when all data wired to the frame is available. New frames are created by right-clicking on the structure border and selecting **Add Frame After** or **Add Frame Before**, as illustrated in Figure 6.28. Data is passed between frames using tunnels in Flat Sequence structures. The data leaves each frame as the frame finishes executing.

6.5.1   Evaluate and Control Timing in a Sequence Structure

It is useful to be able to control and time the execution of VIs. You can use Express VIs or traditional VIs to accomplish the timing task. The Wait (ms) and Tick Count (ms) functions (located in the **Timing** palette of the **Functions≫Programming**

FIGURE 6.28
Removing and adding frames to the Flat Sequence structure

palette) in conjunction with a sequence structure let you accomplish these tasks using traditional VIs. The units of time are given in milliseconds.

You can also accomplish the timing task using Express VIs. The Time Delay Express VI and the Elapsed Time Express VI are located on the same **Timing** palette, as shown in Figure 6.29. When the Time Delay Express VI is placed on the block diagram, a dialog box appears in which you set the time delay value. A similar dialog box opens with the Elapsed Time Express VI. All units are given in seconds.

The Wait (ms) function causes your VI to wait a specified number of milliseconds before it continues execution. The function waits the specified number of milliseconds and then returns the millisecond timer's end value. The Tick Count (ms) function returns the value of the millisecond timer and is commonly used to calculate elapsed time. The base reference time (that is, zero milliseconds) is undefined. Therefore, you cannot convert the millisecond timer output value to a real-world time or date. Note that the value of the millisecond timer wraps from $2^{32} - 1$ to 0.

The internal clock does not have high resolution—about 1 ms on Windows XP/2000 and Macintosh. The resolutions are driven by operating system limitations and not by LabVIEW.

A simple example illustrating the use of the Time Delay Express VI and the Elapsed Time Express VIs is shown in Figure 6.30. Open the VI Timing with Sound Demo located in **Chapter 6** of the **Learning** directory. Set the time between beeps on the slider control and run the VI in **Run Continuously** mode. You should hear a sound approximately every *n* seconds according to the setting on the slider control. The actual time between beeps (to the resolution of the clock) is displayed on the front panel to verify that the desired timing has been achieved. On Windows XP/2000 and Macintosh, if you reduce the time between beeps to 0.001

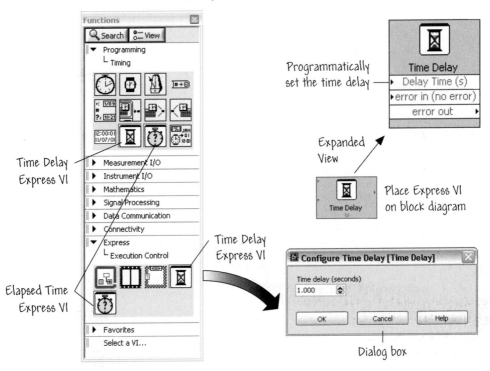

FIGURE 6.29
The Time Delay Express VI and the Elapsed Time Express VI

FIGURE 6.30
Controlling the timing of a VI using the Flat Sequence structure

second (using the **Labeling** tool to type in the value in the digital display of the slider control), you will find that the actual time between beeps is about 1 ms. However, if you continue to reduce the time below 0.001, the actual time between beeps will not reduce accordingly—you have reached the clock resolution of your system.

6.5.2   Avoid the Overuse of Sequence Structures

In general, VIs can operate with a great deal of inherent parallelism. Sequence structures tend to hide parts of the program and to interfere with the natural data flow. The use of Sequence structures prohibits parallel operations, but does guarantee the order of execution. Asynchronous tasks that use I/O devices can run concurrently with other operations if Sequence structures do not prevent them from doing so.

We will be discussing I/O devices (such as GPIB, serial ports, and data acquisition boards) more in subsequent chapters and you will understand better how the use of Sequence structures can inhibit the performance of such devices. The objective here is to alert you to the idea that you should avoid the overuse of Sequence structures. The use of Sequence structures does not negatively impact the computational performance of your program, but it does interrupt the data flow. You should write programs that take advantage of the concept of data flow programming!

6.6   THE FORMULA NODE

The **Formula Node** is a structure that allows you to program one or more algebraic formulas using a syntax similar to most text-based programming languages. It is useful when the equations have many variables or otherwise would require a complex block diagram model for implementation. The Formula Node itself is a resizable box (similar to the Sequence structure, Case structure, For Loop, and While Loop) in which you enter formulas directly in the code, in lieu of creating block diagram subsections.

Consider the equation

$$y = x - e \sin x,$$

where $0 \le e \le 1$. This is a famous equation in astrodynamics, known as Kepler's equation. If you implement this equation using regular LabVIEW arithmetic functions, the block diagram looks like the one in Figure 6.31. You can implement the same equation using a Formula Node as shown in the same figure.

 The Formula Node operates in a similar fashion to the MathScript Node discussed in Chapter 3, but has a more limited set of functions that it can execute.

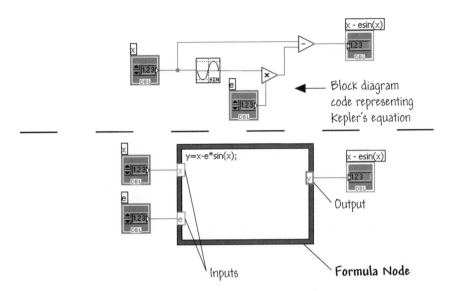

FIGURE 6.31
Implementing formulas in a Formula Node can often simplify the coding

6.6.1   Formula Node Input and Output Variables

You place the Formula Node on the block diagram by selecting it from the **Structures** subpalette of the **Functions≫Programming** palette (see Figure 6.21). You create the input and output terminals of the Formula Node by right-clicking on the border of the node and choosing **Add Input** or **Add Output** from the shortcut menu, as shown in Figure 6.32. Output variables have a thicker border than input variables. You can change an input to an output by selecting **Change to Output** from the shortcut menu, and you can change an output to an input by selecting **Change to Input** from the shortcut menu.

Once the necessary input and/or output terminals are on the Formula Node, enter the input and output variable names in their respective boxes using the **Labeling** tool. Every variable used in the Formula Node must be declared as an input or an output with no two inputs and no two outputs possessing the same name. An output can, however, have the same name as an input. Intermediate variables (that is, variables used in internal calculations) must be declared as outputs, although it is not necessary for them to be wired to external nodes. In Figure 6.32, the y variable and the z variable are both declared as outputs, although the intermediate variable y does not have to be wired to an external node.

6.6.2   Formula Statements

Formula statements use a syntax similar to most text-based programming languages for arithmetic expressions. You can add comments by enclosing them inside a slash-asterisk pair (/*comment*/). Figure 6.33 shows the operators

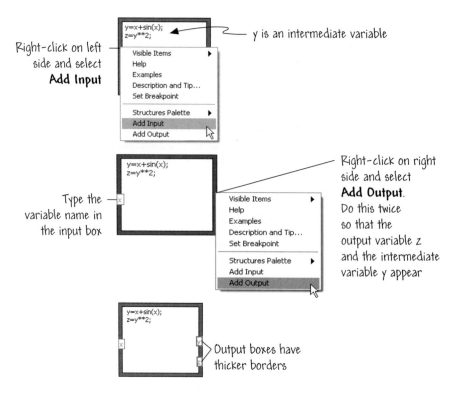

FIGURE 6.32
Formula Node input and output variables

FIGURE 6.33
Functions available in Formula Nodes

and functions that are available inside the Formula Node. You can access the same list of available operators and functions by choosing Help≫Show Context Help and moving the cursor over the Formula Node in the block diagram.

You enter the formulas inside the Formula Node using the **Labeling** tool. Each formula statement must terminate with a semicolon, and variable names are case-sensitive. There is no limit to the number of variables or formulas in a Formula Node. If you have a large number of formulas, you can either enlarge the Formula Node using the **Positioning** tool, or you can right-click in the Formula Node (not on the border) and choose **Scrollbar**. The latter method will put a scrollbar in the Formula Node, and with the **Operating** tool you can scroll down through the list of formulas for viewing.

The following example shows how you can perform conditional branching inside a Formula Node. Consider the following code fragment that computes the ratio of two numbers, x/y:

if $(y \neq 0)$ then
$z = x/y$
else
$z = +\infty$
end if

When $y = 0$, the result is set to ∞. You can implement the code fragment given above using a Formula Node, as illustrated in Figure 6.34.

The Formula Node demonstration depicted in Figure 6.34 can be found in **Chapter 6** *of the* **Learning** *directory and is called* **Formula Code.vi**. *Open the VI and try it out!*

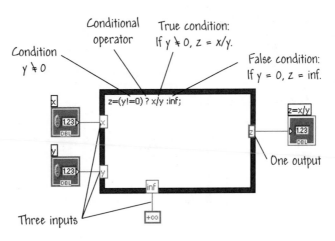

FIGURE 6.34
Implementing the formula $z = x/y$ in a Formula Node

6.7   SOME COMMON PROBLEMS IN WIRING STRUCTURES

When wiring the structures presented in this chapter, you may encounter wiring problems. In this section, we discuss some of the more common wiring errors and present suggestions on how to avoid them. Some common problems with wiring structures are:

- Failing to wire a tunnel in all cases of a Case structure
- Overlapping tunnels
- Wiring underneath rather than through a structure.

6.7.1   Failing to Wire a Tunnel in All Cases of a Case Structure

When you wire from a Case structure to an object outside the structure, you must connect output data from all cases to the object. Failure to do so will result in a bad tunnel, as illustrated in Figure 6.35. This problem is a variation of the no source error. Why? Because at least one case would not provide data to the object outside the structure when that case executed. The problem is easily solved by wiring to the tunnel in all cases. Can you explain why this is not a multiple sources violation? It seems like one object is receiving data from multiple sources. The answer is that only one case executes at a time and produces only one output value per execution of the Case structure. If each case did not output a value, then data flow execution would stop on the cases that did not output a value.

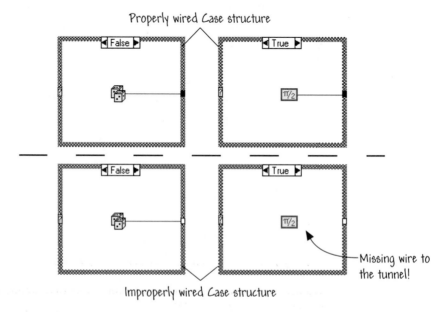

FIGURE 6.35
Failing to wire a tunnel in all cases of a Case structure leads to problems

6.7.2 Overlapping Tunnels

Tunnels are created as you wire, resulting occasionally in tunnels that overlap each other. Overlapping tunnels do not affect the execution of the diagram, so this is not really an error condition. But overlapping tunnels make editing and debugging of the VI more difficult. You should avoid creating overlapping tunnels! The fix is easy—drag one tunnel away with the **Positioning** tool to expose the other tunnels. A problematic situation is illustrated in Figure 6.36.

FIGURE 6.36
Sometimes overlapping tunnels occur when wiring structures

6.7.3 Wiring Underneath Rather Than through a Structure

Suppose that you want to pass a variable through a structure. You accomplish this by clicking either in the interior or on the border of the structure as you are wiring. This process is illustrated in Figure 6.37. If you fail to click in the interior or on the border, the wire will pass underneath the structure, and some segments of the wire may be hidden. This condition is not an error per se, but hidden wires are visually confusing and should be avoided.

As you wire through the structure, you are provided with visual cues for guidance during the wiring process. For example, when the **Wiring** tool crosses the left border of the structure, a highlighted tunnel appears. This lets you know that a tunnel will be created at that location as soon as you click the mouse button. You should click the mouse button! However, if you continue to drag the tool through the structure (without clicking the mouse on the left border) until the tool touches the right border of the structure, a second highlighted tunnel appears. If you continue to drag the **Wiring** tool past the right border of the structure without clicking, both tunnels disappear. When this occurs, the wire passes underneath the structure rather than through it, as depicted in Figure 6.37.

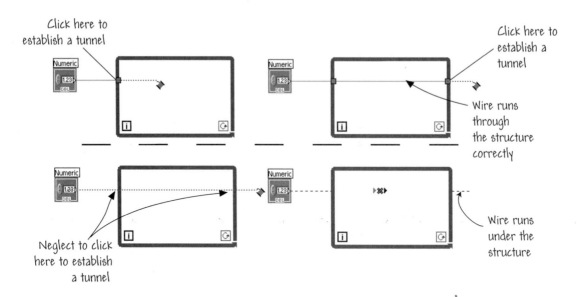

FIGURE 6.37
Wiring underneath rather than through a structure is a common problem

BUILDING BLOCK

6.8   BUILDING BLOCKS: PULSE WIDTH MODULATION

In this Building Block you will create a Pulse Width Modulation VI using a While Loop, a Case structure, and Shift Registers. A new element of the design implementation will be the realization of the VI as a *state machine*. A state machine is an architecture employed by LabVIEW developers to build complex decision-making algorithms quickly. State machines are useful in situations where distinguishable states exist, such as in the Pulse Width Modulation VI where the output is dependent upon the current output value as well as the input values from the Clock and the user controls Period and Duty Cycle.

The VI will have three states, defined by three different cases within a Case structure. To simplify the task, use the Standard State Machine VI Template. On the Getting Started screen, select New≫VI from Template... and create a new VI that uses the **Standard State Machine** template found in From Template≫Frameworks≫Design Patterns. Save the VI as PWM with LED.vi.

Your state machine will have three states: **Update**, **High**, and **Low**. The initial state will be called **Update**. In this state, the most recently entered values for Period and Duty Cycle will be read. These new values will then be stored in shift registers to be read by the other states. At the beginning of each cycle the state machine should return to this state to check for new values for **Period** and/or **Duty Cycle**. The VI is in the **High** state whenever the pulse output is high. Similarly, the VI is in the **Low** state whenever the output is low. In order to determine the transitions between the high and low state, you will need to use the Rising Edge and Falling Edge subVIs created in the previous Building Blocks.

Working versions of the Falling Edge subVI and the Rising Edge subVI can be found in the Building Blocks folder of the Learning directory.

Each state in the template block diagram is represented as a value of an enumeration. This enumeration is a customized control. Right-click on one of the enumerations (for example, **Initialize** on the left-hand side of the block diagram) and select **Open Type Def.** to modify the master copy of the custom control, as illustrated in Figure 6.38. Type definitions are used when you want to place a control of the same data type in many places.

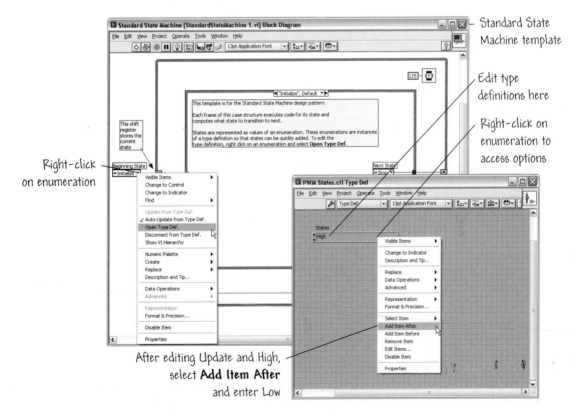

FIGURE 6.38
Editing the enumeration values to include Update, High, and Low

Using the **Labeling** tool, change the values of the enumerations to reflect the three states of your state machine: **Update**, **High**, and **Low**. Edit **Initialize** and change it to **Update**. Similarly, edit **Stop** and change it to **High**. You will need to right-click on the enumeration and select **Add Item After** to add the **Low** state. Then save the control as PWM States.ctl. Note that now all instances of the type definition have been updated with your customized states.

An edited file named PWM States.ctl *can be found in the* Building Blocks *folder of the* Learning *directory.*

Using Figure 6.39, construct a front panel with Duty Cycle and Period as the numeric controls and an LED and Clock numeric indicator as outputs. You want to be able to stop your VI at any time, and since it does not have a final state, you should delete the current input to the **Stop Condition** terminal and create a stop button as the control.

The Case structure currently has only two cases, so you will need to add another case after **High**. Since you saved your new version of the PWM States control, LabVIEW will automatically name the case you add **Low**. Using the block diagrams in Figure 6.40 as a guide, add your subVIs to the Pulse Width Modulation VI along with all other necessary coding to determine state transitions.

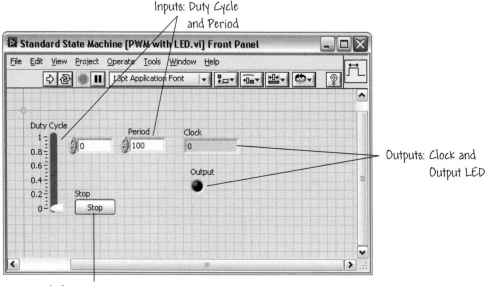

FIGURE 6.39
PWM with LED VI front panel

FIGURE 6.40
PWM with LED VI block diagram

In later chapters we will learn how to graphically display the output wave-form. For now, however, you will convert the output to a Boolean value and display it on the front panel using an LED. The LED should be on whenever the output waveform is high and off when it is low.

When you have completed your VI, experiment with different input values. The larger you make the **Period**, the easier it will be for you to observe the output pattern. You should find that the higher you make the **Duty Cycle**, the longer the LED remains lit for each cycle.

A working version of the PWM with LED VI can be found in the **Building Blocks** *folder in the* **Learning** *directory.*

6.9 RELAXED READING: INVESTIGATING GLOBAL WARMING WITH LABVIEW

The Antarctic ice sheet plays a crucial role in the behavior of the Earth's ice-ocean-atmosphere system because it influences the distribution of heat and mass within that system. Recently, there has been concern that global atmospheric warming could lead to a substantial rise in sea level, and the greatest uncertainty lies in the contribution of melting ice sheets. Improved understanding of this process requires integration of direct measurements of the ice sheet mass balance from satellite measurements and modeling of ice sheet dynamics.

Understanding of the fundamentals of ice flow is not currently sufficiently advanced to make quantitative predictions. Because field measurements cannot define the state of stress when modeling the flow of glaciers and ice sheets, a control system was required to simulate the stress of the flow of glaciers and ice sheets. The objective was to be able to measure the mechanical properties of anisotropic ice from the full depth of a borehole in Antarctica. The control system needed to be flexible and possess a high level of adaptability to measure a large range of physical parameters.

With LabVIEW, it was possible to construct the required multi-axis control system at a lower cost than typical commercial servo-hydraulic turnkey solutions. The system configuration is illustrated in Figure 6.41. A 12-bit PCI-6025 board performs the data acquisition. This board also provides the stepper motor control and timing for the ultrasonic transducer tomography system. It was possible to switch ultrasonic transducer transmitters and receivers to achieve the required pattern of data. In this way, the physical parameter changes in the ice specimen could be measured. Two PCI-7030 RT Series boards and 6030E 16-bit DAQ boards, each with 16 single-ended analog inputs and two analog outputs, made up the multi-axis digital servo hydraulic controller. A LabVIEW Real-Time application provided the error amplifier and valve drive voltage via the analog outputs to drive a precision hydraulic servo valve attached to each of the 60-kN double-acting hydraulic actuators.

The RT Series PCI hardware enabled an interactive solution. The LabVIEW RT application uses a series of layered windows for viewing and setting up of:

- Load or displacement increments, ramp rate, and tomography transducer selection

- Overall test progress of tomography with load and displacement monitoring

- Control loop function (PID and other outer loop controls) for changes in ice sample stiffness

- Mode of control (load or displacement) for individual or paired actuators

- Graphs for stress and strain of each axis

FIGURE 6.41
Schematic of the Control System

- Ultrasonic transducer pulsing levels and digital oscilloscope controls
- Transducer calibration
- Test data capture and replay

The application loads the LabVIEW RT program onto the PCI-7030 control boards after the test parameters are entered. It then awaits the commands from the main PC-based application. With the analog inputs on the RT Series acquisition board, multiple signals can be input from other axes into the closed-loop control system to ensure a stable environment for the tomography component of the experiments. The handshake system between the main control program and the RT Series board control program, combined with soft test limits, ensures a safe shutdown in the event of equipment failures. The LabVIEW RT control program effectively closed the control loop by taking data from the test setup and modifying the valve drive voltage to the actuator systems. The system provides feedback directly to the RT Series board by the conditioned transducer signals, which are connected to the board's analog input channels.

With LabVIEW RT, a low-cost multi-axis control system was constructed in a familiar programming environment with a high level of quality technical support. This project represents a successful application of LabVIEW to the study of the Antarctic ice sheets flows—an important area of research for planetary scientists.

For more information:

Dr. Steve Boon
The Rock and Ice Physics Laboratory
Department of Geological Sciences
University College of London
Gower Street, London, WC1E 6BT
s.boon@ucl.ac.uk

6.10 SUMMARY

In this chapter we studied four structures (For Loops, While Loops, Flat Sequence structures, and Case structures). The While Loop and the For Loop are used to repeat execution of a subdiagram placed inside the border of the resizable loop structure. The While Loop executes as long as the value at the conditional terminal is either TRUE or FALSE. The For Loop executes a specified number of times. Shift registers are variables that transfer values from previous iterations to the beginning of the next iteration.

The Case structure and Sequence structures are utilized to control data flow. Case structures are used to branch to different subdiagrams depending on the values of the selection terminal of the Case structure. Sequence structures are used to execute diagram functions in a specific order. The portion of the diagram to be executed first is placed in the first frame of the Sequence structure, the diagram to be executed second is placed in the second frame, and so on. Use the Flat Sequence structure to avoid using sequence locals and to better document the block diagram.

You can directly enter formulas on the block diagram in the Formula Node. The Formula Node is useful when the function equation has many variables or otherwise would require a complex block diagram model for implementation.

We also discussed the matter of controlling and timing the execution of structures. The Wait (ms) and Tick Count (ms) functions in conjunction with a Sequence structure let you accomplish these tasks using traditional VIs. You can also accomplish the timing task using the Time Delay Express VI and the Elapsed Time Express VI.

KEY TERMS

Case structure: A conditional branching control structure that executes one and only one of its subdiagrams, based on specific inputs. It is similar to If-Then-Else and Case statements in conventional programming languages.

Coercion: The automatic conversion performed in Lab-VIEW to change the numeric representation of a data element.

Coercion dot: Dot that appears where LabVIEW is forced to convert a numeric representation of one terminal to match the numeric representation of another terminal.

Conditional terminal: The terminal of a While Loop that contains a Boolean value that determines whether the loop performs another iteration.

Control flow: Determining the execution order of a program by arranging its elements in a certain sequence.

Count terminal: The terminal of a For Loop whose value determines the number of times the For Loop executes its subdiagram.

Data dependency: The concept that block diagram nodes do not execute until data is available at all the node inputs.

For Loop: Iterative loop structure that executes its subdiagram a set number of times.

Formula Node: Node that executes formulas that you enter as text. Especially useful for lengthy formulas too cumbersome to build in the block diagram.

Frames: Diagrams associated with Sequence structures that look like frames of film and control the execution order of the program.

Iteration terminal: The terminal of a For Loop or While Loop that contains the current number of completed iterations.

Object: Generic term for any item on the front panel or block diagram, including controls, nodes, and wires.

Sequence structure: Program control structure that executes its subdiagram in numeric order. Commonly used to force nodes that are not data-dependent to execute in a desired order.

Shift register: Optional mechanism in loop structures used to pass the value of a variable from one iteration of a loop to a subsequent iteration.

Structure: Program control element, such as a Sequence, Case, For Loop, or While Loop.

Subdiagram: Block diagram within the border of a structure.

Tunnels: Relocatable connection points for wires from outside and inside the structures.

While Loop: Loop structure that repeats a section of code until a Boolean condition is False.

EXERCISES

E6.1 Open the block diagram of Rearrange Cases.vi found in the Exercises&Problems folder in Chapter 6 of the Learning directory. Analyze all the possible temperature values listed in the Case Structure. Notice that the case values are not in ascending numerical order. In order to change the order of case frames, right-click on the **Case** border and select **Rearrange Cases**. This will bring up the **Rearrange Cases** wizard, which displays the current order of the cases. Click the **Sort** button to arrange the cases in ascending numerical order. Click **OK** if you want to leave them in ascending order. Otherwise, if you want the case frames to display in any other order, click and drag the values in the Case List.

E6.2 Open the block diagram of Gray Dots.vi found in the Exercises&Problems folder in Chapter 6 of the Learning directory. Notice there are two gray dots to signify numeric conversion (see Figure E6.2). In general, you should try to eliminate as many gray dots in your VIs as possible because they slow down execution and use more memory for mathematical operations. To eliminate gray dots, right-click on the terminal with the gray dot and select **Representation**. In this VI, right-click on the indicator terminal labeled Data Points and choose the same representation as the input value—which is DBL. To eliminate the other gray dot on the N terminal of the For Loop, change the representation on the control terminal titled Number of Data Points to match the N terminal—which is I32.

E6.3 Open Case Errors.vi found in the Exercises&Problems folder in Chapter 6 of the Learning directory and click on the **Broken Run** arrow. This will display a list of the errors in this VI (see Figure E6.3). Use the error list to determine how to fix the VI so you can run it.

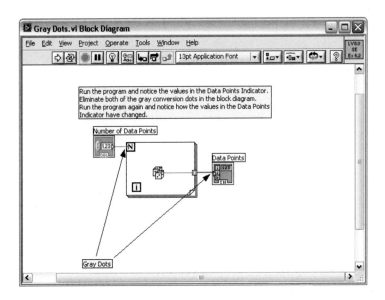

FIGURE E6.2
Eliminating gray dots

FIGURE E6.3
Eliminating Case structure errors

E6.4   Using a single While Loop, construct a VI that executes a loop N times or until the user presses a stop button. Be sure to include the Time Delay Express VI so the user has time to press the **Stop** button.

E6.5   Open the block diagram of the Avoid Sequence.vi found in the Exercises&Problems folder in Chapter 6 of the Learning directory, and gain an understanding of the program. Rewrite this VI without using the Flat Sequence structure. The VI can execute in the correct order without the Sequence structure using data flow. Remember that it is important to use Sequence structures only when it is not possible to wire nodes together to ensure that they execute in the proper order.

E6.6   Open the block diagram of the Loan Calculator.vi found in the Exercises&Problems folder in Chapter 6 of the Learning directory. This program will compute monthly payments required to pay off a loan based on the inputs on the front panel. The user has three choices for how the interest is compounded and two choices for the payment frequency. Click on the **Broken Run** arrow to display the list of errors. You will need to program the following equations into the four Formula Express VIs:

Interest Compounded Monthly	$i = \text{APR}/\text{PymtFreq}$
Interest Compounded Daily	$i = (1 + (\text{APR}/365))^{\text{PymtFreq}} - 1$
Interest Compounded Continuously	$i = e^{(\text{APR}/\text{PymrFreq})} - 1$
Payment Amount	$A = P[i(1+i)^N]/[(1+i)^N - 1]$

E6.7   Open the block diagram of the Compute Equilibrium.vi found in the Exercises&Problems folder in Chapter 6 of the Learning directory. This VI computes the equilibrium force required to keep a beam in equilibrium. Add a Case structure to the program so the user can choose whether the native LabVIEW functions or the Formula Express VIs will be used to calculate the result. The Formula Express VIs required are already programmed for you; you just need to add the Case structure, move them into the appropriate case, and wire them correctly.

     Review: In order for a beam to be at equilibrium, the sum of the moments about any point must be zero. This example uses the sum of the moments about one end of the beam and the sum of the moments about the point where force A is located to determine the C and z values associated with the equilibrium force.

E6.8   Create a VI that uses a Flat Sequence structure and the Elapsed Time Express VI to measure how long it takes LabVIEW to execute Open URL in Default Browser.vi, found on the **Functions≫Programming≫Dialog & User Interface≫Help** palette. Display the execution time on a numeric indicator on the front panel. The elements of the block diagram that you will need to utilize in your VI are depicted in Figure E6.8.

FIGURE E6.8
Elements needed on the block diagram

E6.9 Open the While Loop Exercise VI found in the **Exercises&Problems** folder in **Chapter 6** of the **Learning** directory. This VI is supposed to run continuously until the random number (between 1 and 10) generated outside the While Loop is equal to the user input number (also ranging from 1 to 10).

The uncorrected block diagram is shown in Figure E6.9. Run the VI and verify that the operation of the VI is not as desired. There are two possible modifications that would achieve the desired functionality—both minor. See if you can find both fixes to the VI.

FIGURE E6.9
Block diagram for the While Loop Exercise VI

PROBLEMS

P6.1   Open the block diagram of the **Statistics.vi** found in the **Exercises&Problems** folder in **Chapter 6** of the **Learning** directory. This program will compute the number of combinations possible if you are to select r out of n possible objects. Add Case structures to the code to utilize the inputs **Does Order Matter?** and **Can Samples Repeat?** and display the appropriate result on the front panel. The table below outlines which equation should be used in which case:

	Order Matters	**Order Doesn't Matter**
Samples Repeat	n^r	$(n-1+r)!/(n-1)!r!$
Samples Cannot Repeat	$n!/(n-r)!$	$n!/r!(n-r)!$

P6.2   Use a For Loop to generate 100 random numbers. Determine the most current maximum and minimum number as the random numbers are being generated. This is sometimes referred to as a "running" maximum and minimum. Display the running maximum and minimum values as well as the current random number on the front panel.

Be sure to include the Time Delay Express VI so the user is able to watch the values update as the For Loop executes. Figure P6.2 shows the Time Delay Express VI configuration dialog box (accessed by double-clicking on the VI) with a recommended delay of 0.1 second.

   Place the Time Delay Express VI within the For Loop.

FIGURE P6.2
Time Delay Express VI with configuration dialog box showing 0.1 second delay

P6.3 Construct a VI that displays a random number between 0 and 1 once every second. Also, compute and display the average of the last four random numbers generated. Display the average only after four numbers have been generated; otherwise display a 0. Each time the random number exceeds 0.5, generate a beep sound using the **Beep.vi**.

*The Beep VI can be found on the **Functions≫Programming≫Graphics& Sound** palette.*

P6.4 Construct a VI that has three Round LEDs on the front panel. When you run the program, the first LED should turn on and stay on. After one second, the second LED should turn on and stay on. After two more seconds, the third LED should turn on and stay on. All LEDs should be on for three seconds, and then the program should end.

Use the Flat Sequence structure with three frames and the Time Delay Express VI within each frame.

P6.5 Create a time trial program to compare the average execution times of the Formula Node and the native LabVIEW Math Functions. This program will require a For Loop, a Flat Sequence structure, and a Case structure. The For Loop is required to run the time trial N times and then the results can be averaged using the Statistics Express VI. The Sequence structure is required to sample the Tick Count before and after the code executes. The Case structure is required to determine whether the user would like to execute the Formula Node or the native LabVIEW Math Functions. Your front panel should resemble the one shown in Figure P6.5. To test the timing, use the following formulas:

$$a = X^2/4;$$
$$b = (2 * X) + 1;$$
$$Y = \sin(a + b);$$

where X is the input and Y is the output. Run the time trial for each of the cases. Which method has the fastest execution time? Which method is the easiest to program? Which method is the easiest to understand if someone else were to look at your code?

P6.6 Using a While Loop which iterates every one-hundredth of a second, create a VI that calculates the percentage of the total run-time a push button on the front panel is depressed. The loop should run until a stop button is pressed by the user. There should be an indicator on the front panel displaying how many seconds the VI has been running. Update this indicator with each loop iteration. After

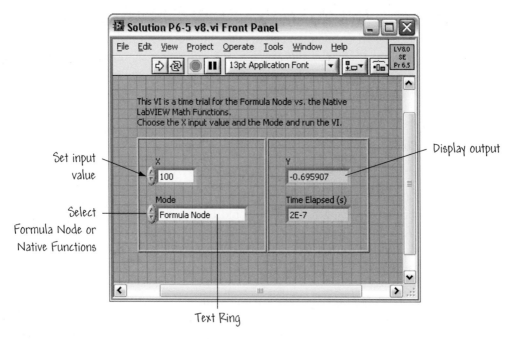

FIGURE P6.5
Suggested front panel for P6.5

FIGURE P6.6
Front panel for the VI designed to compute the run-time a push button is depressed

the stop button is pressed, another indicator should display the ratio of time the button was depressed to total run-time. Your front panel should resemble the one shown in Figure P6.6.

DESIGN PROBLEMS

D6.1   Construct a VI that uses one Formula Node to calculate the following equations:

$$y_1 = 10\sin(a) + b$$
$$y_2 = a^3 + b^2 + 100$$

The inputs are the variables a and b. The outputs are y_1 and y_2.

D6.2   Use a Formula Node to calculate the Body Mass Index (BMI) of an individual (this design problem was originally presented in Chapter 5). The BMI is an internationally used measure of obesity. The subVI will have two numeric inputs: a weight in pounds and a height in inches. Calculate the BMI using the formula

$$\text{BMI} = (703 * W)/H^2$$

where W is the weight in pounds and H is the height in inches. Display the calculated BMI on a numeric indicator.

Classify the body weight according to the Table D6.2 and output a string containing this information. There should also be a "warning" Boolean output that is True for any unhealthy input (i.e., underweight, overweight, or obese).

TABLE D6.2   BMI Classification

<18.5	Underweight
18.5–24.9	Healthy
25–29.9	Overweight
≥ 30	Obese

D6.3   Construct a VI that calculates maximum heart rate using the following formulas:

Gender	Maximum Heart Rate
Male	214 − Age * 0.8
Female	209 − Age * 0.7

On the front panel, place a numeric control to input age and a slide switch to input gender. A numeric indicator should display the maximum heart rate. Use Formula Nodes to solve the heart rate formulas. The front panel should resemble the one depicted in Figure D6.3.

FIGURE D6.3
The Max Heart Rate VI front panel

D6.4 Create a VI that calculates the hyperfactorial of any positive integer n, where the hyperfactorial is equivalent to the value obtained by the operation $1^1 * 2^2 * 3^{3*} \ldots n^n$. If the user inputs a value that is not a positive value, display a message informing the user that the input is not valid and request a new input.

Two important elements of the block diagram will be the One Button Dialog function and the Power of X subVI. The two functions are shown in Figure D6.4.

(a) One Btn Dialog   (b) Power of X

FIGURE D6.4
Use the One Button Dialog and the Power of X subVI

FIGURE 7.5
The terminal of an array shell is black, denoting an undefined data type

> *Remember that you must assign a data object to the empty array shell before using the array on the block diagram. If you do not assign a data type, the array terminal will appear black with an empty bracket.*

7.1.2   Multidimensional Arrays

A two-dimensional (2D) array requires two indices—a row index and a column index—to locate an element. A three-dimensional array requires three indices, and in general, an *n*-dimensional array requires *n* indices. You add dimensions to the array in one of two ways: (a) by using the **Positioning** tool to resize the index display, or (b) by right-clicking on the array index display and choosing **Add Dimension** from the shortcut menu (see Figure 7.6). An additional index display appears for each dimension you add. The example in Figure 7.6 shows a 2D numeric control array.

You can reduce the dimension of an array by resizing the index display appropriately or by selecting **Remove Dimension** from the shortcut menu.

FIGURE 7.6
Adding dimensions to an array

7.2 CREATING ARRAYS WITH LOOPS

With the For Loop and the While Loop you can create arrays automatically with a process known as **auto-indexing**. Figure 7.7(a) shows a For Loop creating a 10-element array using auto-indexing. On each iteration of the For Loop, the next element of the array is created. In this case the loop counter is set to 10; hence a 10-element array is created. If the loop counter was set to 20, then a 20-element array would be created. The array passes out of the loop to the indicator after the loop iterations are complete.

It is possible to pass a value out of a loop without creating an array. This requires that you disable auto-indexing by right-clicking on the tunnel (the square on the loop border through which the data passes) and selecting **Disable Indexing** from the shortcut menu. In Figure 7.7(b), auto-indexing is disabled; thus when the VI is executed, only the last value returned from the Random Number function passes out of the loop.

Open and run Array Auto Index Demo.vi shown in Figure 7.7. It can be found in the Chapter 7 folder in the Learning directory. After running the program, you should find that with auto-indexing enabled, the indicator array will have 10 elements (indexed 0, 1, · · · , 9), and with auto-indexing disabled only the last value of the Random Number function is passed out of the For Loop.

(a) Indexing enabled

(b) Indexing disabled

FIGURE 7.7
Auto-indexing is the ability of For Loops and While Loops to automatically index and accumulate arrays at their boundaries

You can also pass arrays into a loop one element at a time or the entire array at once. With auto-indexing enabled, when you wire an array (of any dimension) from an external node to an input tunnel on the loop border, then elements of the array enter the loop one at a time, starting with the first component. This is illustrated in Figure 7.8(a). With auto-indexing disabled, the entire array passes into the loop at once. This is illustrated in Figure 7.8(b).

*With For Loops auto-indexing is enabled by default. In contrast, with While Loops auto-indexing is disabled by default. If you desire auto-indexing, you need to right-click on the While Loop tunnel and choose **Enable Indexing** from the shortcut menu.*

With a For Loop with auto-indexing enabled, an array entering the loop automatically sets the loop count to the number of elements in the array, thereby eliminating the need to wire a value to the loop count, N. What happens if you explicitly wire a value to the loop count that is different than an array size entering the loop? Or what if you wire two arrays with different numbers of elements to

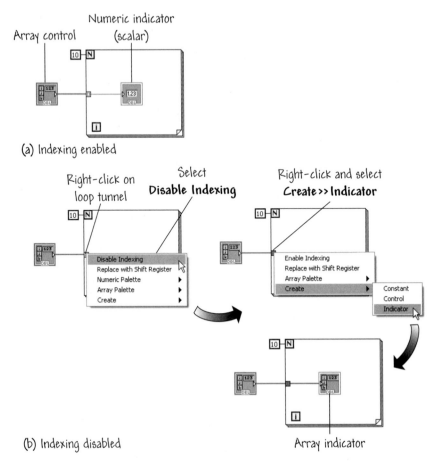

(a) Indexing enabled

(b) Indexing disabled

FIGURE 7.8
Auto-indexing applies when you are wiring arrays into loops

a For Loop? The answer is that if you enable auto-indexing for more than one array entering the same For Loop, or if you set the loop count by wiring a value to N with auto-indexing enabled, the actual loop count becomes the smaller of the two. For example, in Figure 7.9, the array size ($= 4$), and not the loop count N ($= 5$), sets the For Loop count—because the array size is the smaller of the two. The Array Size function used in the VI shown in Figure 7.9 will be discussed in the next section along with other common built-in array functions.

Open and run Array Auto Count Set.vi shown in Figure 7.9. It can be found in the Chapter 7 folder in the Learning directory. After running the program, you should find that only four random numbers are generated, despite the fact that the loop count N has been set to five!

FIGURE 7.9
Automatically setting the For Loop count to the array size

7.2.1   Creating Two-Dimensional Arrays

You can use two nested For Loops (that is, one loop inside the other) to create a 2D array. The outer For Loop creates the row elements, and the inner For Loop creates the columns of the 2D array. Figure 7.10 shows two For Loops creating a 2D array of random numbers using auto-indexing.

Open and run Two Dimensional Array Demo.vi shown in Figure 7.10. It can be found in the Chapter 7 folder in the Learning directory. After running the program, you should find that a 2D array has been created.

7.3   ARRAY FUNCTIONS

Many built-in functions are available to manipulate arrays. Most of the common array functions are found in the **Array** palette of the **Functions≫Programming** palette, as illustrated in Figure 7.11. Several of the heavily used functions are pointed out in Figure 7.11 and discussed in this section.

Row index

Column index

Array dimensions enlarged with the **Positioning** tool to show all array elements

You can change these constants to vary array dimensions

1D array   2D array

FIGURE 7.10
Creating a 2D array using two For Loops and auto-indexing

Select **Programming >> Array**

Array Size

Initialize Array

Build Array

Index Array

Array Subset

FIGURE 7.11
The array functions palette

7.3.1   Array Size

The function Array Size returns the number of elements in the input array, as illustrated in Figure 7.12. If the input is an *n*-dimensional array, the output of the array function Array Size is a one-dimensional array with *n* elements—each element containing a dimension size.

You can run the array size demonstration shown in Figure 7.12 by opening **Array Size Demo.vi** *in the* **Chapter 7** *folder in the* **Learning** *directory. After executing the VI once to verify that the 1D array has length 4 and the 2D array has 2 rows and 4 columns, add an element to the 1D array and run the VI again. What happens? The* **1D Array size** *value increments by 1.*

7.3.2   Initialize Array

The function Initialize Array creates an *n*-dimensional array with elements containing the values that you specify. All elements of the array are initialized to the same value. To create and initialize an array that has more than one dimension, right-click on the lower left side of the Initialize Array node and select **Add**

FIGURE 7.12
The function Array Size returns the number of elements in the input array

FIGURE 7.13
The function Initialize Array creates an array of a certain dimension containing the same value for each element

Dimension or use the **Positioning** tool to grab a resizing handle to enlarge the node. You can remove dimensions by selecting **Remove Dimension** from the function shortcut menu or with the **Resizing** cursor. The Initialize Array is useful for allocating memory for arrays. For instance, if you are using shift registers to pass an array from one iteration to another, you can initialize the shift register using the Initialize Array function.

The input determines the data type and the value of each element of the initialized array. The dimension size input determines the length of the array (see Figure 7.13(a)). For example, if the input value **element** is a double-precision floating-point number with the value of 5.7, and dimension size has a value of 100, the result is a 1D array of 100 double-precision floating-point numbers all set to 5.7, as illustrated in Figure 7.13(b). You can wire inputs to the Initialize Array function from front panel control terminals, from block diagram constants, or from calculations in other parts of the block diagram. One method of obtaining the output array is to right-click on the Initialize Array function and choose Create≫Indicator. Also, you can right-click and choose Create≫Control and Create≫Constant to create inputs for the element and dimension size. Figure 7.13c depicts a 2D array with 3 rows and 2 columns initialized with the long integer value of 0.

You can run the array initialization demonstration shown in Figure 7.13 by opening Array Initialization Demo.vi *located in the* Chapter 7 *folder in the* Learning *directory. Change the initial value of the two arrays on the block diagram, run the VI, and examine the result.*

7.3.3   Build Array

The function Build Array concatenates multiple arrays or adds extra elements to an array. The function accepts two types of inputs—scalars and arrays—therefore it can accommodate both arrays and single-valued elements.

The Build Array function appears with one scalar input when initially placed in the block diagram window, as shown in Figure 7.14.

Array inputs have brackets, while element inputs do not. Array inputs and scalar inputs are not interchangeable! Pay special attention to the inputs of a Build Array function, otherwise you may generate wiring errors that are difficult to detect.

You can add as many inputs as you need to the Build Array function. Each input to the function can be either a scalar or an array. To add more inputs, right-click on the left side of the function and select **Add Input**. You also can enlarge the Build Array node by placing the **Positioning** tool at one corner of the object to grab and drag the resizing handle. You can remove inputs by shrinking the node with the resizing handle or by selecting **Remove Input** from the shortcut menu.

The input type (element or array) is automatically configured when wired to the Build Array function. The Build Array function shown in Figure 7.14 is configured to concatenate an array and one scalar element into a new array.

FIGURE 7.14
Build Array concatenates multiple arrays or appends elements to an array

The Build Array function concatenates the elements or arrays in the order in which they appear in the function, top to bottom. If the inputs are of the same dimension, you can right-click on the function and select **Concatenate Inputs** to concatenate the inputs into a longer array of the same dimension. If you leave the **Concatenate Inputs** option unselected, you will add a dimension to the array.

The Array Build demonstration shown in Figure 7.14 is located in the Chapter 7 folder in the Learning directory and is called Array Build Demo.vi. Open, run, and experiment with the VI.

7.3.4   Array Subset

The function Array Subset returns a portion of an array starting at **index** and containing **length** elements. In the example shown in Figure 7.15 you will find that the array index begins with 0.

*The array subset demonstration shown in Figure 7.15 can be found in the Chapter 7 folder in the Learning directory and is called Array Subset Demo.vi. Open and run the VI in the **Run Continuously** mode. Using the **Operating** tool, change the value of **Index number** and watch the Array Subset function return a portion of the array, starting at the index that you specify and including the three subsequent elements.*

FIGURE 7.15
The function Array Subset returns a portion of an array starting at **index** and containing **length** elements

7.3.5   Index Array

The function Index Array accesses an element of an array. The example shown in Figure 7.16 uses the input **Index number** to specify which element of the array to access. Remember that the index number of the first element is zero.

The index array demonstration shown in Figure 7.16 can be found in the **Chapter 7** *folder of the* **Learning** *directory and is called* **Array Index Demo.vi.** *Open and run the VI in* **Run Continuously** *mode. Using the* **Operating** *tool, change the value of* **Index number** *and watch the Index Array function return a different element based on the index value.*

The Index Array function automatically resizes to match the dimensions of the wired input array. For example, if you wire a 1D array to the Index Array function, a single index input will show. Similarly, if you wire a 2D array to the

FIGURE 7.16
The function Index Array accesses an element of an array

Index Array function, two index inputs will show—one for the row and one for the column.

Once you have wired an input array to the Index Array function, you can access more than one element, or subarray (e.g., a row or a column) using the **Positioning** tool to manually resize the function once placed on the block diagram. When you expand the Index Array function it will expand in increments determined by the dimensions of the array wired to the function. In Figure 7.17, the Index Array function is expanded so that three subarrays can be extracted: a row, a column, and a single element. The index inputs you wire determine the shape of the subarray you want to access or modify. For example, if the input to an Index Array function is a 2D array and you wire only the row input, you extract a complete 1D row of the array. If you wire only the column input, you

Index Array function is stretched to access more than one element or subarray

FIGURE 7.17
Using Index Array to extract a row or column of an array

extract a complete 1D column of the array. If you wire the row input and the column input, you extract a single element of the array. These three cases are illustrated in Figure 7.17. Each input group is independent and can access any portion of any dimension of the array.

*The 2D index array demonstration shown in Figure 7.17 can be found in the **Chapter 7** folder of the **Learning** directory and is called **2D Array Index Demo.vi**. Open and run the VI in **Run Continuously** mode. Using the **Operating** tool, change the **Column number** and **Row number** values and verify the numbers returned in **Column** and **Row**.*

Practice with Arrays

In this exercise you will get more practice using array functions. Open Practice with Arrays.vi located in the Chapter 7 folder of the Learning directory. The front panel and block diagram of the incomplete VI are shown in Figure 7.18. This VI is not completed—you will finish wiring and debugging the VI for practice.

The front panel contains four arrays and one numeric control. The completed VI concatenates the input arrays and the numeric control values to form a new

FIGURE 7.18
Practice with array functions

Display the results on the front panel

Initialize the array with ones

FIGURE 7.19
The completed block diagram for the array practice exercise

array. Using the Array Size function and the Array Initialize function, the VI creates a new array of appropriate dimension and with all elements of the array initialized to 1. In the final computation of the VI, the difference between the two new arrays is computed, and the result is displayed on the front panel. When you compute the difference of two arrays with the same number of elements, the differencing operation subtracts the array values element by element.

The completed block diagram should resemble the one shown in Figure 7.19. Remember to save the VI in the **Users Stuff** folder and close it.

The completed VI for this exercise can be found in the **Chapter 7** *folder of the* **Learning** *directory and is called* **Practice with Arrays Done.vi.** ◆

◆ **Practice with Auto-Indexing**

In this example you will create a VI that computes the final grades in a course using three input arrays containing student scores on three different exams.

The objective of this example is to demonstrate the use of auto-indexing using Arrays and For Loops. To do this, you will compare two different approaches to constructing the VI block diagram. The first approach uses a For Loop with auto-indexing enabled on the input. When auto-indexing is enabled, one element of each array is passed through the input tunnel with each loop iteration (incrementing through the array). The second approach to constructing the VI is to disable auto-indexing on the For Loop and instead use the Index Array function. When auto-indexing is disabled, the entire array is passed through the input tunnel at once—the entire array is available on each iteration. This means that the Index Array function must be used to iterate through each element in the array.

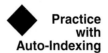

The input to the VI consists of three 1D arrays: **Exam 1 Scores** (weighted as 30% of the final course grade), **Exam 2 Scores** (weighted as 30% of the final course grade), and **Final Exam Scores** (weighted as 40% of the final course grade). You can use the VI front panel shown in Figure 7.20 as a guide. The number of elements in each array is equal to the number of students in the class. The VI should output a 1D array containing the final course grade for each student.

Using the following steps as a guide, you can construct a VI that will compute the average student scores using the Index Array function rather than auto-indexing.

1. Open a new VI and save it as **Array Auto-Indexing.vi** in the **Users Stuff** folder in the **Learning** directory.

2. On the front panel create three numeric arrays and label them **Exam 1 Scores, Exam 2 Scores**, and **Final Exam Scores**. Use Figure 7.20 as a guide.

3. Switch to the block diagram and place a For Loop and wire the three array inputs to the border of the For Loop.

4. The Auto-Indexing is enabled by default for the For Loop. To disable it, right-click each tunnel and select **Disable Indexing**, as illustrated in Figure 7.21(a). This will make the entire array available at each iteration of the For Loop. The Index Array function will then be needed to access the individual scores within each array to be used in the weighted average calculations.

FIGURE 7.20
The Array Auto-Indexing VI front panel

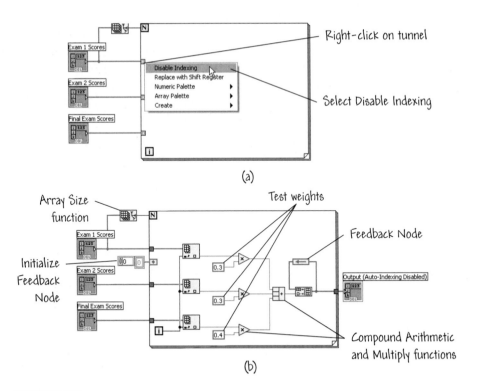

FIGURE 7.21
(a) Select Disable Indexing on the For Loop. (b) Completed block diagram showing Array Size function, Feedback Node, Compound Arithmetic and Multiply functions, and all necessary wiring to complete the diagram

5. You need to specify how many times the For Loop has to iterate. The output array that you are creating should have the same number of elements as the input arrays, that is, one score for each student in the class. To do this, use the Array Size function and pass its output to N, as shown in Figure 7.21(b).

6. The Index Array function is used to extract a single scalar value from each array on every iteration of the For Loop. This way, each iteration of the For Loop will operate on the Exam 1 Scores, Exam 2 Scores, and Final Exam Scores of one specific student to output the final grade for that student. Wire the Loop Iteration terminal to the index input of the Index Array function so that the Index Array function will output the exam scores of the next student on each iteration.

7. Using the Compound Arithmetic and Multiply functions as shown in Figure 7.21(b), perform the necessary operations on the individual array elements to determine the course grade for each student.

8. Use the Build Array function to create a 1D array containing the course grades for the students. The entire output array will be built within the For Loop and then passed out after all calculations have been completed. You

will need to use a Feedback Node to retain grades calculated on previous loop iterations. Continue appending newly calculated grades to the output array as the loop iterates for each student.

9. Create an indicator outside the loop to display the 1D output array. Name the indicator **Output (Auto-Indexing Disabled)**. Note that if you do not disable Auto-Indexing, the tunnel will output a 2D array and a broken wire appears when you connect the tunnel and indicator. Your block diagram should match the block diagram in Figure 7.21(b).

10. Complete the wiring of the block diagram and when ready, run the VI with different input values and verify that it computes the average grades correctly.

Using the following steps as a guide, you can construct a VI that will compute the average student scores using auto-indexing rather than the Index Array function. The objective is to create an identical output array, this time using the auto-indexing capability of the For Loop. This means that a new element from each input array will be passed into the loop upon each iteration, until all elements have been input.

1. Using the same front panel as above, add the **Output (Auto-Indexing Enabled)** indicator, as shown in Figure 7.22.

2. On the same block diagram as in the Array Auto-Indexing VI constructed above, create a new section of the VI that uses the inputs **Exam 1 Scores**, **Exam 2 Scores**, and **Final Exam Scores**. Place the arithmetic functions necessary to calculate the weighted average for each student inside the For

FIGURE 7.22
The Array Auto-Indexing front panel with the additional Output (Auto-Indexing Enabled) indicator

FIGURE 7.23
The completed Array Auto-Indexing block diagram

Loop. These will be the same as before, namely, the Compound Arithmetic and Multiply functions. Use Figure 7.23 as a guide.

3. Wire the exam score arrays directly to the inputs to the Multiply functions. Tunnels will automatically appear on the border of the For Loop with auto-indexing enabled by default.

4. Wire the array indicator **Output (Auto-Indexing Enabled)** to the output of the Compound Arithmetic function. Note that it is not necessary to provide an input to the Count terminal of the For Loop. Because the loop is now using auto-indexing, it will automatically iterate once for each element in the smallest array being input. In this case, the arrays are all the same size, and the loop will iterate once for each student in the class.

5. Your block diagram should resemble the one shown in Figure 7.23. Experiment with different input values and check that the output arrays contain identical values.

6. Remember to save the VI in the **Users Stuff** folder.

The completed VI for this exercise can be found in the **Chapter 7** *folder of the* **Learning** *directory and is called* **Array Auto-Indexing.vi**.   ◆

7.4   POLYMORPHISM

Polymorphism is the ability of certain LabVIEW functions (such as Add, Multiply, and Divide) to accept inputs of different dimensions and representations. Arithmetic functions that possess this capability are polymorphic functions. For example, you can add a scalar to an array or add together two arrays of different lengths. Figure 7.24 shows some of the polymorphic combinations of the Add function.

In the first combination, shown in Figure 7.24(a), the result of adding a scalar and a scalar is another scalar. In the second combination, shown in Figure 7.24(b), the result of adding a scalar to an array is another array. In this situation, the scalar input is added to each element of the input array. In the third combination,

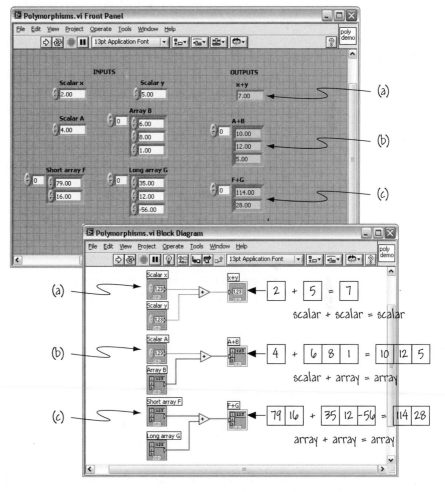

FIGURE 7.24
Polymorphic combinations of the Add function

depicted in Figure 7.24(c), an array of length 2 is added to an array of length 3, resulting in an array of length 2 (the length of the shorter of the two input arrays). Array addition is performed component-wise; that is, each element of one array is added to the corresponding element of the other array.

When two input arrays have different lengths, the output array resulting from some arithmetic operation (such as adding the two arrays) will be the same size as the smaller of the two input arrays. The arithmetic operation applies to corresponding elements in the two input arrays until the shorter array runs out of elements—then the remaining elements in the longer array are ignored!

Consider the VI depicted in Figure 7.25. Each iteration of the For Loop generates a random number that is stored in the array and readied for output once the loop iterations are finished. After the loop finishes execution (after 10 iterations), the Multiply function multiplies each element in the array by the scaling factor 2π. Notice that the Multiply function has two inputs: an array and a scalar. The polymorphic capability of the Multiply function allows the function to take inputs of differing dimension (in this instance, an array of length 10 and a scalar) and produce a sensible output. What happens when you have two array inputs for a Multiply function? In that case, corresponding elements of each array are multiplied.

*The two demonstrations of polymorphism shown in Figures 7.24 and 7.25 can be found in the **Chapter 7** folder in the **Learning** directory and are called*

FIGURE 7.25
Demonstrating the polymorphic capability of the Multiply function

Polymorphisms.vi and *Polymorphism Demo.vi,* respectively. *Open and run each VI. Vary the values of the array elements and examine the results. You can also edit the array lengths and test your knowledge of polymorphisms by first predicting the result and then verifying your prediction by running the VI.*

Practice Using Polymorphism

Open a new VI and recreate the front panel and block diagram shown in Figure 7.26. When you have finished the construction, experiment with the VI. A suggested way to run the VI is in **Run Continuously** mode. Increment and decrement the variable Array length and watch the array length change.

This exercise demonstrates two applications of polymorphism. Referring to Figure 7.26, it can be seen that in one case the Multiply function operates on two inputs of different dimension: an array and a scalar $\pi/2$. In the second illustration of polymorphism, the Subtract function has two array inputs of differing lengths. The resulting array Final Array 2 has length equal to the length of the shorter array (the length is equal to the length of Scaled array). Three array functions are also utilized: Build Array, Array Size, and Initialize Array. Save the VI as Polymorphism Practice.vi and place it in the Users Stuff folder.

FIGURE 7.26
Practicing using polymorphism

*The completed polymorphism practice VI shown in Figure 7.26 can be found in the **Chapter 7** folder in the **Learning** directory and is called **Polymorphism Practice.vi**.*   ◆

7.5 CLUSTERS

A **cluster** is a data structure that, like arrays, groups data. However, clusters and arrays have important differences. One important difference is that clusters can group data of different types, whereas arrays can group only like data types. For example, an array might contain ten numeric indicators, whereas a cluster might contain a numeric control, a toggle switch, and a string control. And although cluster and array elements are both ordered, you access cluster elements by **unbundling** some or all the elements at once rather than indexing one element at a time. Clusters are also different from arrays in that they are of a fixed size.

One similarity between arrays and clusters is that both are made up of either controls or indicators. In other words, clusters cannot contain a mixture of controls and indicators. An example of a cluster is shown in Figure 7.27. The cluster shown has four elements: a numeric control, a horizontal toggle switch, a string control (more on strings in Chapter 10), and a knob. The cluster data type appears frequently when graphing data (as we will see in the next chapter).

Clusters are generally used to group related data elements that appear in multiple places on the block diagram. Because a cluster is represented by only

FIGURE 7.27
A cluster with four elements: numeric control, toggle switch, string control, and knob

one wire in the block diagram, its use has the positive effect of reducing wire clutter and the number of connector terminals needed by subVIs. A cluster may be thought of as a bundle of wires wherein each wire in the cable represents a different element of the cluster. On the block diagram, you can wire cluster terminals only if they have the same type, the same number of elements, and the same order of elements. Polymorphism applies to clusters as long as the clusters have the same number of elements in the same order.

7.6   CREATING CLUSTER CONTROLS AND INDICATORS

Cluster controls and indicators are created by placing a **cluster shell** on the front panel, as shown in Figure 7.28. A new cluster shell has a resizable border and an (optional) label. When you right-click in the empty element area of the cluster shell, the **Controls** palette appears, as illustrated in Figure 7.29. You

Select **Cluster** and move the cluster shell to the front panel

Optional label

Resizable border

FIGURE 7.28
Creating and resizing a cluster

FIGURE 7.29
Creating a cluster by placing objects in the cluster shell

create a cluster by placing any combination of numerics, Booleans, strings, charts, graphs, arrays, or even other clusters from the **Controls** palette into the cluster shell. Remember that a cluster can contain controls *or* indicators, but not both. The cluster becomes a control cluster or indicator cluster depending on the first element you place in the cluster. For example, if the first element placed in a cluster shell is a numeric control, then the cluster becomes a control cluster. Any objects added to the cluster afterwards become control objects. Selecting **Change To Control** or **Change To Indicator** from the shortcut menu of any cluster element changes the cluster and all its elements from indicators to controls or from controls to indicators, respectively. You can also drag existing objects from the front panel into the cluster shell.

7.6.1   Cluster Order

The elements of a cluster are ordered according to when they were placed in the cluster rather than according to their physical position within the cluster shell. The first object placed in the cluster shell is labeled element 0, the second object inserted is element 1, and so on. The order of the remaining elements is automatically adjusted when an element is removed from the cluster. The cluster order determines the order in which the elements appear as terminals on the Bundle and Unbundle functions in the block diagram (more on these cluster functions in the next sections). You must keep track of your cluster order if you want to access individual elements in the cluster, because individual elements in the cluster are accessed by order.

You can examine and change the order of the elements within a cluster by right-clicking on the cluster border on the front panel and choosing the item **Reorder Controls In Cluster** from the shortcut menu, as shown in Figure 7.30. Notice in the figure that a new set of buttons replaces the toolbar, and the cluster appearance changes noticeably. Even the cursor changes to a hand with a # sign above—this is the cluster order cursor.

Two boxes appear side-by-side in the bottom right corner of each element in the cluster. The white boxes indicate the current place of that element in the cluster order. The black boxes indicate the new location in the order, in case you

FIGURE 7.30
Examining and changing the cluster order

have changed the order. The numbers in the white and black boxes are identical until you make a change in the order. Clicking on an element with the cluster order cursor changes the current place of the element in the cluster order to the number displayed in the toolbar at the top of the front panel. When you have completed arranging the elements within the cluster to your satisfaction, click on the **OK** button, or you can revert to the old order by clicking on the **X** button.

A simple example shows the importance of the cluster order. Consider the situation depicted in Figure 7.31(a), where the front panel contains two clusters. In the first cluster, the first component is a numeric control, while in the second cluster, the first component is a numeric indicator. In the associated block diagram, the cluster control properly wires to the cluster indicator. Now, in Figure 7.31(b), the cluster order has been changed so that the string control is the first component of the cluster control. The numeric indicator is still the first component of the cluster indicator. The connecting wire is now broken, and if you attempted to run the VI you would get an error message stating that there is a type conflict.

7.6.2 Using Clusters to Pass Data to and from SubVIs

A subVI connector panel can have a maximum of 28 terminals (as discussed in Section 5.3). In general, you do not want to pass information individually to all 28 terminals when calling a VI or a subVI. A good rule-of-thumb is to use connector passes with 14 or fewer terminals. Otherwise, the terminals are very small and may be difficult to wire. Using clusters, you can group related controls together. One cluster control uses one terminal on the connector, but it can contain many controls. Similarly, you can use a cluster to group indicators.

This allows the subVI to pass multiple outputs using only one terminal. Figure 7.32 illustrates the benefit of using clusters to pass data to and from subVIs.

7.7 CLUSTER FUNCTIONS

Two of the more important cluster functions are the Bundle and Unbundle functions. These functions (and others) are found in the **Cluster & Variant** subpalette of the **Functions≫Programming** palette, as illustrated in Figure 7.33.

7.7.1 The Bundle Function

The Bundle function is used to assemble individual elements into a single new cluster or to replace elements in an existing cluster. When placed on the block diagram, the Bundle function appears with two element input terminals on the left side. You can increase the number of inputs by enlarging the function icon

(a)

(b)

FIGURE 7.31
Example showing the importance of cluster order: (a) Wiring is correct (b) Wiring is incorrect

FIGURE 7.32
Using clusters to pass data to and from subVIs

FIGURE 7.33
The **Cluster** palette

(with the resizing handles) vertically to create as many terminals as you need. You can also increase the number of inputs by right-clicking on the left side of the function and choosing **Add Input** from the menu. Since you must wire all the inputs you create, only enlarge the icon to show the exact desired number of element inputs. When you wire to each input terminal, a symbol representing the data type of the wired element appears in the originally empty terminal, as illustrated in Figure 7.34. The order of the elements in the cluster is equal to the order of inputs to the Bundle function. The order is defined top to bottom, which means that the element you wire to the top terminal becomes element 0, the element you wire to the second terminal becomes element 1, and so on.

The example depicted in Figure 7.34 shows the Bundle function creating a cluster from three inputs: a floating-point real number, an integer, and an array of numbers generated by the For Loop. The output of the new cluster is wired to a waveform graph that displays the random numbers on a graph. Graphs (such as the waveform graph) are covered in Chapter 8.

FIGURE 7.34
The Bundle function

*The example shown in Figure 7.34 can be found in the **Chapter 7** folder in the **Learning** directory and is called **Cluster Bundle Demo.vi**. When you open and run the VI you will see that varying the input x0 changes the x-axis origin.*

In addition to input terminals for elements on the left side, the Bundle function also has an input terminal for clusters in the middle, as shown in Figure 7.35. Sometimes you want to replace or change the value of one or two elements of a cluster without affecting the others. The example in Figure 7.35 shows a convenient way to change the value of two elements of a cluster. Because the cluster input terminal (that is, the middle terminal) of the Bundle function is wired to an existing cluster named **Cluster**, the only element input terminals that must be wired are those that are associated with cluster elements that you want to replace—in the illustration in Figure 7.35 the second and fourth element values are being replaced. To replace an element in the cluster using the Bundle

FIGURE 7.35
Using the Bundle function to replace elements of an existing cluster

function, you first place the Bundle function on the block diagram and then resize the function to show exactly the same number of input terminals as there are elements in the existing cluster that you want to modify. The next step is to wire the existing cluster to the middle input terminal of the Bundle function. Afterwards, any other input terminals on the left side that are wired will replace the corresponding elements in the existing cluster. Remember that if the objective is to create a new cluster rather than modify an existing one, you do not need to wire an input to the center cluster input of the Bundle function.

The example depicted in Figure 7.35 shows the Bundle function being used to change the value of the numeric control and the knob. The corresponding VI can be found in the Chapter 7 folder in the Learning directory and is called Cluster Element Replacement.vi. Open and run the VI in **Run Continuously** mode. Cluster is a control cluster and Cluster Output is an indicator cluster. Toggle the horizontal toggle switch (using the **Operating** tool). You should see that the toggle switch in Cluster Output also switches. Now vary the Knob value away from the default value of 4.8. Notice that the Knob in Cluster Output does not change. This phenomenon occurs because the Knob is the fourth element of the cluster and the value is being replaced by the value set in the numeric control Knob input. To change the value of the Knob in Cluster Output you change the value of Knob input. Try it out!

7.7.2   The Unbundle Function

The Unbundle function extracts the individual components of a cluster. The output components are arranged from top to bottom in the same order as in the cluster. When placed on the block diagram, the Unbundle function appears with two element output terminals on the right side. You adjust the number of terminals with the resizing handles following the same method as with the Bundle function, or you can right-click on the right side of the function and choose **Add Output** from the menu. Element 0 in the cluster order passes to the top output terminal, element 1 passes to the second terminal, and on down the line. The cluster wire remains broken until you create the correct number of output terminals, at which point the wire becomes solid. When you wire an input cluster to the correctly sized Unbundle, the previously blank output terminals will assume the symbols of the data types in the cluster.

While all elements are unbundled using the Unbundle function, the Unbundle By Name function can access one or more elements in a cluster.

The example depicted in Figure 7.36 shows the Unbundle function being used to unpack the elements of the data structure **Cluster**. The cluster has four elements, and each element is split from the cluster and wired to its own individual indicator for viewing on the front panel.

Unbundle function

FIGURE 7.36
The Unbundle function

*The example shown in Figure 7.36 can be found in the **Chapter 7** folder in the **Learning** directory and is called **Cluster Unbundle Demo.vi**. Open and run the VI in **Run Continuously** mode and notice that varying any values in the control cluster immediately changes the values of the various indicators.*

7.7.3   Creating Cluster Constants on the Block Diagram

You can use the same technique you used on the front panel to create a cluster constant on the block diagram. On the block diagram, choose Cluster Constant from the **Functions≫Programming≫Cluster & Variant** palette to create the cluster shell as illustrated in Figure 7.37. Click in the block diagram to place the cluster shell, and place other constants of the appropriate data type within the cluster shell.

If you have a cluster control or indicator on the front panel and want to create a cluster constant containing the same components in the block diagram, you can either drag that cluster from the panel to the diagram or select Create≫Constant from the shortcut menu, as depicted in Figure 7.38.

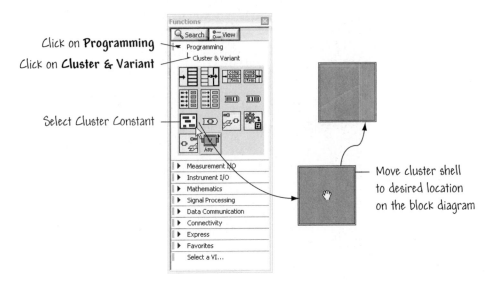

FIGURE 7.37
Creating a cluster constant on the block diagram

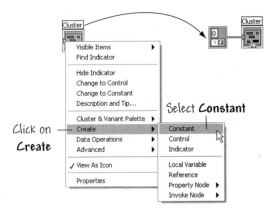

FIGURE 7.38
Create a cluster constant using a cluster on the front panel

7.7.4   Using Polymorphism with Clusters

Since the arithmetic functions are polymorphic, they can be used to perform computations on clusters of numbers. As shown in Figure 7.39, you use the arithmetic functions with clusters in the same way you use them with arrays of numerics. You can also use the string-to-number functions to convert a cluster of numerics to a cluster of strings (strings are covered in Chapter 10).

FIGURE 7.39
Using polymorphism with clusters

The VI in Figure 7.39 can be found in **Chapter 7** *of the* **Learning** *directory. The VI is called* **ClusterScaling.vi.**

BUILDING BLOCK

7.8   BUILDING BLOCKS: PULSE WIDTH MODULATION

The goal in this Building Block is to utilize an array to store the output values of the VI constructed in Chapter 6. You will use the Pulse Measurements VI to confirm that the output has the desired waveform and duty cycle.

To begin, open the Building Blocks VI that you constructed in Chapter 6. It should be called **PWM with LED.vi** and located in the **Users Stuff** folder in the **Learning** directory. Once the VI is open, rename it as **PWM with Measurements.vi** and save the VI in the **Users Stuff** folder.

*In case you do not have a working version of the Building Blocks VI from Chapter 6, a working version of the PWM with LED VI can be found in the **Building Blocks** folder in the **Learning** directory.*

Switch to the block diagram of the PWM with Measurements VI. The changes that you will make are illustrated in Figure 7.40. To begin, add the Boolean To (0,1) function inside the While Loop but outside of the Case structure, as shown in Figure 7.40. The Boolean To (0,1) function is found on the **Functions≫Programming≫Boolean** palette. Wire the input to the **Output** indicator to the Boolean To (0,1) and then wire the output of the Boolean To (0,1) to the While Loop.

FIGURE 7.40
The PWM with Measurements VI

You will need to convert the Boolean outputs to numeric values before creating the array. Recall that by default, auto-indexing is disabled on While Loops and so you will need to enable it before you can output an array from the loop. If you do not use auto-indexing with this While Loop, on each iteration of the While Loop, a single scalar value representing the value of the waveform on the last clock cycle will be output. When auto-indexing is enabled, each iteration of the loop will append a new output value to the end of an array containing all output values. To enable the auto-indexing, right-click on the tunnel and select **Enable Indexing**, as shown in Figure 7.41.

Place the Pulse Measurements VI on your block diagram outside of the While Loop as shown in Figure 7.40. This VI is found on the **Signal Processing≫ Waveform Measurements** palette. You can use **Context Help** to find out more about this VI, including the outputs, which include pulse, pulse duration, and duty cycle. Right-click on the appropriate terminal of the Pulse Measurements VI and add indicators for **Period**, **Pulse Duration**, and **Duty Cycle** that can be displayed on the front panel.

Also, add the Simple Error Handler VI found on the **Programming≫Dialog & User Interface** palette and wire it to the appropriate output of the Pulse Measurements VI. Save your modified VI as PWM with Measurements.vi in the Users Stuff folder in the Learning directory.

What output values would you expect for inputs of Duty Cycle = 0.75 and Period = 100? Input these values on the front panel controls and run the VI. Check the clock before you stop the VI and confirm that it has been running for at least two output cycles so that the Pulse Measurements VI has enough data

FIGURE 7.41
Enable indexing on the Boolean outputs

to make the appropriate calculations. Do your measurements confirm the input values you provided?

The Pulse Measurements VI, like many VIs in LabVIEW, has built-in error handling capabilities. Error handling in LabVIEW follows the data-flow model. Just as data flow through a VI, so can error information. Many nodes in LabVIEW have Error In and Error Out terminals which pass error information through the node. This information is passed in a cluster containing a numeric code identifying the error, a Boolean value reporting whether or not an error occurred, and a string identifying the node in which the error occurred. As the VI runs, LabVIEW tests for errors at each execution node. If LabVIEW does not find any errors, the node executes normally. If LabVIEW detects an error, the node passes the error to the next node without executing. The next node does the same thing, and so on. Now what happens when you run the VI with **Duty Cycle** = 1? Try running the VI with a different value of **Duty Cycle**, but this time, stop execution before two output cycles complete. Observe how these errors in the Pulse Measurements VI are handled.

A working version of the PWM with Measurements VI can be found in the **Building Blocks** *folder of the* **Learning** *directory.*

7.9   RELAXED READING: LABVIEW MONITORS MEDIEVAL BELL TOWER

A long-term investigation is being carried out by the Politecnico di Milano on the Torrazzo cathedral bell tower. The bell tower, shown in Figure 7.42, is the symbol of Cremona, Italy, the city of Stradivarius. The date of construction can be traced back to the 13th century. At a height of 112 m (367 ft), the Torrazzo is the tallest masonry bell tower in Europe. Signs of crack damage and surface deterioration appearing on the tower walls motivated an on-site investigation to understand the tower dynamic and static behavior mainly under the action of wind and thermal gradients. A PC-based system has been set up for continuous measurement, online analysis, and monitoring of slow (such as walls and air temperature and crack openings) and fast (such as wind velocity and direction) signal vibrations.

The challenge was to construct a system for long-term investigations of the static and dynamic behavior of the Torrazzo medieval bell tower. The solution was a hybrid system with high-frequency measurements by means of a DAQ board for vibrations, and low-frequency measurement with a Fieldpoint network for crack monitoring. The LabVIEW-based system is remote-monitored via modem.

Static measurements are mainly related to openings in cracks in the tower. Crack openings depend predominantly on thermal actions during the day (sun radiation) and during the year (seasonal). The air and wall temperatures are measured using 40 RTDs and the crack openings are measured with 16 LVDTs. The

FIGURE 7.42
Torrazzo bell tower

transducers and sensor are placed in various sections of the tower. The main purpose of static measurements is to understand whether the crack openings/closings follow a natural cycle during the year or continually grow larger as time passes.

The dynamic measurements of the complex tower vibrations and their correlation with wind and bell actions provide both direct and diagnostic information. Direct information can be helpful in understanding the risk of structural damage under the action of the wind or bell movements. The direct information can be used for diagnostic purposes: if one notices changes, for example, in the natural frequencies of the tower, then an alarm signal would be sounded that something has changed in the tower's structure, and appropriate action can be taken to avert further problems.

The dynamic signals are measured by nine servoaccelerometers around the tower. Instantaneous speed and direction of the wind are measured by means of four cup anemometers and four wind vanes installed close to the top (85 m) on the four sides of the tower. The dynamics of crack openings are also acquired from LVDTs. The lowest transducer is placed near the base of the tower (about 15 meters), the highest at about 100 meters.

The system development posed many challenges. One of the main challenges was the measurement and data acquisition system, which had strict requirements:

- The signal and supply cables had to be as short as possible, for economic reasons and for measurement reliability.

FIGURE 7.43
Screen capture of data acquisition program (top-left display: stored signal of an accelerometer, bottom left: real-time signal, bottom right: real-time wind measurement)

- A good mass reference had to be given to all signals to avoid ground loops.
- The system had to be flexible and scalable, to allow for component failure or new sensors being added during the measurement campaign.
- Sufficient data had to be acquired to develop a correct and detailed knowledge of the tower behavior without resorting to redundancy and extensive memory requirements.
- The system had to be controlled remotely by modem.

When conducting an unattended long dynamic measurement campaign, it is important to be able to handle signals over a wide range (when signals are high) and a good resolution (when signals are low). The characteristics of the dynamic signals are not completely known *a priori*. To accommodate the range of possible signals, a 16-bit NI PCI-6031 E DAQ board was chosen for dynamic data acquisition.

It is also important to avoid acquiring *too* much data. The LabVIEW software was tasked to acquire data on a ten-minute basis (continuous scan), and store the data to disk only if certain criteria are satisfied. Otherwise, it stores only statistical data for all channels (i.e., mean and RMS). For example, the user can decide to store just ten files for wind coming from direction 0–10° and velocity 5–10 m/s. When these conditions occur, the program stores waveform data until ten files are on the disk. Afterwards only the statistical data are saved. This is repeated in all directions and for different wind velocities. In this way, in a few months' time, enough data will be collected to cover a 360° circle with wind up to 27 m/s, and allow a nearly complete dynamic analysis.

In the case of the static measurement campaign, the sampling rate and, therefore, the amount of data are not a great problem. On the other hand, the high channel count and the distribution of sensors along the tower suggested the installation of a FieldPoint network connected to a PC in a room located at 35 m. The network is made of two FP1001 network modules in two locations at 25 and 75 meters, which communicate with the PC by means of a PCI 485/2 board. The two network modules control analog input modules for the LVDT and RTD signals. This solution allows signals to have short cable paths to the data acquisition and conditioning points and to carry only two serial cables (from the top and the bottom of the tower) instead of hundreds from each sensor to a DAQ board. A program written with LabVIEW controls the FieldPoint network and collects data onto a dedicated PC.

The Torrazzo measurement campaign is being conducted in a hostile and noisy environment with logistic problems and with many transducers of different kinds. The acquisition system is required to be reliable and easy to use, perfectly tailored to the necessities, and at the same time, as cheap as possible. LabVIEW provided the software development environment that made a practical solution possible.

For further information, contact:

Giovanni Moschioni
Politecnico di Milano
Facoltà di Ingegneria di Lecco
Milano, Italy

7.10   SUMMARY

The main topics of the chapter were arrays and clusters. Arrays are a variable-sized collection (or grouping) of data elements that are all the same type, such as a group of floating-point numbers or a group of strings. LabVIEW offers many functions to help you manipulate arrays. In this chapter we discussed the following array functions:

- Array Size
- Initialize Array
- Build Array
- Array Subset
- Index Array

A cluster is a fixed-sized collection of data elements of mixed types, such as a group containing floating-point numbers and strings. As with arrays, Lab-VIEW offers built-in functions to help you manipulate clusters. In this chapter we discussed the following cluster functions:

- Bundle

- Unbundle

The important concept of polymorphism was introduced. Polymorphism is the ability of a function to adjust to input data of different types, dimensions, or representations. For example, you can add a scalar to an array or add two arrays of differing lengths together.

KEY TERMS

Array: Ordered and indexed list of elements of the same type.

Array shell: Front panel object that contains the elements of an array. It consists of an index display, a data object window, and an optional label.

Array Size: Array function that returns the number of elements in the input array.

Array Subset: Array function that returns a portion of an array.

Auto-indexing: Capability of loop structures to assemble and disassemble arrays at their borders.

Build Array: Array function that concatenates multiple arrays or adds extra elements to an array.

Bundle: A cluster function that assembles all the individual input components into a single cluster.

Cluster: A set of ordered, unindexed data elements of any data type, including numeric, Boolean, string, array, or cluster.

Index Array: Array function that accesses elements of arrays.

Initialize Array: Array function that creates an n-dimensional array with elements containing the values that you specify.

> **Polymorphism**: The ability of a function to adjust to input data of different dimension or representation.
>
> **Unbundle**: A cluster function that splits a cluster into each of its individual components.

EXERCISES

E7.1 Open While Loop Indexing.vi found in the Exercises&Problems folder in Chapter 7 of the Learning directory and click on the **Broken Run** button to access the error list. Highlight the error and press the **Show Error** button. This error is a result of an array wired to a terminal that is a scalar value. Recall that While Loops disable indexing on tunnels by default. To eliminate the error, right-click on the loop tunnel and choose **Enable Indexing**. Run the VI. After a few moments, press the **Stop** button on the front panel. An array will appear in the numeric indicator titled Temperature Measurements.

FIGURE E7.1
While Loop Indexing block diagram

E7.2 Open Cluster Order.vi found in the Exercises&Problems folder in Chapter 7 of the Learning directory and click on the **Broken Run** button to access the error list. Highlight the error and press the **Show Error** button. This error is a result of the cluster elements not being in the same order as the elements that are bundled. To eliminate the error, right-click on the cluster border on the front panel and select **Reorder Controls in Cluster**. Change the order so Sinc Data Points is the first element, Over Limit? is the second element, and Limit is the third element. Now you can run the VI.

E7.3 Open Unbundle By Name.vi found in the Exercises&Problems folder in Chapter 7 of the Learning directory and view the block diagram. This VI is the same as the Cluster Unbundle Demo.vi discussed earlier in this chapter. However, in this VI the Unbundle By Name function is used instead of the Unbundle function. The advantage to using the Unbundle By Name is that you can easily identify the cluster values that are unbundled. You can use Unbundle By Name only if every object in the cluster has an owned label associated with it. Using the LabVIEW Help as a source of information, write a short description of the cluster function Unbundle By Name.

FIGURE E7.3
Unbundle By Name VI

E7.4 Open Reversed Array.vi found in the Exercises&Problems folder in Chapter 7 of the Learning directory and view the block diagram. Add code to this VI that will display the data that is in the Array indicator in reversed order in the Reversed Array indicator.

E7.5 Open Rolling Dice.vi found in the Exercises&Problems folder in Chapter 7 of the Learning directory and view the block diagram shown in Figure E7.5. This program uses the random number generator to simulate rolling a six-sided die. The program should keep track of the number of times each value is rolled by incrementing the appropriate value in a 1D array. You will need to wire the Index Array, Increment, and Replace Array Subset nodes appropriately to run this VI.

E7.6 Open Mean.vi found in the Exercises&Problems folder in Chapter 7 of the Learning directory. This VI computes the mean of the values in the input sequence Array. The block diagram is shown in Figure E7.6. Find a function on the **Programming≫Numeric** palette which will allow you to significantly simplify the VI. Then edit the VI so that it uses this function instead of a For Loop.

FIGURE E7.5
Incomplete Rolling Dice VI

FIGURE E7.6
The block diagram for the Mean.vi

E7.7 Construct a VI that outputs a 2D array with 10 rows and 5 columns. The first row should contain five random numbers in the range 0–1, the second row numbers in the range 1–2, and so forth to the tenth row, which should contain five random numbers in the range 9–10.

PROBLEMS

P7.1 Open the VI called **Age.vi** that you created in Chapter 5 in Design Problem 5.1. Currently, the VI has 6 inputs: **Current Year, Current Month, Current Date, Birth Month, Birth Date,** and **Year of Birth.** Create a cluster so that all of the information about an individual date can be carried in one single wire on the block diagram. Then update the Age VI so that it has only 2 inputs: **Current Date** and **Birth Date.**

FIGURE P7.1
The original Age VI and the updated Age with Clusters VI front panels

 If you did not complete Design Problem 5.1, a working version of the Age VI can be found in the Exercises&Problems folder in Chapter 7 of the Learning directory.

P7.2 Create a VI that calculates the dot (inner) product of two n-dimensional vectors. Double check your math by comparing your calculation using arrays and math functions with the results of the **Dot Product.vi** that can be found in the **Functions≫Mathematics≫Linear Algebra** palette.

Review: If the two vectors are denoted by v_1 and v_2, where

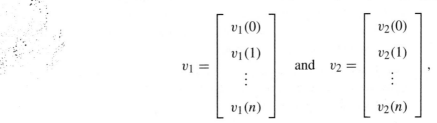

$$v_1 = \begin{bmatrix} v_1(0) \\ v_1(1) \\ \vdots \\ v_1(n) \end{bmatrix} \quad \text{and} \quad v_2 = \begin{bmatrix} v_2(0) \\ v_2(1) \\ \vdots \\ v_2(n) \end{bmatrix},$$

then the dot product is given by

$$v_1 \cdot v_2 = v_1(0)v_2(0) + v_1(1)v_2(1) + \cdots + v_1(n)v_2(n).$$

P7.3   Create a VI that calculates the cross product of two 3-dimensional vectors.
Review: If the two vectors are denoted by v_1 and v_2, where

$$v_1 = \begin{bmatrix} v_1(0) \\ v_1(1) \\ v_1(2) \end{bmatrix} \quad \text{and} \quad v_2 = \begin{bmatrix} v_2(0) \\ v_2(1) \\ v_2(2) \end{bmatrix},$$

then the cross product is given by

$$v_1 \times v_2 = \begin{bmatrix} v_1(1)v_2(2) - v_1(2)v_2(1) \\ v_1(2)v_2(0) - v_1(0)v_2(2) \\ v_1(0)v_2(1) - v_1(1)v_2(0) \end{bmatrix}.$$

P7.4   Create a VI that performs matrix multiplication for two input matrices **A** and **B**. The matrix **A** is an $n \times m$ matrix, and the matrix **B** is an $m \times p$ matrix. The resulting matrix **C** is an $n \times p$ matrix, where **C** = **AB**. Double-check your math by comparing your calculation using arrays and math functions with the results of the AxB.vi found in the **Mathematics≫Linear Algebra** palette.

P7.5   Create a VI that reads 20 temperature measurements using the Demo Temp.vi found in the Exercises&Problems folder in Chapter 7 of the Learning directory. With each temperature measurement, bundle the time (including seconds) and date of the measurement. Include a Time Delay Express VI to slow the loop so it executes four times per second. Run the program and review the time stamps in the output array to be sure there are four samples per second.

P7.6   Build a VI that generates and plots 500 random numbers on a waveform graph indicator. Compute the average of the random numbers and display the result on the front panel. Use the Statistics Express VI found in the **Mathematics≫Probability and Statistics** palette to compute of the average of the random numbers.

P7.7   Create a VI that computes and plots the second order polynomial $y = Ax^2 + Bx + C$. The VI should use controls on the front panel to input the coefficients A, B, and C, and it should also use front panel controls to enter the number of points N to evaluate the polynomial over the interval x_0 to x_{N-1}. Plot y versus x on a waveform graph indicator.

P7.8   Create a VI with a cluster of six buttons labeled Option 1 through Option 6. When executing, the VI should wait for the user to press one of the buttons. When a button is pressed, use the Display Message To User Express VI to indicate which option was selected. Repeat this process until the user presses the **Stop** button. Be sure to include the Time Delay Express VI to give the user time to press the buttons.

Hint 1: This program will require a simplified version of the state machine architecture along with a user menu. A typical state machine in LabVIEW consists of a While Loop, a Case structure, and a Shift register. Each state of the state machine is a separate case in the Case structure. You place VIs and other code that the state should execute within the appropriate case. A Shift register stores the state to be executed upon the next iteration of the loop. This is required in a typical state machine because there are times when the results of the state you are currently in control the state you will go to next. In the program described above, no two states are dependent upon one another, so the shift register is optional.

Hint 2: You can use latched Boolean buttons in a cluster to build a menu for a state machine application. The Cluster To Array function converts the Boolean cluster to a Boolean array, where each button in the cluster represents an element in the array. The Search 1D Array function searches the 1D array of Boolean values created by the Cluster To Array function for a value of True. A True value for any element in the array indicates that the user clicked a button in the cluster. The Search 1D Array function returns the index of the first True value it finds in the array and passes that index value to the selector terminal of a Case structure. If no button is pressed, Search 1D Array returns an index value of −1 and the −1 case executes, which does nothing. The While Loop repeatedly checks the state of the Boolean cluster control until you click the **Stop** button.

DESIGN PROBLEMS

D7.1 Construct a VI that searches the values of an input array of numbers for a specified value. If the specified value is found in the array, the VI should indicate that by turning on an LED and by indicating the corresponding index of the array. Use Figure D7.1 as a guide.

D7.2 Design a VI that produces an array of clusters containing the course average, letter grade, and class rank for each student in a class, given three arrays with the examination scores as input. There is one array of scores for each examination. The course average is determined as follows: 30% from the first exam, 30% from the second exam, and 40% from the final exam. The letter grade is determined from the course average by the criteria listed in Table D7.2.

A Case structure will be useful in determining letter grades.

FIGURE D7.1
The Search Array VI front panel

TABLE D7.2   Course Average Criteria

Course Average	Letter Grade
0%–59%	F
60%–69%	D
70%–79%	C
80%–89%	B
90%–100%	A

Rank the students so that the student with the highest grade has a rank of 1 and the student with the lowest grade in the class has a rank equal to the number of students in the class. You may wish to use the Sort 1D Array and Search 1D Array functions within a loop to establish the student ranks.

Bundle all of the information (course average, letter grade, and class rank) into an array of clusters where each cluster contains all the data for one individual student. Display this array of clusters on the front panel.

Test the VI with different inputs to verify its functionality. Since the array of clusters will become very large for a class with more than a few students, space can be conserved on the front panel by showing only one element of the output array at a time and then using the index display to scroll through the different elements of the cluster array.

D7.3   Construct a VI with one numeric input n that builds an array containing n Fibonacci numbers, beginning with F_1. Fibonacci numbers are the sequence of numbers, denoted by F_n, defined by the relation

$$F_n \equiv F_{n-2} + F_{n-1}$$

Set switch to ON position before clicking on the Run button

Single-plot waveform chart

Multi-plot waveform chart

Temperature data

Random number data

FIGURE 8.4
Examples of single- and multiple-plot waveform charts

the input. In the example shown in Figure 8.5, a new temperature value is plotted on the chart each time the While Loop iterates.

As illustrated in Figure 8.4, waveform charts can display multiple plots. To generate a multi-plot waveform chart, you bundle the data together using the Bundle function or use the Merge Signal function located on the **Functions≫Express Signal Manipulation** palette (see Chapter 7 for a review). In the example in Fig-

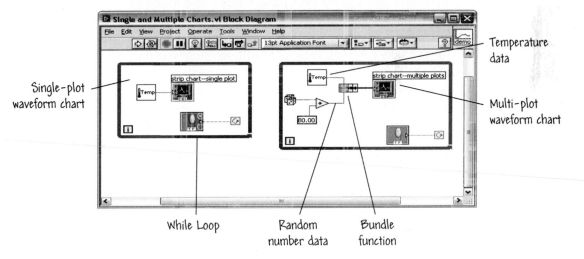

Single-plot waveform chart

Multi-plot waveform chart

Temperature data

While Loop

Random number data

Bundle function

FIGURE 8.5
Block diagrams associated with single- and multiple-plot waveform charts

ure 8.5, the Bundle function groups the output of the two different data sources—Digital Thermometer.vi and the Random Number (0–1) function—that acquire temperature and random number data for plotting on the waveform chart. The random number is biased by the constant 80 so that the plots are easily displayed on the same *y*-axis scale. Remember that Digital Thermometer.vi simulates acquiring temperature data presented in degrees Fahrenheit in the general range of 80°. You can add more plots to the waveform chart by increasing the number of Bundle function input terminals—that is, by resizing the Bundle function.

You can run the single- and multi-plot waveform chart demonstration shown in Figure 8.4 by opening Single and Multiple Charts.vi *located in* Chapter 8 *of the* Learning *directory. Run the VI and examine the single- and multi-plot waveform chart. You can stop the plotting of each chart using the vertical switch button. After stopping the plotting with the vertical switch, remember to reset the switch to the up position before the next execution of the VI.*

Practice with Waveform Charts

In this exercise you will construct a VI to compute and display the circle measurements of circumference, area, and diameter on waveform strip charts. This VI utilizes the Circle Measurements subVI.vi you developed in Chapter 5. Figure 8.6 shows the front panel that you can use as a guide as you construct your VI. The front panel has five items: a numeric control defining the radius of the circle, a push button to stop execution of the VI, and three waveform charts.

Open a new VI front panel and place a numeric control in the window and a push-button Boolean. Label the numeric control r, and add a free label near the push button indicating that pushing the button stops the execution. Place three waveform charts in the front panel window and locate the front panel objects in approximately the same locations as seen in Figure 8.6.

Enter the data r = 1 in the numeric control. In the **Operate** pull-down menu, select **Make Current Values Default**. In future sessions, when the VI is used it will not be necessary to input the default value.

Figure 8.7 shows the block diagram that you can use as a guide for building your VI. The numeric control terminal, waveform charts, and push-button Boolean will automatically appear on the block diagram after you have placed their associated objects on the front panel. The block diagram has two additional items:

- While Loop
- Circle Measurements subVI.vi

Switch to the block diagram window of your VI. First enclose the numeric control, waveform charts, and push-button terminals inside a While Loop. The only other object that is needed is the subVI to compute the circle measurements.

FIGURE 8.6
The front panel for waveform charts to plot circle measurements of circumference, area, and diameter

FIGURE 8.7
The block diagram for waveform charts to plot circle measurements of circumference, area, and diameter

The **Circle Measurements subVI.vi** is located in Chapter 5 of the **Learning** directory.

You can access the subVI by right-clicking on the block diagram and from the **Functions** palette choosing **Select a VI** as illustrated in Figure 8.8. You will need to navigate through the file structure to reach the Chapter 5 directory. Once there, you select the desired subVI—this is demonstrated in Figure 8.8.

Once all the required objects are on the block diagram, wire the diagram as shown in Figure 8.7. As an optional step, you can remove the x-axis scale. Right-click on each waveform chart on the front panel and under the **Visible Items** category choose **X Scale** to deselect the x-axis scale. In this example, the values of the circle measurements are being computed and plotted each iteration of the While Loop—the **X Scale** does not have any physical meaning, other than as a count of the total loop iterations.

Once everything is ready for execution, run the VI. With the default value of $r = 1$, you should find that the computed circumference is $C = 19.74$, the area is $A = 31.01$, and the diameter is $D = 6.28$. Increase the radius and observe how it affects the circle measurements. As you vary the radius you should see a line

FIGURE 8.8
Placing the Circle Measurements subVI.vi on the block diagram

FIGURE 8.14
Two ways to wire multiple-plot waveform graphs

The single-plot waveform graph example shown in Figure 8.13 is called Waveform Graphs demo.vi. *The multiple-plot waveform graph example shown in Figure 8.14 is called* Multiple graphs demo.vi. *Both VIs can be found in* Chapter 8 *of the* Learning *directory.*

Practice with Waveform Graphs

Open a new VI and build the front panel shown in Figure 8.15. Place three objects on the front panel:

- Numeric control labeled No. of points
- Numeric control labeled Rate of growth, r
- Waveform graph labeled Population

We would like the waveform graph y-axis scale to be driven by the data values, rather than remaining at some predetermined fixed values. To modify the y-axis limits of the waveform graph, right-click on the waveform graph and change the **Y Scale** to **AutoScale Y**, as illustrated in Figure 8.16.

Waveform graph

Initial value of population

MathScript Node coding the logistic difference equation

Shift registers

If the population falls below zero the population is extinct

FIGURE 8.15
Front panel and block diagram arrangement to experiment with chaos

Right-click on the waveform graph

Select **AutoScale Y**

FIGURE 8.16
Changing the waveform graph y-axis limits to allow autoscaling

The block diagram that you need to construct is shown in Figure 8.15. You will need to add the following objects:

- For Loop with the loop counter wired to the numeric control No. of points
- Shift registers on the For Loop to transfer data from one iteration to the next
- A MathScript Node containing the equation

$$y = rx(1 - x),$$

where the source of the parameter r is the digital control Rate of growth, r.

The equation in the MathScript Node is known as the *logistic difference equation*, and is given more formally as

$$x_{k+1} = rx_k(1 - x_k)$$

where $k = 0, 1, 2, \cdots$, and x_0 is a given value. In the block diagram (see Figure 8.15), the initial value is wired as $x_0 = 0.02$. The logistic difference equation has been used as a model to study population growth patterns. The model has been scaled so that the values of the population vary between 0 and 1, where 0 represents extinction and 1 represents the maximum conceivable population.

Wire the block diagram as shown in Figure 8.15 and prepare the VI for execution. Set the variable No. of points initially to 30 and the variable Rate of growth, r to 2.7. Execute the VI by clicking on the **Run** button. The value of the population is shown graphically in the waveform graph. You should observe the population reach a steady-state value around 0.63.

Experiment with the VI by changing the values of the Rate of growth, r. For values of $1 < r < 3$, the population will reach a steady-state value. You will find that as $r \to 3$, the population begins to oscillate, and in fact, the "steady-state" value oscillates between two values. When the parameter $r > 4$, the behavior appears erratic. As the parameter increases, the behavior becomes chaotic.[1] Verify that when the parameter $r = 4.1$, the population exceeds the maximum conceivable value (that is, it exceeds 1.0 on the waveform graph) and subsequently becomes extinct! Save your updated VI in the Users Stuff directory and name it Chaos.vi.

The working version of the chaos example shown in Figure 8.15 is called Chaos.vi *and is located in* Chapter 8 *of the* Learning *directory.*   ◆

---

1. For more on chaos, see *Chaos: Making a New Science*, by James Gleick, Penguin Books (New York, 1987).

**Waveform
Charts
and Graphs**

In this example, we will illustrate the difference between charts and graphs in LabVIEW by constructing a VI that plots multiple signals. The VI will plot multiple signals on a Waveform Chart and plot those same signals on a Waveform Graph.

To get started, open a new VI and place Triangle Wave PtByPt.vi and Sine Wave PtByPt.vi on the block diagram. The signals will be generated using these two VIs found on the **Signal Processing≫Point By Point≫Signal Generation PtByPt** palette, as illustrated in Figure 8.17. The inputs to the sine and triangle wave point-by-point VIs are amplitude, frequency (f), phase, and time. The output is the corresponding output signal (sine or triangle wave) at the requested input time. The use of these signals in a While Loop simulates point-by-point sampling of an actual input signal.

Use Figure 8.18 as a guide to constructing the VI. Add six numeric controls to the block diagram and label them amplitude, amplitude 2, frequency, frequency 2, phase, and phase 2. Wire three controls (amplitude, frequency, and phase) to the Triangle Wave PtByPt VI and the remaining three to the Sine Wave PtByPt VI.

Notice that the iteration terminal of the While Loop is scaled (by 500) and used to provide a time input signal to the point by point signal generation VIs.

FIGURE 8.17
Accessing the SineWave PtByPt VI and Triangle Wave PtByPt VI

FIGURE 8.18
Generating and plotting a sine wave and a triangle wave on a waveform graph and a waveform chart

Adding the delay of 10 ms to the loop using the Wait Until Next ms Multiple function slows down the data generation and makes the display in the waveform chart easier to follow.

To display the data as it is generated, place a waveform chart within the While Loop. The chart is updated on each loop iteration with the most recently generated data point from each point by point signal generation VI. In order to display both signals as separate plots on the same chart, you need to bundle the signals into one cluster to input to the chart. The Bundle function is shown in Figure 8.18. The waveform chart will be updated with two points (one from each signal generation VI) upon each loop iteration. On the front panel, right-click on the waveform chart and select Visible Items≫Plot Legend. Rename the two plots Triangle Wave and Sine Wave.

Place a waveform graph on the block diagram outside of the While Loop. The graph will be used to display both the sine wave and the triangle wave, as well as the sum of these two signals (refer to the block diagram in Figure 8.18). The waveform graph will display the acquired signals after all data points have been generated.

When displaying multiple plots on a waveform graph you must use the Build Array function to combine the signals to be displayed on the graph. Place the Build Array function on the block diagram outside of the While Loop. Right-click on the Build Array function and verify that **Concatenate Inputs** is not selected. With **Concatenate Inputs** deselected, the function will produce a 2D array with a row for each of the input arrays, as desired. Otherwise, the function will build a 1D array by concatenating all of the input arrays, leading to a single signal on the graph.

In order to display an additional signal, use the Add function to sum the sine and triangle signals. Then create tunnels on the While Loop for the sine wave signal, the triangle wave signal, and the sum of the two signals. Wire the three signals appropriately to the Build Array function, as depicted in Figure 8.18. On the front panel, right-click on the waveform graph and select Visible Items≫Plot Legend. Resize the **plot legend** and rename each plot to accurately reflect the signal it represents.

Complete the wiring of the block diagram. Then run the VI and check that your graph plots three different waveforms. You should find that as the While Loop iterates, the chart indicator is continuously updated with new data points. The data generated appears all at once on the graph indicator only when the loop is stopped and data acquisition is complete.

Vary the inputs and observe the effects on the sine and triangle signals. Save the VI as Multiple Plots.vi in the Users Stuff folder when you are finished.

You can find a completed VI for the waveform chart and graph demonstration shown in Figure 8.18 in Chapter 8 *of the* Learning *directory. The VI is called* Multiple Plots.vi.   ◆

8.3   XY GRAPHS

As previously mentioned, waveform graphs are ideal for plotting evenly sampled waveforms. But what if your data is available at irregular intervals, or what if you want to plot two dependent variables against each other (e.g., x versus y)? The **XY graph** is well-suited for use in situations where you want to specify points using their (x, y) coordinates. The XY graph is a general-purpose Cartesian graph that can also plot multivalued functions—such as circles and ellipsoids. The XY Graph VI is located on the **Graph** subpalette of the **Controls≫Modern** palette.

A single-plot XY graph and its corresponding block diagram are shown in Figure 8.19. The Merge Signal function is used to combine the **X** and **Y** arrays

Data to plot the unit circle

Data to plot the function $e^{-0.1x}$ versus x

Merge Signal function

FIGURE 8.19
Examples of the XY graphs: single and multiple plots

into a single output for use with the XY Graph. The Express XY Graph VI makes creating XY graphs simple. The Express XY Graph VI is also located on the **Graph Indicators** subpalette of the **Controls≫Express** palette, as shown in Figure 8.1. When placed on the front panel, the Express XY Graph VI places a corresponding terminal on the block diagram and an Express VI for quick configuration of your X and Y inputs. Wire the data arrays into the **X Input** and **Y Input** and they will automatically be configured for the graph, as shown in Figure 8.19.

*The XY graph example shown in Figure 8.19 is called **XY Graphs demo.vi** and is located in **Chapter 8** of the **Learning** directory. The single XY graph produces a unit circle, and the multiple XY graph produces a unit circle and a plot of $e^{-0.1x}$ for $0 \le x \le 41$.*

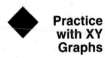

Practice with XY Graphs

Suppose that you need to borrow \$1000 from a bank and are given the option of choosing between simple interest and compound interest. The annual interest rate is 10% and, in the case of compound interest, the interest accrues annually. In either case, you pay the entire loan off in one lump sum at the end of the loan period. Construct the VI shown in Figure 8.20 to compute and graph the amount due at the end of each loan period, where the loan period varies from $N = 1$ year to $N = 20$ years in increments of 1 year.

Create a multiple-plot XY graph showing the final payments due for both the simple interest and the compound interest situations.

FIGURE 8.20
A VI to compute and plot the time value of money

Define the following variables: F = final payment due, P = amount borrowed, i = interest rate, N = number of interest periods. The relevant formulas are:

- Simple interest

$$F = P(1 + iN)$$

- Compound interest

$$F = P(1 + i)^N$$

Open a new front panel and place the following three items on it:

- Numeric control labeled P
- Numeric control labeled i
- XY Graph

The XY graph in Figure 8.20 has been edited somewhat. Right-click on the first plot in the legend and select **Common Plots** and choose the scatter plot (top row, center). Then right-click in the same place again and select a **Point Style**. Repeat this process for the second plot listed in the legend.

One solution to coding the VI is shown in Figure 8.20. In this block diagram, the formulas are coded using MathScript Nodes. Can you think of other ways to code the same equations?

Once your VI is working, you can use it to experiment with different values of annual interest and different initial loan amounts. Based on your investigations, would you prefer to have simple interest or compound interest?

The XY graph example shown in Figure 8.20 is called Time Value of Money.vi *and is located in* Chapter 8 *of the* Learning *directory.*   ◆

8.4   CUSTOMIZING CHARTS AND GRAPHS

Charts and graphs have editing features that allow you to customize your plots. This section covers how to configure some of the more important customization features. In particular, the following items are discussed:

- Autoscaling the chart and graph x- and y-axes

- Using the Plot Legend

- Using the Graph Palette

- Using the Scale Legend

- Chart customizing features, including the scrollbar and the digital display

By default, charts and graphs have the **plot legend** showing when first placed on the front panel. Using the **Positioning** tool, you can move the scale and plot legends and the graph palette anywhere relative to the chart or graph. Figure 8.21 shows some of the more important components of the chart and graph customization objects.

8.4.1   Axes Scaling

The x- and y-axes of both charts and graphs can be set to automatically adjust their scales to reflect the range of the plot data. The **autoscaling** feature is enabled or disabled using the **AutoScale X** and **AutoScale Y** options from the **X Scale** and **Y Scale** submenus of the shortcut menu, as shown in Figure 8.22.

FIGURE 8.21
The chart and graph customization objects

FIGURE 8.22
Using the **Autoscaling** feature of charts and graphs

The use of autoscaling may cause the chart or graph to update more slowly.

The X and Y scales can be varied manually using the **Operating** or **Labeling** tools—just in case you do not want to use the autoscale feature. For instance, if the graph x-axis end marker has the value 10, you can use the **Operating** tool to change that value to 1 (or whatever other value you want!). The graph x-axis marker spacing will then change to reflect the new maximum value. In typical situations, the scales are set to fit the exact range of the data. You can use the **Loose Fit** option (see Figure 8.22) if you want to round the scale to multiple values of the scale increment—in other words, with a loose fit the scale marker numbers are rounded to a multiple of the increment used for the scale. For example, if the scale markers increment by 5, then with a loose fit, the minimum and maximum scale values are set to multiples of 5.

On both the **X Scale** or **Y Scale** submenus, you will find the **Formatting** option. Choosing the option opens up a dialog box, as shown in Figure 8.23, that allows you to configure many components of the chart or graph. This dialog can also be opened by right-clicking and selecting **Properties** to change the properties of the graph. You can modify the x- and y-axis characteristics individually. The following choices are available:

- The **Appearance** tab is used to specify which elements of the object are visible. This provides a way to choose to show the graph palette, plot legend, scrollbar, scale legend, and cursor legend. You can also select these items on the **Visible Items** pull-down menu by right-clicking on the graph.

- Within the **Format and Precision** area, you change the format and precision of numeric objects. You can choose the number of digits of precision for the scale marker numbers, as well as the notation. You can choose among floating point, scientific, SI notation, hexadecimal, octal, binary, absolute time, and relative time. The choices depend on the format choice.

- The **Plots** tab is used to configure the appearance of plots on a graph or chart. You can select the line type (e.g., solid versus dashed), point markers, and scale titles.

- Under the **Scales** tab you format scales and grids on graphs and charts. For example, you can set the axis origin $X0$ (or $Y0$) and the axis marker increment ΔX (or ΔY). The grid style and colors provide control over the gridlines: no gridlines, gridlines only at major tick mark locations, or gridlines at both major and minor tick marks. This option also allows you to change the color of the gridlines. Here you can also specify the scale as either a linear or a logarithmic scale. The **Scales** tab lets you select major and minor tick marks for the scale, or none at all. A major tick mark corresponds to a scale label,

Format and Precision

Scale marker precision can be selected here

Precision Formatting
(a)

Scales

Change the color of the gridlines by clicking here

No gridlines

Scale Formatting
(b)

FIGURE 8.23
Scale formatting

while a minor tick mark denotes an interior point between labels. You have to click on the **Scale Style** icon to view the choices. With this menu, you can make the axes markers invisible.

- The **Cursors** tab is used to add cursors to a graph or chart and to configure the appearance of the cursors.

- Use the **Documentation** tab to describe the purpose of the object and to give users instructions for using the object.

8.4.2   The Plot Legend

The plot legend provides a way to customize the plots on your charts and graphs. This is where you choose the data point style, the plot style, the line style, width and color, and other characteristics of the appearance of the chart or graph. For example, on multiple-plot graphs you may want one curve to be a solid line and the other curve to be a dashed line or one curve to be red and the other blue. An example of a plot legend is shown in Figure 8.24. When you move the chart or graph around on the front panel, the plot legend moves with it. You can change the position of the plot legend relative to the graph by dragging the plot legend with the **Positioning** tool to the desired location.

The plot legend can be visible or hidden—and you use the **Visible Items** submenu of the chart or graph shortcut menu (see Figure 8.24) to choose. After you customize the plot characteristics, the plot retains those settings regardless of whether the plot legend is visible or not. If your chart or graph receives more plots than are defined in the plot legend, they will have the default characteristics.

FIGURE 8.24
The plot legend for charts and graphs

When the plot legend first appears, it is sized to accommodate only a single plot (named Plot 0, by default). If you have a multiple-plot chart or graph, you will need to show more plot labels by dragging down a corner of the legend with the cursor to accommodate the total number of curves. Each curve in a multiple-plot chart or graph is labeled as Plot 0, Plot 1, Plot 2, and so forth. You can modify the default labels by assigning a name to each plot in the legend with the **Labeling** tool. Choosing names that reflect the physical value of the data depicted in the chart makes good sense—for instance, if the curve represents the velocity of an automobile you might use the label Automobile velocity in km/hr. The legend can be resized on the left to give the plot labels more room or on the right to give the plot samples more room. This is useful when you assign long names to the various curves on a multiple-plot chart or graph. You can right-click on each plot in the legend and change the plot style, line, color, and point styles of the plot. The shortcut menu is shown in Figure 8.21.

8.4.3   The Graph Palette and Scale Legend

The graph palette and scale legend allow you access to various aspects of the chart and graph appearances while the VI is executing. You can autoscale the axes and control the display format of the axes scales. To aid in the analysis of the data presented on the charts and graphs, buttons allow you to pan and zoom in and out. The graph palette and scale legend shown in Figure 8.25 can be displayed using the **Visible Items** submenu of the chart or graph shortcut menu.

On the right of the scale legend are two buttons for each axis that control the autoscaling of the axes. Clicking on the autoscaling button (either the x- or y-axis buttons) will cause the chart or graph to autoscale the axis. When you press the lock button, the chart or graph will autoscale either of the axes scales continuously.

The x-axis and y-axis labels can also be modified by entering the desired labels (using the **Labeling** tool) into the text area provided on the left side of the scale legend. The upper text is for the x-axis.

The scale format buttons (on the right of the scale legend) provide run-time control over the format and precision of the x and y scale markers, respectively. As shown in Figure 8.25, you can control the **Format** of the axes markers, the **Precision** of the marker numbers (from 0 to 6 digits after the decimal point), and the **Mapping Mode**. Remember that in the Edit Mode, you can control the format, precision, and mapping mode through shortcut menus on the chart or graph. The chart or graph scale legend gives you the added feature of being able to control the format and precision of the x and y scale markers while the VI is executing.

The three buttons on the graph palette allow you to control the operational mode of the chart or graph. You are in the standard mode under normal circumstances where the button with the crosshair (left side of the palette) is selected.

FIGURE 8.25
The graph palette and scale legend

This indicates that you can click in the chart or graph area to move cursors around. Other operational modes include panning and zooming. When you press the pan button (depicted in Figure 8.25), you can scroll through the visible data by clicking and dragging sections of the graph with the pan cursor. Selecting the zoom button gives you access to a shortcut menu that lets you choose from several methods of zooming.

As depicted in Figure 8.25, you have several zoom options:

- **Zoom by Rectangle**. In this mode, you use the cursor to draw a rectangle around the area you want to zoom in and when you release the mouse button, the axes will rescale to zoom in on the desired area.

- **Zoom by Rectangle**—with zooming restricted to X data. This is similar in function to the **Zoom by Rectangle** mode, but the Y scale remains unchanged.

- **Zoom by Rectangle**—with zooming restricted to Y data. This is similar in function to the **Zoom by Rectangle** mode, but the X scale remains unchanged.

- **Zoom to Fit**. This autoscales all x- and y-scales on a graph or chart.

- **Zoom in about a Point**. With this option you hold down the mouse on a specific point on the chart or graph to continuously zoom in until you release the mouse button.

- **Zoom out about a Point**. With this option you hold down the mouse on a specific point and the graph will continuously zoom out until you release the mouse button.

For the zoom in and zoom out about a point modes, <Shift>-clicking will zoom in the other direction.

8.4.4   Special Chart Customization Features

Charts have the same customizing features as graphs—with two additional options: **Scrollbars** and **Digital Displays**. The **Visible Items** submenu (of the chart shortcut menu) is used to show or hide the optional digital display(s) and a scrollbar.

Charts have scrollbars that can be used to display older data that has scrolled off the chart. The chart scrollbar is depicted in Figure 8.21. The scrollbar can be made visible by right-clicking on the chart and selecting the scrollbar from the **Visible Items** submenu, as illustrated in Figure 8.26.

There is one digital display per plot that displays the latest value being plotted on the chart. You can place a digital display on the chart by selecting **Digital Display** from the **Visible Items** submenu, as illustrated in Figure 8.26. The digital display can be moved around relative to the chart using the **Positioning** tool. The last value passed to the chart from the block diagram is the latest value for the chart and the value that is displayed on the digital display. If you want to view past values contained in the chart data you can navigate through the older data using the scrollbar. There is a limit to the amount of data the chart can hold in the buffer. By default, a chart can store up to 1024 data points. When the chart reaches the limit, the oldest point(s) are discarded to make room for new data. If you want to change the size of the chart data buffer, select **Chart History Length** from the shortcut menu and specify a new value of up to 100,000 points.

When you are not running the VI, you can clear the chart by right-clicking on the chart and selecting **Clear Chart** from the **Data Operations** menu, as shown in Figure 8.26. However, if you are running the VI (you are in the Run Mode),

FIGURE 8.26
The scrollbar and the digital display for charts

then **Clear Chart** becomes a shortcut menu option and it will automatically appear in the chart shortcut menu after clicking on the **Run** button. This allows you to clear the chart while the VI is executing.

If you have a multiple-plot chart, you can choose between overlaid plot or stacked plot. When you display all plots on the same set of scales, you have an overlaid plot. When you give each plot its own scale, you have a stacked plot. You can select **Stack Plots** or **Overlay Plots** from the chart shortcut menu (see Figure 8.26) to toggle between stacked and overlaid modes. Figure 8.27 illustrates the difference between stacked and overlaid plots.

Display multiple charts stacked one above the other

By default the chart displays multiple plots on a single chart.

FIGURE 8.27
Stacked and overlaid charts

*The multiple-plot chart example shown in Figure 8.27 is called **Chart Customizing.vi** and is located in **Chapter 8** of the **Learning** directory. The two charts on the block diagram display the same data, except that the chart on the left uses the stacked chart, and the chart on the right uses the overlaid chart.*

8.4.5 Special Graph Customization Features: Cursor Legend

To add cursors to a graph, you first bring the **Cursor Legend** to the front panel graph by right-clicking on the graph and selecting Visible Items≫Cursor Legend from the shortcut menu, as illustrated in Figure 8.28. The cursor legend is shown in Figure 8.29. To add a cursor to the graph, right click in the **Cursor Legend** and select **Create** Cursor. You will be given three options to choose

FIGURE 8.28
Placing the **Cursor Legend** on the front panel

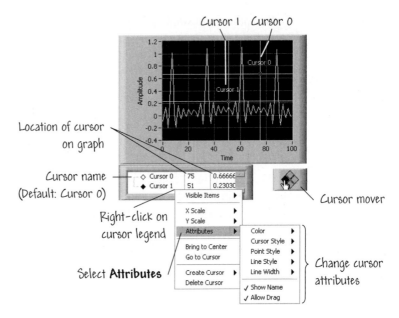

FIGURE 8.29
The **Cursor Legend** is associated with graphs

from: **Free**, **Single-Plot**, and **Multi-Plot**. Choosing **Free** creates a cursor that moves freely within the plot area. Choosing **Single-Plot** creates a cursor that is positioned on a single plot associated with that cursor. If you have one plot on the graph, the cursor will move only along that plot. If you have multiple plots on a single graph, then choosing **Single-Plot** will associate the cursor with the first plot listed in the bf Plot Legend. Repeating the process and selecting **Single-Plot** again will associate the second cursor with the second plot on the graph as listed in the **Plot Legend**. To modify the association of the cursors with their respective plots, right-click the **Cursor Legend** on the row of the desired cursor and select **Snap To** from the shortcut menu to associate one or all plots with the cursor. The third option, **Multi-Plot**, positions the cursor only on a specific data point in the plot area. This mode is only valid for mixed-signal graphs.

Various attributes of the cursors can be modified and personalized. As illustrated in Figure 8.29, right-click on the **Cursor Legend** on the row of the cursor that you wish to modify. For example, in Figure 8.29 the attributes of Cursor 1 are being modified. The attributes that can be modified include: **Color**, **Cursor Style**, **Point Style**, **Line Style**, **Line Width**, **Show Name**, and **Allow Drag**. Selecting **Color** displays the color picker so you can select the color of the cursor. **Cursor Style** provides various cursor styles, including combinations of dots and dashes in an assortment of configurations. **Point Style** provides various point styles for the intersection of the cursor, including circles, filled circles, squares, filled squares, crosses, and a host of other choices. **Line Style** and **Line Width** offer various solid- and dotted-line styles and line widths, respectively.

Two other important options appearing in the **Attributes** menu include **Show Name** and **Allow Drag**. **Show Name** displays the name of the cursor on the graph and **Allow Drag** lets the mouse drag the cursor.

You can control the location of the cursor using the **Bring to Center** and **Go to Cursor** options found on the **Cursor Legend** pull-down menu. Selecting **Bring to Center** results in centering the cursor on the graph without changing the *x*- and *y*-scales. When the cursor mode is **Single-Plot** or **Multi-Plot**, this option centers the cursor on the plot on which the cursor is currently positioned and updates the cursor coordinates in the cursor legend. When the cursor mode is **Free**, this option centers the cursor in the plot area and updates the cursor coordinates in the cursor legend. The **Go to Cursor** option changes the *x*- and *y*-scales to show the cursor at the center of the graph.

You cannot change the mode of a cursor after you create it. You must delete the cursor and create another.

8.4.6   Using Graph Annotations

It is often very useful to add annotations to graphs, especially when the graphs are in their final form and are being used in reports or presentations. In LabVIEW you can use annotations to highlight important data points in the plot area. The annotation includes a label and an arrow that identifies the annotation and data point. A graph can have any number of annotations. Figure 8.30 shows an example of a graph using annotations.

To add an annotation to a graph, right-click the graph and select Data Operations≫Create Annotation from the shortcut menu to display the **Create Annotation** dialog box, as illustrated in Figure 8.30. You can assign the annotation name and you can specify how the annotation snaps to plots in the plot area. To specify the annotation name, just type the desired name or phrase in the **Annotation Name** field.

To specify how the annotation snaps to plots in the plot area, use the **Lock Style** pull-down menu. The **Lock Style** component includes the following options:

- **Free**—Allows you to move the annotation anywhere in the plot area.
- **Snap to All Plots**—Allows you to move the annotation to the nearest data point along any plot in the plot area.
- **Snap to One Plot**—Allows you to move the annotation only along the specified plot.

You can customize the behavior and appearance of the annotation. You can hide or show the annotation name or arrow in the plot area, specify the color of

FIGURE 8.30
Annotating a graph

the annotation, and specify line, point, and annotation style. To accomplish this, right-click the annotation and select options from the shortcut menu to customize the annotation.

8.4.7   Exporting Images of Graphs, Charts, and Tables

You can export images of graphs, charts, tables, digital data, and digital waveform controls and indicators into the following formats:

- Windows: .emf, .bmp, and .eps files
- Mac OS: .pict, .bmp, and .eps files

To export an image, right-click the control and select Data Operations≫ Export Simplified Image from the shortcut menu. In the **Export Simplified Image** dialog box, select the image format and whether you want to save the image to the clipboard or to disk. You also can use the **Export Image** method to export an image programmatically.

8.4.8   Using Context Help

If you are new to LabVIEW, it can often be confusing when you try to wire data to charts and graphs. Do you use a Build Array function, a Bundle function, or both? What order do the input terminals use? You can find valuable information in the **Context Help** window. For example, if you select **Show Context Help** from the **Help** menu and put your cursor over a waveform graph terminal in the diagram, you will see the information shown in Figure 8.31. The **Context Help** window shows you what data types to wire to the waveform graph, how to specify point spacing with the Bundle function, and which example to use when you want to see the different ways you can use a waveform graph. The **Context Help** window also shows similar information for XY graphs and waveform charts.

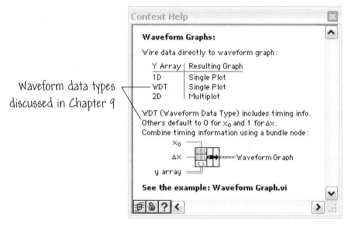

FIGURE 8.31
Using Context Help to determine which data types to wire to the waveform graph

BUILDING BLOCK

8.5   BUILDING BLOCKS: PULSE WIDTH MODULATION

The goal of this Building Block is to plot the waveform output of a pulse width modulation VI using a waveform chart. To begin, open the Building Blocks VI that you constructed in Chapter 7. It should be called **PWM with Measure-ments.vi** and located in the **Users Stuff** folder in the **Learning** directory. Once the VI is open, rename it **PWM with Chart.vi** and save the VI in the **Users Stuff** folder.

In case you do not have a working version of the Building Blocks VI from Chapter 7, a working version of the PWM with Measurements VI can be found in the Building Blocks folder in the Learning directory.

In the PWM with Measurements VI, the output was displayed using an LED indicator. You should replace that LED with a waveform chart. To accomplish this, in the block diagram of the PWM with Chart VI, select and delete the Boolean LED named Output. Then add a waveform chart and wire the chart to the output of the Boolean to (0,1) function, as illustrated in Figure 8.32. Clear all broken wires. Rename the waveform chart Pulse Width Modulated Signal, as illustrated in Figure 8.32.

When you have added and named the waveform chart on your VI, experiment with different input values. By default, the **Chart History Buffer** keeps the most recent 1024 data points in memory. Thus you may find that, for large periods, you will need to increase this value in order to view a significant portion of the

FIGURE 8.32
The PWM with Chart VI showing the use of the waveform chart.

output signal. This can be accomplished by right-clicking on the plot, selecting **Chart History Length**, and replacing the default **Chart History Length** with a larger number—preferably one that is several times the size of the period. For example, you might choose to keep 5000 data points in memory.

We have not specifically stated units for either the *x*-axis or the *y*-axis of the chart. Since this signal is not being used to communicate with any physical devices, explicit units of measurement are not critical. The important information that the signal carries is how the magnitude varies with respect to time. As long as the units of magnitude and time remain consistent in this simulation, it is not necessary to expressly relate them to physical units. We denote the high and low values of the signal 1 and 0.

Observe what the output looks like for various input values. Change the update mode and watch how the data is plotted for all three update modes. To change the update mode, right-click on the chart and select Advanced≫Update Mode. The three options are **Strip Chart**, **Scope Chart**, and **Sweep Chart**. When you are finished, save the VI as PWM with Chart.vi.

A working version of the PWM with Chart VI can be found in the Building Blocks *folder of the* Learning *directory.*

8.6   RELAXED READING: SPACECRAFT ATTITUDE DETERMINATION ON THE WEB

Real-time spacecraft attitude determination involves estimation of the orientation of the spacecraft in space using the sensor data as it is being telemetered to the ground from the spacecraft. Attitude determination is very important in the early phase of a space mission because it is essential that the spacecraft be oriented in the desired way and that all attitude sensors be operational. During this phase, the spacecraft analysts monitor the real-time attitude determination process occurring in the mission control center.

In the 1990s, a prototype for a real-time spacecraft attitude determination system (RTADS) existed in LabVIEW. The LabVIEW real-time attitude determination system was investigated as potential flight dynamics functionality. The current version, built specific to X-ray Timing Explorer (XTE) spacecraft, is capable of handling magnetometer, sun, and star tracker data and of displaying the estimated real-time attitude data and its comparison with the on-board estimated and telemetered attitude data in text, as well as real-time graphical plots.

With the support of the Computer Sciences Corporation (CSC) Leading Edge Forum (LEF), researchers began working in the area of real-time data rendering over the internet. The existing RTADS was web-enabled using the National Instruments internet Developers Toolkit. The feasibility of interactive monitoring by spacecraft analysts via third-party browsers was firmly established.

FIGURE 8.33
Front panel of Real-time Attitude on the web being run

The web-enabled prototype, called RAW (Real-time Attitude on the Web), shown in Figure 8.33, in its current form reads spacecraft telemetry data from a file on hard disk on the workstation on which it resides. The major functional steps of the RTADS prototype include extraction of relevant sensor data from among magnetometer, digital sun sensor, and two star trackers, conversion to engineering units, and generation of "observation vectors" and transformation to body coordinates. If the star tracker is a selected sensor, star identification is carried out using *a priori* attitude. Reference vectors are generated, and in the final step, the attitude determination process is performed and the results are displayed on a LabVIEW front panel as text and as graphical plots.

The web interface is accomplished using the G web server and its built-in functions in conjunction with HTML documents. The key VI provided by NI was the **VI-CTRL.vi** designed to control custom VIs. This VI was very helpful in minimizing the CGI programming effort involved in web-enabling. Most of the browser-based input controls were accomplished using this VI in conjunction with image maps. The web document also embeds video help, audio help, and e-mail buttons activated using JavaScript.

Some of the features of the LabVIEW real-time attitude determination system on the web include:

- **Restricted access**: Access to the front panels and chosen subdirectories is restricted using usernames and passwords, which were configured without any programming effort using a Windows-like interface. IP-address-based restrictions are also incorporated in the same way.

- **Browser-based inputs**: Several user-selectable browser-based inputs are provided, such as mode of TLM/Ephemeris data, type of estimation method, sensor selection and standard deviation, mode of *a priori* attitude, *a priori* attitude, and plot size. Additionally, browser-based controls are provided for stopping, closing the front panel, and restarting the program.

- **Visual and e-mail alarms**: Visual and e-mail alarms are provided, which are activated when the error between estimated attitude angles and the telemetered on-board attitude angles exceeds a pre-set threshold.

- **Multiple component types**: The system has both LabVIEW and MATLAB components. Existing MATLAB components were easily incorporated using MATLAB script nodes without having to re-code.

- **Online visual and audio help files**: Audio and video help files to aid browser-input interface embedded into the HTML document and activated using JavaScript.

To sum up, some of the advantages of this approach are:

- Offers off-site monitoring

- Requires minimal programming effort

- Legacy modules in other programming languages can be incorporated and web enabled

- Can be served from workstations on which LabVIEW is not installed if a stand-alone executable is built and shipped.

Some of the scenarios where such systems can be put to use are for spacecraft monitoring and control for low-cost constellation missions that cannot afford spacecraft-specific control centers and for remote access by payload scientists at universities and international institutions.

For further information, contact:

Sheela Belur
Computer Sciences Corporation
bvsheela@yahoo.com

8.7   SUMMARY

You can display plots of data using charts or graphs. Three types of charts were discussed—strip, scope, and sweep—and two types of graphs—waveform and XY. Both charts and graphs can display multiple plots at a time. Charts append new data to old data by plotting one point at a time. Graphs display a full block of data. The waveform graph plots only single-valued functions with points that are evenly distributed. Conversely, the XY graph is a general-purpose Cartesian graph that lets you plot unevenly spaced, multivalued functions. You can customize the appearance of your charts and graphs using the legend and the palette. The format and precision of the X and Y scale markers and zoom in and out are controlled with the palette even while the VI is executing.

KEY TERMS

Autoscaling: The ability of graphs and charts to adjust automatically to the range of plotted values.

Legend: An object owned by a chart or graph that displays the names and styles of the plots.

Plot: A graphical representation of data shown on a chart or graph.

Palette: An object owned by a chart or graph from which you can change the scaling and format options while the VI is running.

Scope chart: A chart modeled after the operation of an oscilloscope.

Strip chart: A chart modeled after a paper strip chart recorder, which scrolls as it plots data.

Sweep chart: Similar to a scope chart, except that a line sweeps across the display to separate new data from old data.

Waveform chart: Displays one point at a time on one or more plots.

Waveform graph: Plots single-valued functions or array(s) of values with points evenly distributed along the x-axis.

XY graph: A general-purpose, Cartesian graph used to plot multivalued functions, with points evenly sampled or not.

EXERCISES

E8.1 Open Multiple Graphs Demo.vi found in the Exercises&Problems folder in Chapter 8 of the Learning directory. Right-click on each graph and select Visible Items≫Cursor Legend. Right-click on the **Cursor Legend** and create a single-plot cursor. In the **Attributes** pull-down menu change the line style, point style, and color of the cursor. Also, in the **Attributes** menu, select **Allow Drag**. Again, right-click on the **Cursor Legend** and select Snap To≫Plot0. Add a second cursor for **Plot1** and repeat the editing process described above. When ready, practice moving the cursors to different values on the graph.

E8.2 Create a new VI where temperature data, created with the Digital Thermometer.vi in the Activity directory, is displayed on a graph. At the beginning of each execution of the VI, clear the graph.

E8.3   Create a VI that plots an ellipse

$$r^2 = \frac{A^2 B^2}{A^2 \sin^2 \phi + B^2 \cos^2 \phi}$$

where r, A, and B are input parameters and $0 \le \phi \le 2\pi$.

FIGURE E8.3
The ellipse with $r = 2$, $A = 3$, and $B = 1$

E8.4   Open the NI Example Finder and open Waveform Graph.vi found in the Fundamentals≫Charts and Graphs folder. Investigate the block diagram and locate the Build Array and Bundle functions. Run the VI. Note how the Bundle function is used to specify x parameters other than the defaults of $x_0 = 0$ and $\Delta x = 1$. The Build Array function is used to create graphs with multiple plots. Close the VI without saving it when you are finished with your observations.

E8.5   Open the NI Example Finder and go to the **Search** tab. Type in "charts." Open Charts.vi. Run the VI and examine the different types of update modes, specifically the differences between the strip chart, scope chart, and sweep chart. When you are done examining the VI, stop and close the VI. Do not save any changes.

E8.6   Open the NI Example Finder and go to the **Browse** tab. Select browsing according to **Task**. Open Building User Interfaces≫Displaying Data≫Graphs and Charts≫3D Graph Properties-Torus.vi. Run the VI and change the **Plot Style**. Use your mouse to click on the 3D graph and rotate the position of the graph. Experiment with changing the **Transparency** setting. Stop the VI. Switch to the block diagram. The 3D Parametric Surface Graph is a special type of LabVIEW graph that uses ActiveX to generate the 3D image. This graph and other types of 3D graphs are available in the **Controls≫Modern≫Graph** palette.

PROBLEMS

P8.1 Create a new VI to plot a circle using an XY Graph.

P8.2 Create a new VI where temperature data, created with the Digital Thermome-
ter.vi in the Activity directory, is displayed on a strip chart. Compute and display
the running average of the temperature data.

P8.3 Open a blank VI and place a Simulate Signal Express VI on the block diagram.
Configure the Express VI to generate a 50-Hz sine wave. Click **OK** to exit
the Express VI configuration page. Right-click on the **Sine** terminal and select
Create≫Graph Indicator. Run the VI.

P8.4 Create a VI that graphs the function $\sin x$ where $x = 0 \ldots n\pi$ and the integral
$y = \int_0^{n\pi} \sin x \, dx$. The value of n should be an input on the front panel.

P8.5 Open and run Chart Mistake.vi which can be found in the Exercises&Problems
folder in Chapter 8 of the Learning directory. The VI generates two signals: a
sawtooth waveform and a triangle waveform. Both signals have their amplitude
controlled by front panel inputs. Whenever their sum exceeds 2, an LED lights
up.

Take a look at the block diagram (shown in Figure P8.5) after you have run
the VI. The designer of this VI intended the Waveform Graph to display the
sawtooth and triangle waveforms being generated, with the graph continuously
updating on each loop iteration. Find the mistake in the implementation of this
goal. Why do the time values on the graph never increment? How can you display
two separate plots? Make the necessary adjustments so that this VI does what its
designer intended.

FIGURE P8.5
The Chart Mistake VI block diagram

P8.6   Open the NI Example Finder and navigate to the Temperature Analysis VI found in the Fundamentals≫Charts and Graphs folder. Run the VI and note the maximum value displayed on the Max indicator on the right. Create an annotation on the Temperature Graph indicator marking the maximum point and specifying the value.

Once the graph is annotated, change the color of the annotation from the default yellow to blue. This color will show up better when you export the image to a file, where the graph has a white, rather than black, background. As a final step, export the image of the graph to an encapsulated postscript file (.eps) file. When viewed with an external viewer, the .eps file will appear similar to the graphic in Figure P8.6. Close the Temperature Analysis VI without saving any changes.

FIGURE P8.6
An encapsulated postscript (.eps) image of the Temperature Graph annotated with the maximum temperature

DESIGN PROBLEMS

D8.1   In this problem, you will design a VI which normalizes a waveform of 50 randomly generated points. The completed VI should have a waveform chart and a waveform graph on the front panel. On the waveform chart, display the random points as they are being generated.

To improve the visualization of the waveform chart, use a delay in the loop of 50 milliseconds. Place controls on the front panel so that the offset and scale of the waveform can be adjusted each time the VI is run. When all 50 points have been generated, perform the necessary scaling and shifting on this waveform so that it is centered on the x-axis and has a peak-to-peak amplitude of 1.

Plot both the original and the normalized waveforms on a single waveform graph. Make the **Cursor Legend** and **Graph Palette** visible on the graph. Create two labeled cursors and lock one to the original waveform and the other to the normalized waveform.

Run your VI and use cursors and the zooming tools to verify that your VI is producing the expected output. As a test case, set the offset to 0 and the scale to 2. Then confirm that the original signal varies between 0 and 2, while the normalized signal varies between -1 and 1.

D8.2 Employ the Sine Wave PtByPt.vi to generate a sine wave continuously until a user presses a stop button on the front panel. Allow the user to control the frequency, phase, and amplitude of the sine wave with numeric controls on the front panel. The sine wave should be displayed on a waveform chart and updated with each new data point generated. Then use the Collector Express VI to display the last 1000 points generated on a waveform graph when the user presses the stop button.

*The Collector Express VI can be found on the **Functions≫Express≫Signal Manipulation** palette.*

D8.3 Create a VI that produces 20 random integers between 0 and 100 and then plots them on a waveform graph. Display the **Cursor Legend** and use cursors to determine the coordinates of the minimum and maximum generated values. Using the Array Max & Min function, display the coordinates of the maximum and minimum on the front panel. Run the VI and compare the values output by the Array Max & Min function to the values you find with the cursors. Verify that they are equivalent.

CHAPTER 9

Data Acquisition

Data Acquisition (DAQ) is the measurement or generation of physical signals. This chapter focuses on the use of the DAQ Assistant and the basic notions associated with analog and digital input and output (I/O). Some common terms and concepts associated with data acquisition are also discussed, including the components of a DAQ system, signal conditioning, and the types of signals encountered.

GOALS

1. Review some basic notions of signals and signal acquisition.
2. Introduce the organization of the DAQ VIs.
3. Understand the basics of analog and digital input and output using the DAQ Assistant.

9.1   COMPONENTS OF A DAQ SYSTEM

Some think that having a plug-in DAQ device properly installed in a personal computer is equivalent to having a complete DAQ system. This is not the case at all. In fact, the plug-in DAQ device is only one component of the system. A DAQ system generally has (in addition to plug-in DAQ devices) sensors and transducers, signal conditioning, and a suite of software for acquiring, manipulating, analyzing, storing, and displaying the raw acquired data. Two options for constructing a DAQ system are illustrated in Figure 9.1.

In Option A (see Figure 9.1), the plug-in DAQ device resides in the computer. This computer can be a tower, a desktop model, or a laptop with PCMCIA slots. In Option B, the DAQ device is external to the computer. With the latter approach you can build DAQ systems using computers without available plug-in slots (such as some laptops), and the computer and DAQ module communicate through various buses, such as the USB port. Option B systems are usually more practical for remote data acquisition and control applications where you want to bring the DAQ system into the field.

Before a computer-based system can measure a physical signal, a sensor (or transducer) must convert the physical signal into an electrical signal (such as voltage or current). Generally, you cannot connect signals directly to a plug-in DAQ device in a computer, as you can with most stand-alone instruments. In many cases, the measured physical signal is very low-voltage and susceptible to noise. In these situations, the measured signal may need to be amplified and filtered before conversion to a digital format for use in the computer. A signal-conditioning

FIGURE 9.1
Two options for a DAQ system

accessory conditions measured signals before the plug-in DAQ device converts them to digital information. More will be said on signal conditioning later in the chapter. Software controls the DAQ system—acquiring the raw data, analyzing the data, and presenting the results.

The subject of data acquisition cannot be adequately covered in one chapter, although the fundamental ideas can be introduced and discussed in enough detail to generate enthusiasm for pursuing other sources of information.[1] The most effective way to learn about data acquisition is by doing it. Reading about it is not enough. At its most basic level, the task of a DAQ system is the measurement or generation of physical signals.

9.2   TYPES OF SIGNALS

The concepts of signals and systems arise in a wide variety of fields, including science, engineering, and economics. In an effort to develop an analytic framework for studying certain natural phenomena, the notion of an input–output representation has arisen, and is illustrated in Figure 9.2. In the input–output representation, the input signals are operated on by the DAQ system to produce the output signals. Signals are physical quantities that are functions of an independent variable (such as time) and contain information about a natural phenomenon. For example, the input signals might be degraded and weak video signals received from

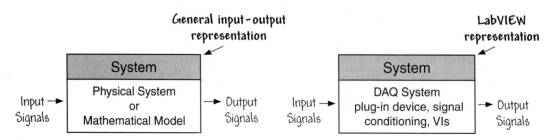

FIGURE 9.2
Modeling physical phenomena with input-output representations

a spacecraft approaching a distant asteroid. The DAQ system includes the hardware and software to acquire, process, and enhance the spacecraft signals (e.g., the spacecraft images of the asteroid can be enhanced to show features and colors more clearly), and the output signals would be the enhanced asteroid images. In

---

1. A good source of information for the beginner is the *Data Acquisition Basics Manual*, available from National Instruments, Inc. Part Number 320997E-0. For the advanced student, see Chapters 14 to 17 of *LabVIEW Graphical Programming, 3rd ed.,* by Gary W. Johnson and Richard Jennings (McGraw-Hill, New York, 2001).

general, physical signals are converted to electrical signals (such as voltage or current) before being amplified and filtered by the signal conditioning and acquired by the DAQ hardware. This conversion is accomplished by some type of transducer. The list of common transducers includes microphone (sound), camera (images), thermocouple (temperature), strain gauge (bending strain), thermistor (temperature), accelerometer (acceleration), antenna (radio waves), and encoder (rotation or distance). Once the physical signals are measured and converted to electrical signals by the sensor, the information contained in them may be extracted and analyzed.

Five common classes of information that can be extracted from signals are illustrated in Figure 9.3. The signals evolve either continuously or only at discrete points in time and are known as continuous- or discrete-time signals, respectively. Room temperature is an example of a continuous-time signal (although we may discretize the temperature signal by recording the temperature only at specific times or sampling points). The end-of-day closing Dow Jones stock index is an example of a discrete signal, taking on a new value once per working day.

For purposes of discussing data acquisition, we will use the following signal classifications:

- For digital signals, we have two types:
  - On–Off
  - Pulse Train
- For analog signals, we have three types:
  - DC–Static or Slow Changing Signals
  - AC–Fast Changing Signals
  - Frequency-Domain–Frequency Analysis of AC and DC Signals

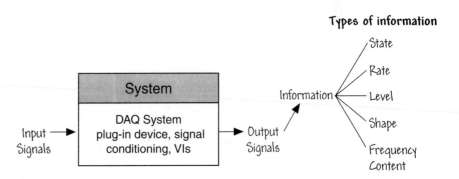

FIGURE 9.3
Measuring and analyzing signals provides information: state, rate, level, shape, and frequency content

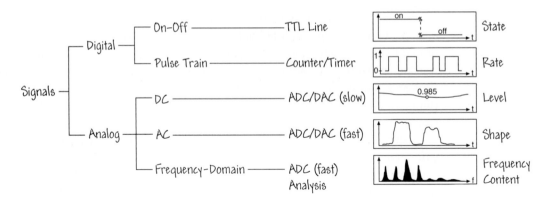

FIGURE 9.4
Signal types (ADC → analog-to-digital converter, DAC → digital-to-analog converter, TTL → transistor-transistor logic)

The five signal types just listed are classified as analog or digital according to how they convey information. A digital signal has only two possible discrete levels: high level (on) or low level (off). An analog signal, on the other hand, contains information that varies continuously with respect to time.

A schematic with the various signal types is shown in Figure 9.4. Each signal type is unique in the information conveyed, and the five signal types closely parallel the five basic types of signal information: state, rate, level, shape, and frequency content.

9.2.1　Digital Signals

The two types of digital signals that we will consider here are the on–off switch and the pulse train signals. The on–off signal illustrated in Figure 9.5 conveys information concerning the immediate digital *state* of the signal. A simple digital state detector is used to measure this signal type. An example of a digital on-off signal is the output of a transistor-transistor logic (TTL) switch.

The second type of digital signal is the pulse train signal illustrated in Figure 9.6. This signal consists of a series of state transitions, and information is contained in the number of state transitions, the *rate* at which the transitions occur, and the time between one or more state transitions. The output signal of an optical encoder mounted on the shaft of a motor is an example of a digital pulse train signal.

9.2.2　Analog DC Signals

Analog DC signals are static (or slowly varying) analog signals that convey information in the *level* (or amplitude) of the signal at a given instant. The analog DC signal is depicted in Figure 9.7 in the context of the input–output representation. Since the analog DC signal is static or varies slowly, the accuracy of the

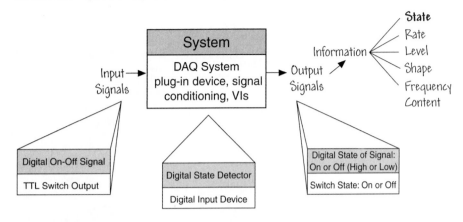

FIGURE 9.5
The on–off signal

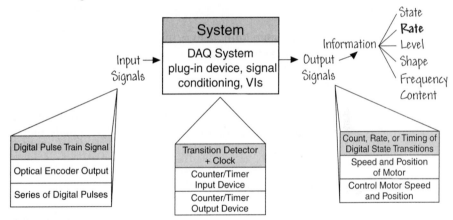

FIGURE 9.6
The pulse train signal

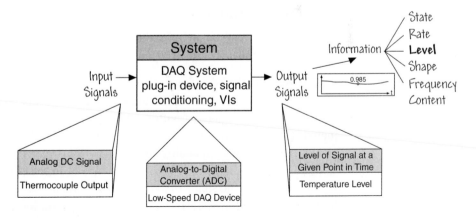

FIGURE 9.7
Analog DC signals are static or slowly varying analog signals

FIGURE 9.8
Common examples of DC signals include temperature, battery voltage, flow rate, pressure, strain-gauge output, and fluid level

measured level is of more concern than the time or rate at which you take the measurement. Common examples of DC signals include temperature, battery voltage, flow rate, pressure, strain-gauge output, and fluid level, as illustrated in Figure 9.8. The information contained in the signals is often displayed on meters, gauges, strip charts, and numerical readouts for the easiest interpretation by the user.

The DAQ system will return a single value indicating the magnitude of the signal at the requested time. For the measurement to be an accurate representation of the signal, the DAQ system must possess adequate accuracy and resolution capability and adequate sampling rates (usually slow).

9.2.3   Analog AC Signals

Analog AC signals differ from other signals in that they convey useful information in the signal level (or amplitude) *and* the way this level varies with time. The information associated with an AC signal (also referred to as a waveform) includes such information as time to peak, peak magnitude, time to settle, slope, and shapes of peaks. An analog AC signal is illustrated in Figure 9.9.

You must take a precisely timed sequence of individual amplitude measurements to measure the shape of a time-domain signal. DAQ systems that are used to measure time-domain signals usually have an **analog-to-digital conversion** (ADC) component, a sample clock, and a trigger.

The physical signal must be sampled and measured at a rate that adequately represents the shape of the signal. Therefore, when acquiring analog time-domain

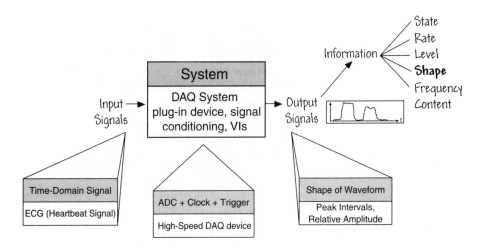

FIGURE 9.9
Analog AC signals convey information in the signal level and in the way this level varies with time

signals, DAQ systems need a high-bandwidth (fast) ADC to sample the signal at high rates. The sampled value must be measured accurately without significant loss of precision.

Generally the signal sampling needs to start at a specfic time to guarantee that an interesting segment of the signal is acquired. Once started, the sample clock accurately times the occurrence of each sample, defined as a single analog-to-digital (A/D) conversion. In many situations, triggering is necessary to initiate the sampling process at a precise time. The trigger starts the acquisition at the proper time based on either the state of the signal or some external condition specified by the user. For example, the acquisition of a sine wave or cosine wave looks the same with a 90-degree phase shift (1/4-period shift in time). A sine wave is triggered when the signal passes through zero with positive slope and the cosine is triggered when the value reaches the first peak. There are an unlimited number of different time-varying signals, some of which are shown in Figure 9.10.

9.2.4   Analog Frequency-Domain Signals

Analog frequency-domain signals are similar to AC time-domain signals in that they also convey information in the way the signals vary with time. However, the information extracted from a frequency-domain signal is based on the signal frequency content, as opposed to the shape of the signal. An analog frequency-domain signal is shown in Figure 9.11.

Frequency-domain signals are converted by performing mathematical operations on their time-domain counterparts. As with DAQ systems that measure time-domain signals, a system used to measure a frequency-domain signal must

FIGURE 9.10
Six different time-varying signals

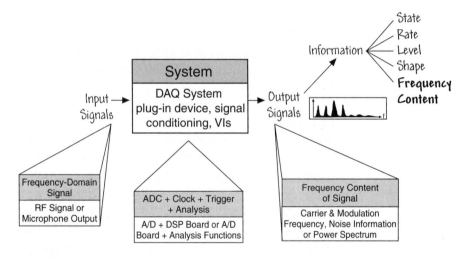

FIGURE 9.11
The information extracted from a frequency-domain signal is based on the signal frequency content

also include the same fundamental hardware components, including an ADC, a sample clock, and a trigger to accurately capture the time-domain waveform. Additionally, the DAQ system must include the necessary analysis capabilities to extract frequency information from the time-domain signal. You can perform this type of digital signal processing (DSP) using application software or special DSP hardware designed to analyze the signal quickly and efficiently.

In short, a DAQ system that acquires analog time-domain signals and can compute their frequency-domain representation, should also possess a high-bandwidth ADC capability to sample signals at high rates and an accurate clock to take samples at precise time intervals. Also, it is common to use triggers to initiate the measurement process precisely at prespecified times. Finally, a complete DAQ system should provide a library of analysis functions, including a function to convert time-domain information to frequency-domain information. Chapter 11 discusses some of the LabVIEW analysis functions, but time and space limitations make it impossible to cover all analysis functions available to VI developers.

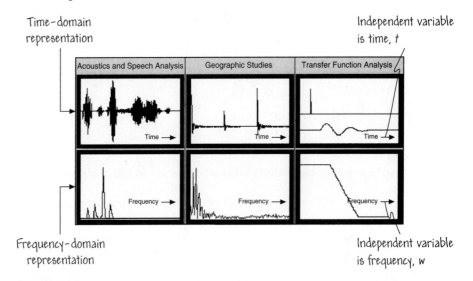

Time–domain representation

Independent variable is time, t

Frequency–domain representation

Independent variable is frequency, w

FIGURE 9.12
Three examples of frequency-domain signals: speech and acoustics analysis, geophysical signals, and transfer function frequency response

Figure 9.12 shows some examples of frequency-domain signals. Each example in the figure includes a graph of the originally measured signal as it varies with respect to time, as well as a graph of the signal frequency spectrum. While you can analyze any signal in the frequency domain, certain signals and application areas lend themselves especially to this type of analysis. Among these areas are speech and acoustics analysis, geophysical signals, vibration, and studies of system transfer functions.

9.2.5   One Signal—Five Measurement Perspectives

The five classifications of signals presented in the previous discussions are not mutually exclusive. A particular signal may convey more than one type of information, and you can classify a signal as more than one type and measure it in more than one way. In fact, you can use simpler measurement techniques with the digital on–off, pulse train, and DC signals, because they are just simpler cases of the analog time-domain signals.

The measurement technique you choose depends on the information you want to extract from the signal. In many cases you can measure the same signal

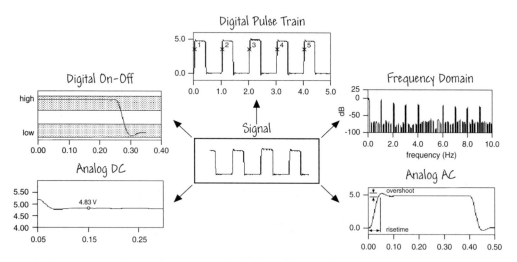

FIGURE 9.13
A series of voltage pulses can provide information for all five signal classes

with different types of systems, ranging from a simple digital input board to a sophisticated frequency analysis system. Figure 9.13 demonstrates how a series of voltage pulses can provide information for all five signal classes.

9.3 COMMON TRANSDUCERS AND SIGNAL CONDITIONING

When measuring a physical phenomenon, a transducer must convert this phenomenon (such as temperature or force) into a measurable electrical signal (such as voltage or current). Table 9.1 lists some common transducers used to convert physical phenomena into measurable quantities.

All transducers output an electrical signal that is often not well suited for a direct measurement by the ADC (Analog-to-Digital Converter). For example, the output voltage of most thermocouples is very small and susceptible to noise, and often needs to be amplified and filtered before measuring. Figure 9.14 shows some common types of transducer and signal pairs and the required signal conditioning. A highly expandable signal conditioning system—dubbed Signal Conditioning eXtensions for Instrumentation (**SCXI**) and made by National Instruments, Inc.—conditions low-level signals in a noisy environment within an external chassis located near the sensor. The close proximity improves the signal-to-noise ratio (measure of signal quality) of the signals reaching the DAQ device. LabVIEW works with SCXI and many other types of hardware signal conditioning systems.

Some types of **signal conditioning** (such as linearization and scaling) can be performed in the software. LabVIEW provides several VIs for such purposes. The remainder of this section is devoted to short descriptions of some of the basic ideas involved in signal conditioning. Further information on signal conditioning

TABLE 9.1   Phenomena and Transducers

Phenomenon	Transducer
Temperature	Thermocouples
	Resistance temperature detectors (RTDs)
	Thermistors
	Integrated circuit sensor
Light	Vacuum tube photosensors
	Photoconductive cells
Sound	Microphone
Force and pressure	Strain gauges
	Piezoelectric transducers
	Load cells
Position (displacement)	Potentiometers
	Linear voltage differential transformer (LVDT)
	Optical encoder
Fluid flow	Head meters
	Rotational flowmeters
	Ultrasonic flowmeters
pH	pH electrodes

and SCXI hardware (i.e., setup procedures for SCXI hardware, hardware operating modes, and programming considerations for SCXI) can be found in other LabVIEW documentation (check the NI website).

Some common types of signal conditioning follow:

- **Transducer excitation**: Certain transducers (such as strain gauges or microphones) require external voltages or currents to excite their own circuitry in a process known as transducer excitation. The process is similar to a television needing power to receive and decode video and audio signals. The necessary excitation in a DAQ system can be provided by the plug-in DAQ devices and the signal-conditioning peripherals, and sometimes by external instruments.

- **Linearization**: Common transducers (such as strain gauges, thermistors, RTDs, and thermocouples) generate voltages that are nonlinear with respect to the phenomena they represent. The NI-DAQ driver and LabVIEW software can perform **software linearization** to scale a transducer voltage to the correct units of strain or temperature.

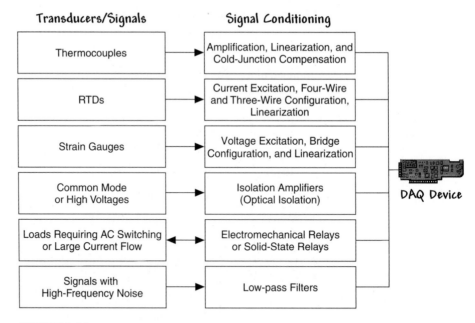

FIGURE 9.14
Common types of transducers/signals and the required signal conditioning

- **Isolation**: Another common use for signal conditioning is to isolate the transducer signals from the computer and other transducers. For example, when the signal being monitored contains large voltage spikes that could damage a computer or harm a person, you should not connect the signal directly to a DAQ device without some type of isolation. Figure 9.15 shows two common methods for isolating signals.

FIGURE 9.15
Two common methods for isolating signals

- **Filtering**: Another form of signal conditioning is the filtering of unwanted signals from the desired signal. Notch filters are commonly used to reduce 60 Hz AC powerline noise present in many signals. Other well-known types of filters include low-pass, high-pass, and band-pass filters. Some DAQ devices and signal-conditioning devices have built-in filters.

- **Amplification**: This is the most common type of signal conditioning. Amplification maximizes the use of the available voltage range to increase the accuracy of the digitized signal and to increase the signal-to-noise ratio (SNR). Low-level signals should be amplified at the DAQ device or at an external signal-conditioning peripheral positioned near the source of the signal, as shown in Figure 9.16. One reason to amplify low-level signals close to the signal source, instead of at the DAQ device, is to increase the signal-to-noise ratio. Consider the case where you amplify the signal only at the DAQ device. Then the DAQ device will also measure and digitize any noise that enters the lead wires along the path as the signal travels from the source to the DAQ device. On the other hand, the ratio of signal voltage to noise voltage that enters the lead wires is larger if you amplify the signal close to the signal source. Table 9.2 shows how the SNR changes with the location of amplification.

 You can minimize the effects of external noise on the measured signal by using shielded or twisted-pair cables and by minimizing the cable length. Keeping cables away from AC power cables and computer monitors will also help minimize 50/60 Hz noise.

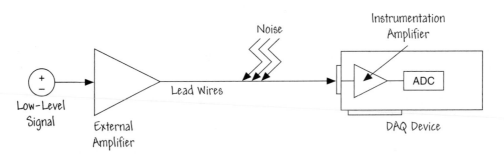

FIGURE 9.16
Low-level signals should be amplified at the DAQ device or at an external signal-conditioning peripheral positioned near the source of the signal

TABLE 9.2   Effects of Amplification on Signal-to-Noise Ratio (S.C. → Signal-Conditioning Peripheral)

	Signal Voltage	S.C. Amplification	Noise in Lead Wires	DAQ Device Amplification	Digitized Voltage	SNR*
Amplify only at DAQ device	0.01 V	None	0.001 V	×100	1.1 V	10
Amplify at S.C. and DAQ device	0.01 V	×10	0.001 V	×10	1.01 V	100
Amplify only at S.C.	0.01V	×100	0.001 V	None	1.001 V	1000

$$*\text{SNR (dB)} = 20 \log_{10} \frac{\text{Amplitude}_{\text{signal}}}{\text{Amplitude}_{\text{noise}}}$$

9.4   SIGNAL GROUNDING AND MEASUREMENTS

Up to this point, we have discussed three components of the DAQ system: transducers, signals, and signal conditioning. Now you might be tempted to think that all that remains is to wire the signal source to the DAQ device and begin acquiring data. However, a few important items must be considered:

- The nature of the signal source (grounded or floating)
- The grounding configuration of the amplifier on the signal-conditioning hardware or DAQ device
- The cabling scheme to connect all the components together

A DAQ system is depicted in Figure 9.17 highlighting the signals and the cabling.

9.4.1   Signal Source Reference Configuration

Signal sources come in two forms: referenced and nonreferenced. Referenced sources are usually called **grounded signals**, and nonreferenced sources are called **floating signals**. A schematic of a grounded signal source is shown in Figure 9.18(a).

Grounded signal sources have voltage signals that are referenced to a system ground, such as earth or a building ground. Devices that plug into a building ground through wall outlets, such as signal generators and power supplies, are

FIGURE 9.17
A DAQ system highlighting the signals and the cabling

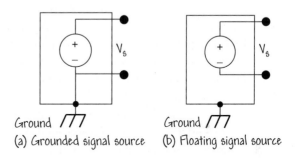

FIGURE 9.18
Grounded signal sources have voltage signals that are referenced to a system ground; floating signal sources contain a signal that is not connected to an absolute reference

the most common examples of grounded signal sources. Grounded signal sources share a common ground with the DAQ device and the computer.

Floating signal sources contain a signal (e.g., a voltage) that is not connected to an absolute reference, such as earth or a building ground. Some common examples of floating signals are batteries, battery-powered sources, thermocouples, transformers, isolation amplifiers, and any instrument that explicitly floats its output signal. As illustrated in Figure 9.18(b), neither terminal of the floating source is connected to the electrical outlet ground.

9.4.2   Measurement System

A schematic of a measurement system is depicted in Figure 9.19. A measurement system can be placed in one of three categories:

- Differential
- Referenced single-ended (RSE)
- Nonreferenced single-ended (NRSE)

FIGURE 9.19
Types of signal sources and measurement systems

In a **differential** measurement system, neither the positive or negative terminal of the source or transducer is connected to a ground. National Instruments DAQ devices have advanced instrumentation amplifiers that can be configured as differential measurement systems. Figure 9.20 depicts the 8-channel differential measurement system used in an NI Multifunction DAQ device. A **channel** is a pin or wire lead where analog or digital signals enter or leave a DAQ device. For this configuration, the analog input ground pin labeled AIGND is the measurement system ground. The analog multiplexers (labeled MUX in the figure) increase the number of available measurement channels while still using a single instrumentation amplifier.

An ideal differential measurement system, shown in Figure 9.21, reads only the potential difference between its two terminals—the (+) and (−) inputs. It completely rejects any other voltages present in the system. In other words, an ideal differential measurement system completely rejects the common-mode voltage (voltage between the measurement and system ground).

A **referenced single-ended** (RSE) measurement system measures a signal with respect to the system ground and is often called a ground referenced measurement. Figure 9.22 depicts a 16-channel version of an RSE measurement system.

DAQ devices often use a **nonreferenced single-ended** (NRSE) measurement system, which is a variation of the RSE measurement system. In an NRSE measurement system, all measurements are made with respect to a common reference, because often many transducers share a common power source and ground. Figure 9.23 depicts an NRSE measurement system. AISENSE is the common reference for taking measurements and all signals in the system share this common reference. Note that the signals are not referenced to the system ground, AIGND.

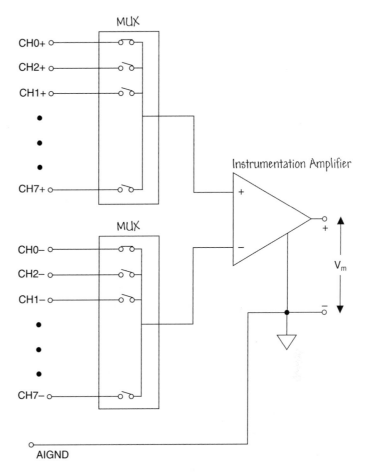

FIGURE 9.20
An 8-channel differential measurement system

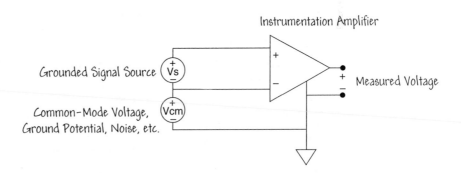

FIGURE 9.21
An ideal differential measurement system completely rejects common-mode voltage

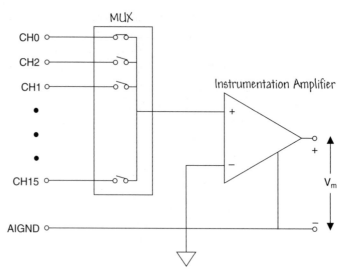

FIGURE 9.22
A 16-channel RSE measurement system

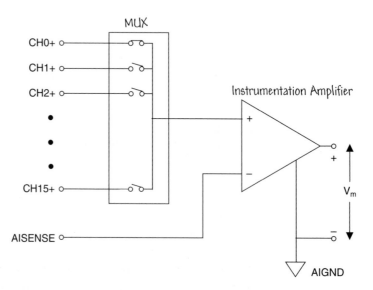

FIGURE 9.23
An NRSE measurement system

Differential measurements offer the best immunity to noise when measuring grounded and floating signal sources. NRSE offers the next best noise immunity, where all signals share the same ground reference point. An RSE measurement system can be used with a grounded signal source if the signal levels are high and the cabling has a low impedance. RSE measurements are subject to noise and ground loops (ground is not actually zero volts for all transducers). In general, there is a trade-off between noise immunity and the number of channels. Differential measurements require more connections than NRSE measurements because differential measurements require a single pin for the $(-)$ terminal of each signal whereas (N)RSE measurements share the $(-)$ terminal for each signal. When higher channel counts are needed, you can use single-ended measurement systems when all input signals meet the following criteria:

1. High-level signals (normally, greater than 1 V).

2. Short or properly shielded cabling traveling through a noise-free environment (normally less than 15 ft).

3. All signals can share a common reference signal at the source.

A summary of analog input connections is given in Figure 9.24.

9.5 ANALOG TO DIGITAL CONVERSION CONSIDERATIONS

When preparing to configure a DAQ device you need to consider the quality of the analog-to-digital conversion. There are many questions that arise when making analog signal measurements with your DAQ system. For example, what are the signal magnitude limits? How fast does the signal vary with time? The latter question is important because the sample rate determines how often the A/D conversions take place. The four parameters of concern are

- Resolution
- Device range
- Signal input range
- Sampling rate

Depending on the type of DAQ device you have, these four parameters will be either set in the hardware (using dip switches and jumpers on the board) or set using software (with the **Measurement and Automation Explorer** or in LabVIEW as discussed later in this chapter).

The number of bits used to represent an analog signal determines the **ADC resolution**. You can compare the resolution on a DAQ device to the number of divisions on a ruler. For a fixed ruler length, the more divisions you have on the ruler, the more precise the measurements can be made. A ruler marked off in millimeters can be read more accurately than a ruler marked off in centimeters. Similarly, the higher the ADC resolution, the higher the number of divisions of the ADC range and, therefore, the more accurately the analog signal can be represented. For example, a 3-bit ADC divides the signal range into 2^3 divisions ($2^3 = 8$), with each division represented by a binary or digital code between 000

FIGURE 9.24
Summary of analog input connections

and 111 (0 and 7, respectively). The ADC then translates each measurement of the analog signal to one of the digital divisions (rounding to the nearest digital division). Figure 9.25 shows a sine wave digital image obtained by a 3-bit ADC. Clearly, the digital signal does not represent the original signal adequately because there are too few divisions to represent the varying voltages of the analog signal. By increasing the resolution to 16 bits, however, the ADC's number of divisions increases from 2^3 to 2^{16} (8 to 65,536). The ADC can now provide an adequate representation of the analog signal.

The **device range** refers to the minimum and maximum analog signal levels that the ADC can handle. You should attempt to match the range to that of the analog input signal to take best advantage of the available resolution. Fortunately,

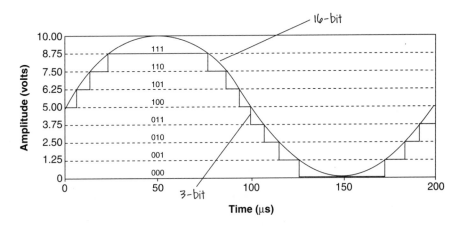

FIGURE 9.25
16-bit versus 3-bit resolution (5 kHz sine wave)

many DAQ devices feature selectable ranges. Consider the example shown in Figure 9.26(a), where the 3-bit ADC has eight digital divisions in the range from 0 to 10 volts. If you select a range of −10.00 to +10.00 volts, as shown in Figure 9.26(b), the same ADC now separates a 20-volt range into eight divisions. The smallest detectable voltage increases from 1.25 to 2.50 volts, and you now have a much less accurate representation of the signal.

The **signal input range** is the maximum and minimum value of the signal you are measuring. The closer the signal input range is to the incoming analog signal maximum and minimum, the more digital divisions will be available to the ADC to represent the signal. Using a 3-bit ADC and a range setting of 0.00 to 10.00 volts, we see in Figure 9.27 the effects of a signal input range between 0 and 5 volts and 0 and 10 volts. With a signal input range of 0 to 10 volts, the

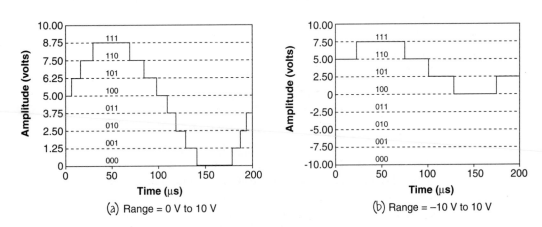

FIGURE 9.26
The 3-bit ADC in (a) has eight digital divisions in the range from 0 to 10 volts—if you select a range of −10.00 to 10.00 volts as in (b), the same ADC now separates a 20-volt range into eight divisions

(a) Signal input range: 0 V to 10 V   (b) Signal input range: 0 V to 5 V

FIGURE 9.27
Precise signal input range selection allows the ADC to use more digital divisions to represent the signal

ADC uses only four of the eight divisions in the conversion; with a signal input range of 0 to 5 volts, the ADC now has access to all eight digital divisions. This makes the digital representation of the signal more accurate.

The resolution and range of a DAQ device and the signal input range determine the smallest detectable change in the input voltage. This change in voltage represents 1 **least significant bit** (LSB) of the digital value and is often called the **code width**. The smallest code width is calculated with the following formula:

$$V_{CW} = \frac{\text{range}}{2^{\text{resolution}}},$$

where the resolution is given in bits. For example, a 12-bit DAQ device with a 0 to 10 V range detects a 2.4-mV change. This is calculated as follows:

$$V_{CW} = \frac{\text{range}}{2^{\text{resolution}}} = \frac{10}{2^{12}} = 2.4 \text{ mV},$$

while the same device with a -10 to 10 V range detects only a change of 4.8 mV:

$$V_{CW} = \frac{\text{range}}{2^{\text{resolution}}} = \frac{20}{2^{12}} = 4.8 \text{ mV}.$$

A high-resolution ADC provides a smaller code width for a given range. For example, consider the two preceding examples, except that the resolution is now 16-bit. A 16-bit DAQ device with a 0 to 10 V range detects a 0.15-mV change:

$$V_{CW} = \frac{\text{range}}{2^{\text{resolution}}} = \frac{10}{2^{16}} = 0.15 \text{ mV},$$

while the same device with a -10 to 10-V range detects a change of 0.3 mV:

$$V_{cw} = \frac{\text{range}}{2^{\text{resolution}}} = \frac{20}{2^{16}} = 0.3 \text{ mV}.$$

You also need to determine whether your signal is unipolar or bipolar, as this affects the code width. **Unipolar** signals range from 0 V to a positive value (e.g.,

0 V to 10 V). **Bipolar** signals range from a negative value to a positive value (e.g., −5 V to 5 V). A smaller code width is obtained by specifying the range to be unipolar, if indeed the signal is unipolar; and conversely, specifying the range as bipolar if the signal is bipolar. If the maximum and minimum variation of the analog signal is smaller than the value of the range, you will need to set the signal input range to more accurately reflect the analog signal variation. If the smallest range on the DAQ devices does not provide enough resolution, signal conditioning or a higher accuracy DAQ device should be considered.

Some confusion may exist about selecting the signal input range rather than selecting the gain. *When you set the signal input range, you are effectively selecting the gain for that signal. Signal input range settings automatically magnify the magnitude of the signal to create more precise analog-to-digital conversions.*

The **sampling rate** is the rate at which the DAQ device samples an incoming analog signal. Figure 9.28 shows an adequately sampled signal as well as the effects of undersampling. The sampling rate determines how often an analog-to-digital (A/D) conversion takes place. Computing the proper sampling rate requires knowledge of the maximum frequency of the incoming signal and the accuracy required of the digital representation of the analog signal. It also requires some knowledge of the noise affecting the incoming signal and the capabilities of your hardware. A fast sampling rate acquires more points in a given time, allowing, in general, a better representation of the original signal than a slow sampling rate would allow. In fact, sampling too slowly may result in a misrepresentation of the incoming analog signal.

The effect of undersampling is that the signal appears as if it has a different shape in the time domain, thus appearing to have a different frequency than it truly does. This misrepresentation is called an *alias*. To prevent undersampling, you must sample greater than twice the rate of the maximum frequency com-

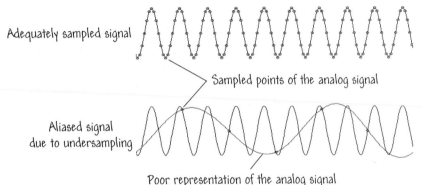

FIGURE 9.28
Sampling too slowly may result in a poor representation of the analog signal

ponent of the incoming signal. One way to deal with aliasing is to use low-pass filters that attenuate any frequency components in the incoming signal above the Nyquist frequency (defined to be one-half the sampling frequency). These filters are known as *antialiasing filters* and these filters must be applied in hardware before the ADC.

Signal Input Range Selection

The kind of calculations required to select the signal input range for a DAQ application are illustrated in this example. Keep in mind that the objective is to minimize the code width while making sure that the entire signal fits within the allowable device range.

Assume your transducer output is a sine wave with an amplitude of ± 30 mV and an offset of 10 mV. Your device has signal input ranges of 0 to +10 V, ± 10 V, and ± 5 V. What signal input range would you select for maximum precision if you use a DAQ device with 12-bit resolution?

Table 9.3 shows how the code width of a hypothetical 12-bit DAQ device varies with device range and signal input range. The values in Table 9.3 depend on the hardware, and you should consult your DAQ hardware documentation to determine the available device voltage ranges for your device.

Step 1: Select a device range. Since the signal input is bipolar (the sine wave contains both positive and negative voltages), we must choose one of the bipolar ADC ranges. For our device , there are two, so to start we choose ± 5 V, for a range of 10 V.

Step 2: Determine the code width. Using the formula for determining the code width, we compute

$$V_{\text{cw}} = \frac{\text{range}}{2^{\text{resolution}}} = \frac{10}{2^{12}} = 2.4 \text{ mV}.$$

Step 3: Choose a signal input range. Since our signal has a maximum magnitude of +30 mV, we must choose signal input ranges that allow us to read ± 30 mV. Referring to Table 9.3, we find that (for a range of -5 to 5 V) a signal input range of -50 mV to 50 mV will cover the signal variation of ± 30 mV. Choosing this range yields a code width (or precision) of 24.4 μV.

Step 4: Repeat steps 1 through 3 for a range of ± 10 V. We must also try the other range of ± 10 V to see if it yields a smaller code width. Repeating steps 1 through 3 gives us a signal input range of -0.1 V to 0.1 V and a code width of 48.8 μV. This does not improve our accuracy. Thus, we choose the settings as follows:

- Device Range $= -5$ to 5 V
- Signal Input Range $= -50$ mV to 50 mV

TABLE 9.3   Code Width for Various Device Ranges and Signal Input Ranges

Device Range	Signal Input Range	Code Width*
0 to 10 V	0 to 10 V	2.44 mV
	0 to 5 V	1.22 mV
	0 to 2.5 V	610 μV
	0 to 1.25 V	305 μV
	0 to 1 V	244 μV
	0 to 0.1 V	24.4 μV
	0 to 20 mV	4.88 μV
−5 to 5 V	−5 to 5 V	2.44 mV
	−2.5 to 2.5 V	1.22 mV
	−1.25 to 1.25 V	610 μV
	−0.625 to 0.625 V	305 μV
	−0.5 to 0.5 V	244 μV
	−50 to 50 mV	24.4 μV
	−10 to 10 mV	4.88 μV
−10 to 10 V	−10 to 10 V	4.88 mV
	−5 to 5 V	2.44 mV
	−2.5 to 2.5 V	1.22 mV
	−1.25 to 1.25 V	610 μV
	−1 to 1 V	488 μV
	−0.1 to 0.1 V	48.8 μV
	−20 to 20 mV	9.76 μV

* Code width is the smallest measurable difference in voltage

9.6   DAQ VI ORGANIZATION

The LabVIEW DAQ VIs are organized in the NI-DAQmx palette, as illustrated in Figure 9.29. These VIs are device drivers for the National Instruments DAQ hardware line. NI-DAQmx, originally released in NI-DAQmx 7.0, is a next-generation DAQ driver architecture that incorporates significant improvements over previous NI-DAQ drivers in both performance and ease of use. For students learning about data acquisition for the first time, the special advantage of NI-DAQmx over previous versions of NI-DAQ is the inclusion of the DAQ Assistant

FIGURE 9.29
The **DAQmx-Data Acquisition** palette

for configuring channels and measurement tasks. The task of writing VIs to acquire data is significantly simplified with the DAQ Assistant, as we shall learn later in this chapter. For these reasons, the remainder of this chapter will cover NI-DAQmx.

The NI-DAQmx VIs are a special type of VI called Polymorphic VIs. The result is a core set of VIs that can adapt to different DAQ functionality, such as analog input, analog output, digital I/O, etc. You access the palette by opening the **Functions** palette and clicking on **Measurement I/O≫DAQmx-data Acquisition**, shown in Figure 9.29. In this book, we will focus on the functionality of these three DAQ operations: **Analog Input**, **Analog Output**, and **Digital I/O**.

9.7   CHOOSING YOUR DATA ACQUISITION DEVICE

In this chapter we introduce the NI-DAQmx driver to communicate with the National Instruments PCI-6251 M Series data acquisition device in LabVIEW. The PCI-6251 M Series is but one of many high-speed, general-purpose data acquisition options available to users of LabVIEW. National Instruments offers plug-in data acquisition devices that can measure signals in frequency ranges from DC to Gigahertz with digital accuracies of 8 to 24 bits. Many common interfaces for PC connectivity are available including PCI, PCIe, PXI, USB, Firewire, and Ethernet. Engineers at National Instruments designed the NI-DAQmx driver so that the same VIs can be used to communicate with nearly all NI data acquisition devices and even allow many of those devices to be simulated.

9.7.1   M Series Data Acquisition Devices

The M Series multifunction data acquisition devices from National Instruments are optimized for higher accuracy at fast sampling rates for a broad range of applications. These devices are ideal for laboratory environments where a versatile, high-speed, high-quality data acquisition device is needed. M Series devices have 16, 32, or 80 analog inputs with a resolution of 16 or 18 bits and acquisition rates up to 1.25 MS/s. They also include up to 48 digital I/O lines, up to four analog outputs, two 32-bit, 80 MHz counter/timers, up to seven programmable input ranges, and digital and analog triggering (analog triggering not available on all models). M Series devices are ideal for a wide range of applications including test, control, and design.

9.7.2   Low Cost Data Acquisition for Students

The USB-6008 and USB-6009 multifunction data acquisition devices from National Instruments provide reliable data acquisition at prices affordable to most students. Although the USB-6008 and USB-6009 devices cannot match the performance of an M Series device, employing them offers a low-cost solution to data acquisition that is well suited for home and laboratory use. Most of the exercises presented in this chapter can be easily completed using USB data acquisition. With plug-and-play USB connectivity, these devices are simple enough for quick measurements, but versatile enough for more complex measurement applications. The USB-6008 and USB-6009 have the following characteristics:

- Small, portable multifunction data acquisition devices
- 12- or 14-bit input resolution, at up to 48 kS/s
- Built-in, removable connectors for easier and more cost-effective connectivity
- 2 true DAC analog outputs for accurate output signals
- 12 digital I/O lines (TTL/LVTTL/CMOS)
- 32-bit event counter
- Student kits available

You must install the NI-DAQmx device driver before completing the rest of this chapter. Visit the Learning with LabVIEW companion website for additional information.

9.7.3   Simulated Data Acquisition

You can create NI-DAQmx simulated devices in NI-DAQmx 7.5 and later. Using NI-DAQmx simulated devices, you can craft VIs for use with data acquisition devices without actually having the hardware. No time is lost when you later plug-in or move to a computer with the hardware you simulated, because you can import the NI-DAQmx simulated device configuration to the physical device using the MAX Portable Configuration Wizard. You can prototype and test your data acquisition applications before you get to the laboratory, knowing you are ready to acquire real-world signals. If you do not have access to a data acquisition device, follow the steps outlined below to complete the exercises and homework problems presented in the rest of the chapter using a simulated data acquisition device.

Simulated data acquisition is currently only available on the Windows plat-form.

Creating NI-DAQmx Simulated Devices

Since this book uses the National Instruments PCI-6251 data acquisition device for the exercises, we will add that device as an NI-DAQmx simulated device. To accomplish this task, follow these steps:

1. Open the **Measurement and Automation Explorer (MAX)** by clicking on the **Measurement & Automation** icon on your desktop (see section 9.8.1 for more information on MAX).

2. Right-click on **Devices and Interfaces** and select **Create New**

3. A dialog box prompts you to select a device to add. Select **NI-DAQmx Simulated Device** and click **Finish**.

4. In the **Choose Device** dialog box, select the family of devices for the device you want to simulate. In this case, we will choose **M Series DAQ**.

5. Select **PCI-6251** and click **OK**.

9.7.4 Macintosh, Linux, Palm OS, Windows Mobile for Pocket PC, and Select Windows CE OS Devices

NI-DAQmx Base driver software offers a variety of other OS users a programming interface similar to the polymorphic VIs available in the NI-DAQmx driver software as well as increased ease of use in connecting with high-performance NI data acquisition hardware. NI-DAQmx Base offers a clean, concise programming interface, programmatic channel and task creation, and other features previously available only with NI-DAQmx for Windows. The new driver software also includes ready-to-use National Instruments LabVIEW VIs and C function features similar to those included in the full-featured version of NI-DAQmx.

National Instruments built NI-DAQmx Base driver software for these OSs using the NI Measurement Hardware Driver Development Kit and developed the software almost entirely in the multiplatform NI LabVIEW graphical development environment.

For more information on updates to DAQ for the Mac, refer to www.ni.com/ mac. For more information on updates to DAQ for Linux, refer to www.ni.com/ linux. For more information on updates to DAQ for PDAs, refer to www.ni.com/ pda.

9.8   DAQ HARDWARE CONFIGURATION

LabVIEW provides utilities designed to help you define which signals are connected to which channels on your data acquisition device and to easily group those into combined tasks. In previous years, significant amounts of time were spent defining the signal types and connections and converting voltage to engineering units—and all this before beginning development and programming of the actual DAQ system! For example, if you are using thermocouples, you must perform cold-junction compensation (CJC) calculations and apply appropriate scaling factors to convert raw measured voltages into actual temperature readings. This process is now one of entering the necessary information in dialog boxes, or through wizards, to define an input signal, the type of transducer being used, any scaling factors required, CJC values, and the conversion factors, as well as the timing and triggering information you desire. You can then reference the channel name or task (that you assign) for the input signal(s) and the conversion from voltage to physical units is performed automatically (and transparently). You can save different configuration files for different settings or systems. Not only can you assign **channel** and **task names**, sensor types, engineering units, and scaling information, as well as timing and triggering information, to each channel using the LabVIEW utility, but you can also define the physical quantities you are measuring on each DAQ hardware channel. Once the software has been properly configured, the hardware will be configured correctly to make the measurement for each channel in terms of the physical quantity.

9.8.1   Windows

LabVIEW for Windows installs a configuration utility for establishing all device and channel configuration parameters. This utility is known as the **Measurement & Automation Explorer**, or MAX for short. MAX is also helpful in troubleshooting and self-calibration. The MAX utility reads the information the Device Manager records in the Windows registry and assigns a logical device name to each National Instruments DAQ device in your system. You use the device name to refer to the device in LabVIEW. Figure 9.30 shows the relationship between the DAQ device and MAX. The Windows Configuration Manager keeps track of all the hardware installed in your system, including National Instruments DAQ devices.

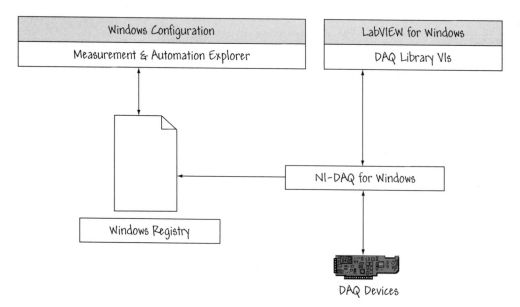

FIGURE 9.30
Windows configuration management

If your National Instruments DAQ device does not show up in MAX it may not be configured correctly or it may not be supported in the version of NI-DAQ that is installed. The installation guide for your DAQ device can be found on www.ni.com/manuals.

You can check the Windows Configuration by accessing the Device Manager on your computer. You will find **Data Acquisition Devices**, which lists all DAQ devices installed in your computer, as shown in Figure 9.31. Highlight a DAQ device and select **Properties** or double-click on the device, and you see a dialog window with tabbed pages. **General** displays overall information regarding the device. You use **Resources** to specify the system resources to the device such as interrupt levels, DMA, and base address for software configurable devices. **Driver** specifies the driver version and location for the DAQ device.

LabVIEW for Windows DAQ VIs access the National Instruments standard NI-DAQ for Windows 32-bit dynamic link library (DLL). The LabVIEW setup program installs the NI-DAQ DLL in the System32 folder of the Windows directory. NI-DAQ for Windows supports all National Instruments DAQ devices and SCXI.

The nidaq32.dll file, the high-level interface to your device, is loaded into the System32 folder of the Windows directory. The nidaq32.dll file then interfaces with the Windows Registry to obtain the configuration parameters defined by MAX. You access MAX either by double-clicking on its icon on the desktop or selecting **Measurement & Automation Explorer** from the **Tools** menu in LabVIEW, as illustrated in Figure 9.32.

FIGURE 9.31
Checking the Windows Configuration by accessing the Device Manager

Using MAX (Windows)

The objective of this exercise is to use MAX to examine the configuration for the DAQ device in your computer and to configure one NI-DAQmx task.

Start MAX by double-clicking on the desktop icon or by selecting **Measurement & Automation Explorer** from the **Tools** menu in LabVIEW, as illustrated in Figure 9.32.

Expand the **Devices and Interfaces** section as seen in Figure 9.33. The figure shows what MAX looks like for a PCI-6251 and a PCI-GPIB. The PCI-6251 is an NI-DAQmx device, so open the **NI-DAQmx Devices** section as well. MAX shows the National Instruments devices and software in your system. Note the device name indicated in the quotes after the DAQ device. The LabVIEW DAQ VIs use the device name to determine which device performs DAQ operations.

FIGURE 9.32
Accessing the Measurement & Automation Explorer (MAX)

In Figure 9.34, we see that the DAQ device PCI-6251 is "Dev1." You may have a different device installed and some of the options shown may be different.

You can get more information about your DAQ device configuration by examining the properties using the **Attributes**, **Device Routes**, and **Calibration** tabs. With the DAQ device highlighted, MAX displays the attributes of the device, such as the system resources that are being used by the device, the serial number, socket number, and memory range, as shown in Figure 9.33. Clicking the **Device Routes** tab provides detailed information about the internal signals that can be routed to other destinations on the device, as shown in Figure 9.34. This is a powerful resource that gives you a visual representation of the signals that are available to provide timing and synchronization with components that

FIGURE 9.33
Checking the properties of the DAQ devices

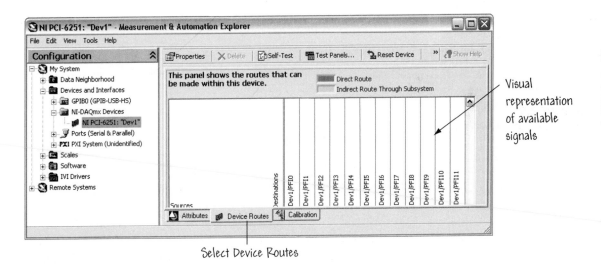

FIGURE 9.34
Accessing information about the DAQ device configuration

are on the device and shared with other external devices. The **Calibration** tab provides information about the last time the device was calibrated both internally and externally. If you right-click the NI-DAQmx device in the configuration tree and select **Self Calibrate**, the DAQ device will be calibrated using a precision voltage reference source. Once the device has been calibrated, the **Self Calibration** information updates in the **Calibration** tab.

*Press the **Show/Hide** button in the top right corner of the MAX window to hide the online help.*

Next, we want to create and configure an NI-DAQmx task. An NI-DAQmx task is a shortcut to configuring a data acquisition task that allows you to save a configuration complete with channel, timing, and triggering information. Right-click on the **Data Neighborhood** icon and choose **Create New**, as shown in Figure 9.35. The window shown in Figure 9.36 will appear. In the window, select **NI-DAQmx Task** and press the **Next** button, as depicted in Figure 9.36. You will now configure a channel to take a reading from a temperature sensor (Analog Input).

After pressing the **Next** button, a window for configuring the input will appear, and you can select the input as shown in Figure 9.36. We will select the Analog Input≫Temperature≫Thermocouple as shown in Figure 9.37. Once we do so, a new dialog box opens up to permit us to assign a new local channel to the task. In this case, we select **ai0** associated with Dev1—our PCI-6251 DAQ device. Once you have completed the job of creating the local channel, select **Next** to proceed.

FIGURE 9.35
Configuring a new virtual channel

FIGURE 9.36
Creating a new virtual channel configuration

The next dialog box that appears allows you to name your task, as shown in Figure 9.38. By default, the name **MyTemperatureTask** is provided—we will accept this name this time around. You can change the name to something more descriptive depending on your application. Once you have named the NI-DAQmx task, select **Finish**.

*As you proceed through the configuration process, you can always go back to a previous page using the **Back** button.*

When you are finished defining the channel, MAX entries should appear as shown in Figure 9.39. In future sessions, you can edit the configurations through MAX by highlighting the desired task and right-clicking on **Temperature**. MAX provides important configuration information. For example, as seen in Figure 9.39, the temperature minimum and maximum values are shown to be 0°C and 100°C, respectively. These values can be readily changed by editing the screen directly. Also, if you click on the double arrow ≫ under the **Channel List**, the screen will expand to show the device type and physical channel. You can right-click on **Temperature** to open a dialog box in which the channel assignment can be changed should the need arise in future sessions. When you are finished configuring the NI-DAQmx task, close the MAX window to exit. ◆

FIGURE 9.37
Selecting the measurement type and creating a new local channel

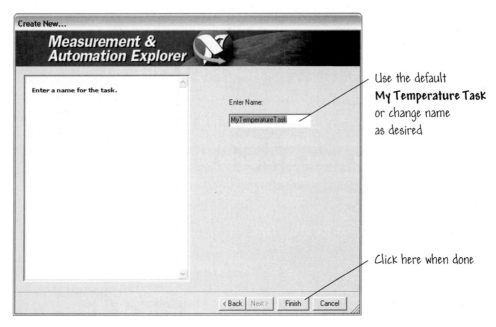

FIGURE 9.38
Naming the task and finishing the task configuration

9.8.2   Channels and Tasks

In Section 9.8.1, we created an NI-DAQmx task to acquire temperature readings from a thermocouple using an NI PCI-6251 DAQ device. In the sections following this one, we will use the DAQ Assistant to create other NI-DAQmx tasks. But what exactly is an NI-DAQmx task? And what are NI-DAQmx channels?

If you are familiar with LabVIEW 6i (or earlier versions), you know that when developing DAQ applications, you configured a set of *virtual channels* using MAX. This configuration was a collection of property settings that included a physical channel, the type of measurement specified in the channel name, and scaling information. In the traditional NI-DAQ (prior to LabVIEW 7 Express), virtual channels were a simple method to remember which channels were used for different measurements. NI-DAQmx *channels* are similar to the virtual channels of traditional NI-DAQ channels and must be configured accordingly. In NI-DAQmx (introduced with LabVIEW 7 Express), you can configure virtual channels either in MAX or using the DAQ Assistant—in both cases the configuration process is almost identical.

An NI-DAQmx *task* is a collection of one or more channels, timing, triggering, and other properties that apply to the task itself. For example, a task might represent a measurement you want to perform. In the previous section, we created a task to measure temperature from one channel on a DAQ device.

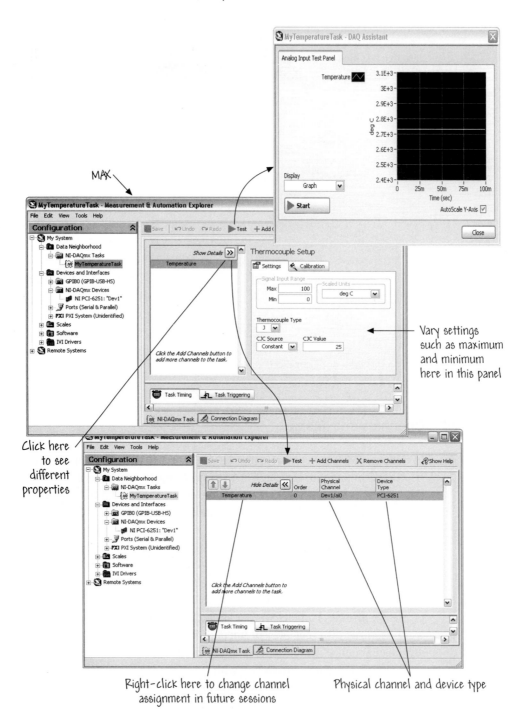

FIGURE 9.39
The MAX final configuration

To complete the exercises in the next section using the DAQ Assistant, you must have installed NI-DAQmx and have installed an NI-DAQmx-supported device, or created an NI-DAQmx simulated device in MAX.

9.9   USING THE DAQ ASSISTANT

In this section, we introduce the DAQ Assistant. The DAQ Assistant is a graphical interface that we can use to configure measurement tasks and channels (described in Section 9.8).

The DAQ Assistant is located on the **Functions≫Measurement I/O≫ DAQmx-Data Acquisition** palette. To launch the DAQ Assistant, place it on the block diagram, as illustrated in Figure 9.40. When the DAQ Assistant is placed on the block diagram, a DAQ Assistant dialog box will automatically appear. Once the DAQ Assistant is open, the steps required to configure the NI-DAQmx task are basically the same as described in Section 9.8 using MAX.

FIGURE 9.40
Locating and placing the DAQ Assistant on the block diagram

The general process of constructing a data acquisition VI using the DAQ Assistant is as follows:

- Open a new VI.

- Place the DAQ Assistant on the block diagram.

- A DAQ Assistant dialog box appears to help you configure the measurement task.

- Configure, name, and test the NI-DAQmx task.

- Click the **OK** button to return to the block diagram.

- Edit the front panel and block diagram to complete the VI.

- If desired, generate an NI-DAQmx Task Name control so that the task can be used in other applications.

Practice with the DAQ Assistant

In this example, you will construct an NI-DAQmx task to acquire voltage readings from a measurement device attached to a PCI-6251 DAQ device.

The first step is to open a new VI and place the DAQ Assistant on the block diagram. The DAQ Assistant will automatically launch to begin the process of creating and configuring the NI-DAQmx task. In the first dialog box, click the **Analog Input** button to display the **Analog Input** options, as illustrated in Figure 9.41. Select **Voltage** to create a new voltage analog input task. The dialog box then displays a list of channels on each DAQ device installed. The number of channels listed depends on the number of channels you have on the DAQ device. In our case, we find a list **ai0, ai1, ...** Select **ai0** and click the **Finish** button.

At this point, the DAQ Assistant opens a new window, shown in Figure 9.42, which displays the options for configuring the selected channels. In the **Signal Input Range** section of the **Settings** tab, enter 10 for the **Max** value and enter −10 for the **Min** value. In the **Task Timing** tab, select the **N Samples** option. Enter a value of 100 in the **Samples To Read** input box.

At this point, the NI-DAQmx task is ready to test. You can test the task to verify that you correctly configured the channel. Click the **Test** button, as shown in Figure 9.42, to access the **Analog Input Test Panel** dialog box that will automatically open. Testing the channel is easy—click the **Start** button once or twice to confirm that you are acquiring data. Once you have verified the proper functioning and configuration, click the **Close** button to return to the DAQ Assistant. Once there, click the **OK** button to return to the block diagram. The DAQ Assistant appears on the block diagram ready for integration into the VI application.

We are now ready to add a graph to the VI to plot the voltage data acquired from the DAQ device. On the block diagram, right-click the **data** output on the DAQ Assistant and select Create≫Graph Indicator, as illustrated in Figure 9.43. The waveform graph will be added and automatically wired. Switch the view to the front panel. After running the VI, notice that the waveform graph

FIGURE 9.41
Using the DAQ Assistant to configure an analog input channel

legend displays the channel name **Voltage**. The DAQmx task is now ready to acquire voltage data and to display the output on the waveform graph.

It is possible to rename the channels configured using the DAQ Assistant. You might, for instance, decide later that a more descriptive name would be helpful in organizing your configured channels. To accomplish this, return to the block diagram, right-click the DAQ Assistant, and select **Properties** to rename the

FIGURE 9.42
Configuring the channel settings and testing the DAQmx task

channel, as shown in Figure 9.44. Right-click **Voltage** in the **Channel List** box and select **Rename** to display the **Rename a channel or channels** dialog box. In the **New Name** text box, enter **Extreme Voltage**, then click the **OK** button. You are done—the channel has been renamed. To verify the name change, click the **OK** button and return to the block diagram. Then, switch to the front panel and run the VI. After running the VI, notice that **Extreme Voltage** appears in the waveform graph plot legend.

FIGURE 9.43
Adding a waveform graph to the DAQ Assistant

*You also can select the name of the channel and press the <F2> key to access the **Rename a channel or channels** dialog box.*

Save the VI as Read Voltage.vi in the Users Stuff folder in the Learning directory.   ◆

9.9.1   DAQmx Task Name Constant

If you configure an NI-DAQmx task using the DAQ Assistant, then the task is a *local task*; hence, it cannot be saved to MAX for use in other applications. If you want to make the task available to other applications, you can use the DAQ

FIGURE 9.44
Renaming a channel

Assistant to generate an NI-DAQmx Task Name control so that the task can be saved to MAX and used in other applications. This is readily accomplished by right clicking on the DAQ Assistant in the block diagram and selecting **Convert to Task Name Control** in the pull-down menu, as illustrated in Figure 9.45. The DAQmx Task Name Constant is a LabVIEW data type used by the DAQ VIs for communicating to the DAQ devices. After converting the DAQ Assistant to a DAQmx Task Name Constant, the DAQ Assistant will appear so that you can re-configure the task, if needed. After exiting the DAQ Assistant, the DAQmx task is available in MAX and hence ready for use in other applications. You will probably want to rename the task.

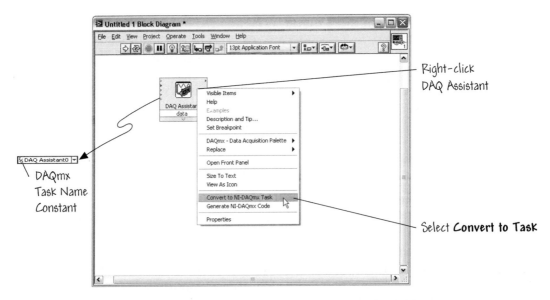

FIGURE 9.45
Create a Task Name Constant to make a DAQmx task available to other applications through MAX

If you want to use one of the DAQmx tasks from MAX in an application, you first place a DAQmx Task Name Constant on the block diagram. You can find the DAQmx Task Name Constant on the **Functions≫Measurement I/O≫NI-DAQmx Data Acquisition** palette, as shown in Figure 9.46. The task names can

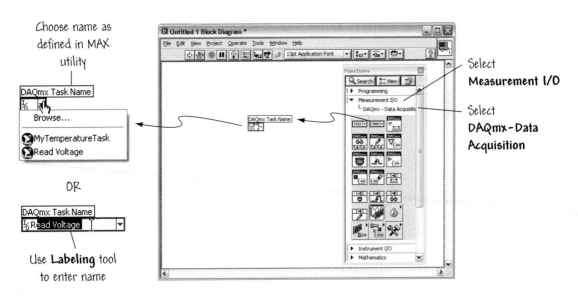

FIGURE 9.46
Entering DAQmx task names

be entered in two different ways. You can use the **Operating** tool to click on the DAQmx Task Name Constant and choose the desired task name found in the MAX utility, as shown in Figure 9.46. Another way to enter the DAQ task name is to use the **Labeling** tool to enter the task channel number. The task name must refer to one of the tasks configured using MAX, as discussed in the previous sections.

9.10   ANALOG INPUT

On the first screen of the DAQ Assistant (see Figure 9.41) you choose the measurement type. The available types are analog input, analog output, counter input, counter output, and digital I/O. In this section we focus on analog input. In subsequent sections we will cover analog output and digital I/O.

The analog input is used to perform analog-to-digital (A/D) conversions. In the DAQ Assistant, clicking on **Analog Input** opens the screen that lists the available analog input measurement types: voltage, temperature, strain, current, resistance, frequency, position, acceleration, custom voltage with excitation, and sound pressure. In the *Practice with the DAQ Assistant* example in Section 9.9, the voltage measurement type was selected. Each measurement type has its own characteristics necessary for turning the acquired voltage to useful data, such as the resistor value for a current measurement or a strain gauge parameter for a strain measurement.

Once you have selected the channels to add to your task, the DAQ Assistant settings and testing screen opens. The DAQ Assistant is shown in Figure 9.42. In the lower section of the DAQ Assistant are two tabs: **Task Timing** and **Task Triggering**. After setting the signal input range, terminal configuration, and scaling for each channel, the next step is to configure the timing and triggering.

Only one analog input task can run at a time on a single Multifunction DAQ device, although a single task can acquire data on multiple channels.

9.10.1   Task Timing

When performing analog input, the task can be timed to acquire a single sample (on demand), a single sample (hardware timed), *n* samples, or to acquire data continuously. When **1 Sample (On Demand)** is selected (see Figure 9.42), NI-DAQmx acquires one value from an input channel and immediately returns the value to your application. This operation does not require any buffering or hardware timing. Timing is defined by the rate at which the DAQ Assistant runs. For example, if you periodically monitor the fluid level in a tank, you would acquire single data points. You can connect the transducer that produces a voltage representing the fluid level to a single channel on the measurement device and initiate a single-channel, single-point acquisition when you want to know the fluid level.

It is possible to acquire multiple samples for the channels in your task by executing the VI in a repetitive manner. This process is only suggested for measuring static or slow changing signals and point-by-point control applications because the time between samples may vary or it may take too much time to accurately capture your signal. The preferable way to acquire multiple samples is to use hardware timing. This method employs a buffer on the DAQ device and in computer memory to acquire data more efficiently and with very accurate timing. For these types of applications, set the sample mode to **N Samples**. With NI-DAQmx, you can acquire multiple samples on a single channel or multiple samples on multiple channels. For example, you can monitor both the fluid level and the temperature in a tank using two transducers connected to two channels on the device. Each time you execute the DAQ Assistant, *n* samples of data are acquired at the specified rate.

If you want to view, process, or log large amounts of data for more than one iteration of *n* samples, you need to continually acquire samples. For these types of applications, set the sample mode to **Continuous**. In this configuration the DAQ Assistant would typically be placed inside a loop and would transfer small buffers of data back to the computer each time the loop iterate, with no lost data between buffers.

9.10.2   Task Triggering

So far we have discussed the acquisition of data that begins immediately after the DAQ Assistant is run in software. Often it is desirable for the acquisition of data to begin after a stimulus such as detection of a defined voltage level or change in a digital signal. The stimulus used to begin a task is called a **trigger**. Both analog input and output tasks can be triggered when using the PCI-6251. The start trigger is used to define when a task starts. The DAQ Assistant provides three start triggers: analog edge, analog window, and digital edge. The reference trigger establishes the reference point (time zero) in a set of input samples. A reference trigger can be set up to acquire data both before and after the trigger is received. Data acquired up to the reference point is known as **pretrigger data**. Data acquired after the reference point is known as **posttrigger data**. Not all data acquisition devices support triggering. When using the PCI-6251, these triggers are implemented in hardware.

*Triggers are only supported when Task Timing is configured to **N Samples**.*

**Practice
with
Analog Input**

In this example we will construct a VI that acquires an analog signal and displays the output on a meter. For the DAQ device, we are using the DAQ Signal Accessory from National Instruments to produce the temperature measurements. The sensor outputs a voltage proportional to the temperature and is hard-wired to channel 0 of the DAQ device.

If you do not have an NI-DAQ Accessory, you can still follow this example up to the point of actually acquiring temperature data from the device.

To begin the development, open a blank VI and place a Meter on the front panel. The Meter is located on the **Controls≫Modern≫Numeric** palette. Adjust the scales on the meter to read in the range 0.0 to 0.4. Then place a Vertical Toggle Switch on the front panel and name it **Power**. The Vertical Toggle Switch is located on the **Controls≫Modern≫Boolean** palette. Configure the toggle switch to a default value of False and a mechanical action of **Latch When Pressed**. Create two free labels titled **Off** and **On**. The process of building the front panel is illustrated in Figure 9.47. The needed controls and indicators should be on the front panel before continuing to the block diagram. When the front panel is ready, switch to the block diagram to continue.

On the block diagram, place a While Loop and enlarge the loop to accommodate several programming elements. Then place a DAQ Assistant (see Section 9.9 for instructions on using the DAQ Assistant) inside the loop. When the DAQ As-

FIGURE 9.47
Configuring the front panel of the temperature acquisition VI

sistant appears on the screen, proceed to configure a DAQmx task to read an analog input channel and return the voltage. The steps to accomplish this are illustrated in Figure 9.48. In brief, the steps are as follows:

1. Select Analog Input≫Voltage as the measurement type.

2. Select Dev1≫ai0 as the physical channel.

3. Click the **Finish** button.

4. When the **Analog Input Voltage Task** dialog box appears, configure the **Task Timing** to **1 Sample (On Demand)**.

5. Click the **OK** button to close the **Analog Input Voltage Task Configuration** dialog box.

The settings specified for the task are now saved in the DAQ Assistant. At this point, we need to add several programming elements to the block diagram to complete the VI. First, we want the loop to execute every 100 ms. To accomplish this goal, place a Time Delay function on the block diagram. This function is located on the **Functions≫Programming≫Timing** palette. When the Time Delay function is placed on the block diagram, a dialog box will appear in which the time delay value is set. The default value is a 1 s delay—change the delay to 0.1 s.

It is recommended that you include the capability to stop the program execution in the case where the DAQ Assistant produces an error. Fortunately, the DAQ Assistant outputs include a status indicator that is accessible by the Unbundle by Name function, located on the **Functions≫Programming≫Cluster & Variant** palette. This function is used to access the **status** from the error cluster when wired to the DAQ Assistant as shown in Figure 9.49.

You will also want to be able to halt the program execution using the toggle switch on the front panel. Actually, you will need to be able to halt program execution if an error occurs *or* if the user clicks the power switch (i.e., the toggle switch). To accomplish this task, place the Or function on the block diagram inside the While Loop. The Or function is located on the **Functions≫Programming≫ Boolean** palette. Wire the block diagram using Figure 9.49 as a guide.

*If you located the While Loop on the **Functions≫Express≫Execution Control** palette, the **loop condition** will be automatically wired to a **Stop** button. Delete the **Stop** button and wire the toggle switch instead.*

When you have finished wiring the block diagram and the **Run** button indicates that the program is ready for execution, save the VI as AI Single Point.vi in the Users Stuff folder in the Learning directory.

Return to the front panel and run the VI. The meter will display the voltage and the temperature sensor outputs. Stop the execution of the VI by clicking the power switch. ◆

FIGURE 9.48
Configuring the DAQmx task using the DAQ Assistant

Temperature
measurement

FIGURE 9.49
Acquiring analog input one point at a time

9.11   ANALOG OUTPUT

Analog output is used to perform digital-to-analog (D/A) conversions. The first screen of the DAQ Assistant is where you choose the analog output measurement type. Refer to Section 9.9 for an introduction to using the DAQ Assistant. You can also configure a DAQmx analog output task using MAX as described in Section 9.8. We will focus on the use of the DAQ Assistant in this section, but actually the process of configuring the DAQmx task is almost the same using the DAQ Assistant as it is with MAX.

Configuring a DAQmx task to perform analog output consists of the same basic steps as for the analog input (see Section 9.10). In the DAQ Assistant, clicking on **Analog Output** opens a list that shows the available analog output signal generation types: voltage and current.

A compatible device must be installed that can generate a voltage or current task.

FIGURE 9.50
DAQ Assistant for the analog output DAQmx task

Once you have selected the channels to add to your task, the DAQ Assistant settings and testing screen opens, as illustrated in Figure 9.50. In the lower section of the DAQ Assistant are two tabs: **Task Timing** and **Task Triggering**. After setting the Signal Output Range, the next step in configuring the DAQmx task is to configure the timing and triggering.

Only one analog output task can run at a time on a single Multifunction DAQ device.

9.11.1   Task Timing

The analog output task can be timed to generate a single sample, *n* samples, or continuous samples. How do you decide when to use single-sample timing? If the signal level is more important than the sample generation rate, then choose **1**

Sample (On Demand). Each time the DAQ Assistant runs, the output value on the analog output channel(s) will update. This method is used when a constant signal, or slowly changing signal is needed. You can implement timing in LabVIEW to control the time at which signal value(s) are updated. Generating a single sample does not require any buffering or hardware timing.

You should use **N Samples** if you want to generate a *finite* time-varying signal, such as a 5-second AC sine wave. As discussed in Section 9.10 for analog input, one way to generate multiple samples (representing, for example, a time-varying signal) for one or more channels is to generate single samples in a repetitive manner. Just as before, **1 Sample** is only suggested for generating static or slowly changing signals and point-by-point control applications because the time between samples may vary or take too much time to generate to actually recreate your signal. The best way to generate a time-varying signal is to use hardware timing that uses a buffer in computer and DAQ device memory to generate samples timed more accurately. With hardware timing, a TTL signal, such as a clock on the device, controls the rate of generation. A hardware clock can run much faster and is more accurate than a software loop. As with other functions, you can generate multiple samples on a single channel or multiple channels.

Some devices do not support hardware timing for analog output, such as the NI USB-6008 and USB-6009. Consult the device documentation if you are unsure if the device supports hardware timing.

Continuous sample generation is similar to generating *n* samples, except that multiple buffers or arrays of data can be output without missing any samples between buffers. When performing continuous sample generation, an event must occur to stop the sample generation. If you want to continuously generate signals, such as generating a non-finite signal, set the timing mode to **Continuous**.

9.11.2 Task Triggering

As discussed in Section 9.11.1, two very common output tasks are generating *n* samples and generating continuous samples. Just as with analog input, these tasks can be initiated by triggers. The start trigger starts the generation. The DAQ Assistant provides three start triggers: analog edge, analog window, and digital edge. The reference trigger is not supported for analog output tasks.

**Practice
with
Analog Output**

In this example, we develop a VI that outputs an analog voltage using a DAQ device. The VI will output the voltage from 0 to 9.5 V in 0.5 V increments. For the DAQ device, we are using the DAQ Signal Accessory from National Instruments. The goal is to use the VI developed in this example to output an analog signal to the DAQ device, and to use the analog input VI developed in Section 9.10 to read the data and display the result on a meter.

*If you have a DAQ Signal Accessory, connect **Analog Out CH0** to **Analog In CH1**. If you do not have a DAQ Signal Accessory, you can still follow this example up to the point of actually generating analog output to the DAQ device.*

In this example, we begin the process of designing a VI to output analog data by starting with a partially completed VI. Open the Voltage Output VI located in the **Chapter 9** folder in the **Learning** directory. You will find the front panel and block diagram shown in Figure 9.51.

On the block diagram there are two interesting functions: the Time Delay function and the Select function. The Time Delay function is configured to cause the For Loop to execute every 500 ms. The Select function checks whether the

FIGURE 9.51
Generating a voltage output from 0 to 9.5 V in 0.5 V increments

FIGURE 9.52
Placing the DAQ Assistant on the block diagram

loop is in its last iteration, and if so, then the DAQ device outputs 0 volts. It is a good idea to reset the output voltage to a known level to prevent damage to connected devices.

On the block diagram, place the DAQ Assistant, as illustrated in Figure 9.52. When the DAQ Assistant appears, configure the DAQmx task to generate an analog output voltage. The following steps (shown in Figure 9.53) will produce the desired task:

- Select Analog Output≫Voltage on the first DAQ Assistant dialog box.

- Select Dev1≫ao0 as the physical channel and click the **Finish** button.

- In the **Analog Output Voltage Task Configuration** dialog box that appears, configure the **Task Timing** to **1 Sample (On Demand)** and change the output range minimum to 0 and maximum to 10.

- Click the **OK** button to close the **Analog Output Voltage Task Configuration** dialog box. This saves the settings specified for the task in the DAQ Assistant.

FIGURE 9.53
Configuring the analog output DAQmx task

Once you are back on the block diagram, wire the DAQ Assistant as shown in Figure 9.54. Save the VI as Voltage Output Done.vi in the Users Stuff folder in the Learning directory.

Open the AI Single Point VI that you completed in Section 9.10. Configure the meter scale minimum to 0.0 and maximum to 10.0. On the block diagram of the AI Single Point VI, double-click the DAQ Assistant to open the **Analog Input Voltage Task Configuration** dialog box. Right-click **Voltage** in the channel list section and select **Change Physical Channel** as illustrated in Figure 9.55. Select **ai1** for the channel because you wired the DAQ Signal Accessory to output a voltage on **Analog Out CH0** and acquire the voltage from **Analog In CH1**. Change the voltage range to 0 to 10 and then click the **OK** button to close the dialog box.

Add wires

Select
**Disable
Indexing**

FIGURE 9.54
Wiring the DAQ Assistant

Right-click **Voltage**   Select **Change Physical Channel**

FIGURE 9.55
Reconfiguring the DAQmx task to read from channel ai1

Analog Input

Acquire 3 V

Analog Output

Generates 3 V

FIGURE 9.56
Generating and acquiring analog signals

Run the AI Single Point VI. Recall that this VI will run until there is an error or the toggle switch is used to stop the execution of the program. Once the AI Single Point VI is executing, run the Voltage Output Done VI. Notice that the Voltage Output Done VI outputs the voltage in 0.5 V increments from 0 to 9.5 V as displayed in the digital indicator in Figure 9.56. At the same time, the meter in the AI Single Point VI displays the voltage readings over the same range. When the For Loop executes its last iteration, the VI outputs 0 V to reset the analog output channel, as desired.

What have we accomplished? The Voltage Output Done VI is generating an analog signal output to the DAQ Accessory. This is a digital-to-analog conversion. The AI Single Point VI is acquiring the signal on the DAQ Accessory and displaying the data on its meter. This is an analog-to-digital conversion.   ◆

9.12   DIGITAL INPUT AND OUTPUT

Most multifunction DAQ devices have digital inputs and outputs (I/O). A digital channel or line can generally be configured as an input or output before or while the line is being used. Digital I/O can generally be thought of as analog I/O with only two possible voltage values. Digital I/O hardware generates or accepts binary on/off signals. For example, TTL is one standard of binary signal where on = 5 volts and off = 0 volts. In addition to on/off binary values, you may also find 1/0, high/low, and 5 v/0 v are used to define the binary signal interchangeably.

These signals are often used to control processes, generate patterns for testing, and communicate with peripheral equipment.

The binary value of individual digital lines can be written or read programmatically using the DAQ Assistant. Each line corresponds to a channel in the task. Digital lines are often used in groups of four or eight and are referred to as a port. All lines in a port are generally configured for the same function, either input or output. Since a port contains multiple digital lines, by writing to or reading from a port, you can set or retrieve the states of multiple lines simultaneously.

The DAQ Assistant can also be used to read or write port values using an 8-bit integer or an array of Booleans. When using *polled* I/O, software timing is used to decide when to read or write a value to a line or port by executing the DAQ Assistant. Each port corresponds to a channel in the task.

Most M Series devices and other advanced digital I/O devices are capable of hardware-timed digital I/O signal generation and acquisition. These devices can create arbitrary digital waveforms and acquire digital signals at very high speeds.

Practice with Digital I/O

In this example, we plan to control the digital I/O lines on the DAQ device. As in previous examples in Section 9.10 and 9.11, we are using the DAQ Signal Accessory from National Instruments. Our final VI will turn on the LEDs of port 0 on the DAQ Signal Accessory based on the digital value set on the front panel. Each LED is wired to a digital line on the DAQ device. The lines are numbered 0, 1, 2, and 3, starting with the LED on the right.

The LEDs on the DAQ Signal Accessory use negative logic. That is, writing a 1 to the LED digital line turns off the LED. Writing a 0 to the LED digital line turns on the LED.

Open the Digital Example VI, located in the Chapter 9 folder in the Learning directory. You will find the block diagram shown in Figure 9.57.

Place the DAQ Assistant on the block diagram within the While Loop. When the DAQ Assistant dialog box appears, configure the DAQmx task for Digital I/O. The following steps (shown in Figure 9.58) will produce the desired task:

- Select Digital I/O≫Port Output.
- Select Dev1≫port0 for the physical channel and click the **Finish** button.
- In the **Digital Port Output Setup** dialog box that appears, select **Invert All Lines In Port** because the LEDs use negative logic.
- Click the **OK** button to close the configuration dialog box.

All of the settings specified for the task are saved internally in the DAQ Assistant. The Boolean buttons on the front panel are stored in an array to simplify the code. The Array Subset function extracts only the first four elements in the array. The output of the array subset needs to be reversed since element 0 of the

Boolean buttons stored in array

Extracts first four elements of the array

While Loop

Stop execution if error occurs or stop button is pressed

FIGURE 9.57
Digital I/O example block diagram

Select **Port Output**

Select **Digital I/O**

Select **Invert All Lines In Port**

Select **Dev1 >> port0**

FIGURE 9.58
Configuring the DAQmx task for digital I/O

FIGURE 9.59
The Digital Example VI completed

array is the most significant bit. The array is then converted to a number with the Boolean Array To Number function, which passes the output to the DAQ Assistant to write that value to the port.

Once you are back on the block diagram, wire the DAQ Assistant as shown in Figure 9.59. When the VI is ready, click the **Run** button on the front panel. Turn the Boolean LEDs on and off on the front panel and observe that the LEDs on the DAQ Signal Accessory turn on and off at the same time and in the same order. You are now performing digital I/O!

Save the VI as Digital Example Done.vi in the Users Stuff folder in the Learning directory. ◆

More Practice with Digital I/O

In this example our goal is to create a VI that reads in a digital input signal then displays it in binary on LEDs. We will generate the input signal using a counter, which is a digital timing device. Counters are typically used for event counting, frequency measurement, period measurement, position measurement, and pulse generation. A counter contains the following four main components:

- **Count Register**—Stores the current count of the counter. You can query the count register with software.

- **Source**—An input signal that can change the current count stored in the count register. The counter looks for rising or falling edges on the source signal. Whether a rising or falling edge changes the count is software selectable. The type of edge selected is referred to as the active edge of the signal. When an active edge is received on the source signal, the count changes. Whether an active edge increments or decrements is also software selectable.

- **Gate**—An input signal that determines if an active edge on the source will change the count. Counting can occur when the gate is high, low, or between various combinations of rising and falling edges. Gate settings are made in software.

- **Output**—An output signal that generates pulses or a series of pulses, otherwise known as a pulse train.

As in the previous example, we will again use the DAQ Signal Accessory. The Signal Accessory has a quadrature encoder which produces pulses as the knob is rotated. Start by connecting the output of Quadrature Encoder A to the source of Counter 0. Now open a blank VI and place the DAQ Assistant on your block diagram.

When the configuration window appears, select Counter Input≫Edge Count so that the DAQ Assistant will read the value stored in the counter representing the number of edges on the signal from the quadrature encoder. Select **ctr0** as the channel, since you connected the quadrature encoder to Counter 0 on the Signal Accessory. The default settings for task timing and edge count setup are fine. Press **OK** to save these settings. When LabVIEW prompts you to confirm auto loop creation, select **Yes**. This will automatically create a loop that will continue reading the counter output until the user presses Stop on the front panel.

Create a numeric indicator on the **data** output of the DAQ Assistant and display it on the front panel. Your block diagram should look like the one in Figure 9.60.

Run the VI and turn the knob on the Signal Accessory. You should observe that the counter increments each time you turn the knob. Save the VI as Digital IO 1.vi in the Users Stuff folder in the Learning directory.

Return to the block diagram and place a node to convert the waveform data output to a double. Since there are 4 LEDs on the Signal Accessory, we can only display 2^4, or 16, values using the LEDs. Thus we will divide the output by 16 and display the remainder of this operation on the LEDs. Utilize the Quotient & Remainder function to perform this task. As the counter continues incrementing by one, the result of this division will increment from 0 up to 15 and then wrap back around to 0. Use the Build Array function to convert the remainder from the division to an array so that it can be a data input to the DAQ Assistant.

FIGURE 9.60
The Digital 1 I/O VI block diagram

Now place a second DAQ Assistant on the block diagram. This time, select Digital I/O≫Port Output as the measurement type. Choose **port0** as the channel, since this is the port the LEDs are on. The default configuration settings are fine and do not need to be updated. Connect the array containing the output value to be displayed on the LEDs to the data input on the second DAQ Assistant. Rewire the error cluster so that it passes through the second DAQ Assistant before it reaches the stop terminal. Your block diagram should now resemble the one seen in Figure 9.61.

FIGURE 9.61
The Digital I/O 2 VI block diagram

Now run your VI and turn the knob on the Signal Accessory. The LEDs on digital port 0 should count from 0-15 in binary. You have now created a VI that performs digital input (from the counter) as well as digital output (to the LEDs).

Save the VI as Digital IO 2.vi in the Users Stuff folder in the Learning directory.

Working versions of Digital IO 1.vi *and* Digital IO 2.vi *can be found in* Chapter 9 *of the* Learning *directory.*   ◆

BUILDING BLOCK

9.13   BUILDING BLOCKS: PULSE WIDTH MODULATION

In this exercise, you will develop a VI that outputs a pulse width modulated digital signal to a data acquisition device using the DAQ Assistant. You will also investigate a more robust approach to generating a pulse width modulated output on a DAQ device using an existing LabVIEW example. To begin, open the VI you saved as PWM with Chart.vi in Chapter 8 and save a new copy of the VI as PWM with Digital Output.vi in the Users Stuff folder in the Learning directory.

In case you do not have a working version of the Building Blocks VI from Chapter 8, a working version of the PWM with Chart VI can be found in the Building Blocks *folder in the* Learning *directory.*

The Boolean values are currently being written to the waveform chart. You should also output the Boolean values to a single digital output line. To accomplish this, add the DAQ Assistant inside the While Loop and configure it as Digital I/O≫Line Output on **port0/line0**. Next, wire the Boolean data to the DAQ Assistant. When performing digital output, the DAQ Assistant expects an array of Boolean values. Therefore, you will need to add a Build Array function to convert the single Boolean value to an array of Boolean values. Use the block diagram in Figure 9.62 as a guide to constructing your VI.

9.13.1   Generating Pulse Width Modulated Signals with Hardware Counters

In these past exercises, the sampling rate of the signal has been 100 Hz (10 milliseconds) and is determined by software timing. The maximum possible software timed rate in LabVIEW is 1000 Hz (1 millisecond). This is a limitation

Several versions of this system have operated successfully for more than a year. Researchers now are conducting complex and lengthy experiments with a minimum of user intervention. It is estimated that the automation systems using National Instruments products have cumulatively saved in excess of 10,000 man hours of manual control, monitoring, and data collection.

For more information, contact:

Robert O. Hamburger, Principal Engineer
Bloomy Controls Inc.
bob.hamburger@bloomy.com

9.15 SUMMARY

DAQ systems consist of the following elements:

- Signals
- Transducers
- Signal-conditioning hardware
- DAQ device or module
- Application software

There are two types of signals:

- Analog—provides level, shape, or frequency content information
- Digital—provides state or rate information

With many transducers it is necessary to provide for signal conditioning. Some types of signal conditioning are

- Amplification
- Transducer excitation
- Linearization
- Isolation
- Filtering

There are two types of signal sources:

- Grounded sources—devices that plug into the building ground
- Floating sources—isolated from the building ground system

There are three types of measurement systems:

- Differential—use this type whenever possible!
- Referenced single-ended—use single-ended types if you require more channels
- Nonreferenced single-ended

Multifunction DAQ devices typically include:

- Analog-to-digital converters (ADCs)
- Digital-to-analog converters (DACs)
- Digital I/O ports
- Counter/timer circuits

When configuring the DAQ device, you should consider how the following parameters will affect the quality of the digitized signal:

- Resolution: Increasing resolution increases the precision of the ADC.
- Device Range: Decreasing range increases precision.
- Signal Input Range: Changing signal input range to reflect the actual signal range increases precision.

The Measurement & Automation Explorer (MAX) is a utility that helps configure the channels on the DAQ device according to the sensors to which they are connected.

The LabVIEW DAQ VIs are organized into palettes corresponding to the type of operation involved—analog input, analog output, counter operations, or digital I/O. Specifically, the DAQ VIs are organized into six palettes (we covered the first three topics in this chapter):

- **Analog Input**
- **Analog Output**
- **Digital I/O**
- **Counter**
- **Calibration and Configuration**
- **Signal Conditioning**

KEY TERMS

ADC resolution: The resolution of the ADC measured in bits. An ADC with 16 bits has a higher resolution (and thus a higher degree of accuracy) than a 12-bit ADC.

Analog-to-digital converter (ADC): An electronic device (often an integrated circuit) that converts an analog voltage to a digital number.

Bipolar: A signal range that includes both positive and negative values (e.g., −5 V to 5 V).

Channel: Pin or wire lead where analog or digital signals enter or leave a data acquisition device.

Channel name: A unique name given to a channel configuration in the Measurement & Automation Explorer or DAQ Assistant.

Code width: The smallest detectable change in an input voltage of a DAQ device.

Data acquisition (DAQ): Process of acquiring data from plug-in devices.

Device Range: The minimum and maximum analog signal levels that the analog-to-digital converter can digitize.

Differential measurement system: A method of configuring your device to read signals in which you do not connect inputs to a fixed reference (such as the earth or a building ground).

Digital-to-analog converter (DAC): An electronic device (often an integrated circuit) that converts a digital number to an analog voltage or current.

Floating signal sources: Signal sources with voltage signals that are not connected to an absolute reference or system ground. *Also* called nonreferenced signal sources.

Grounded signal sources: Signal sources with voltage signals that are referenced to system ground, such as the earth or building ground. *Also* called referenced signal sources.

Handshaked digital I/O: A type of digital I/O where a device accepts or transfers data after a signal pulse has been received. *Also* called latched digital I/O.

Immediate digital I/O: A type of digital I/O where the digital line or port is updated immediately or returns the digital value of an input line. *Also* called nonlatched digital I/O.

Input/output (I/O): The transfer of data to or from a computer system involving communication channels and DAQ interfaces.

Linearization: A type of signal conditioning in which the voltage levels from transducers are linearized, so that the voltages can be scaled to measure physical phenomena.

LSB: Least significant bit.

Measurement & Automation Explorer: Provides access to all National Instruments DAQ and GPIB devices.

Nonreferenced single-ended (NRSE) measurement system: All measurements are made with respect to a common reference, but the voltage at this reference may vary with respect to the measurement system ground.

Referenced single-ended (RSE) measurement system: All measurements are made with respect to a common reference or ground. *Also called a grounded measurement system.*

Sampling rate: The rate at which the DAQ device samples an incoming signal.

SCXI: Signal Conditioning eXtensions for Instrumentation. The National Instruments product line for conditional low-level signals within an external chassis near the sensors so that only high-level signals in a noisy environment are sent to the DAQ device.

Signal conditioning: The manipulation of signals to prepare them for digitizing.

Signal Input Range: The range that you specify as the maximum and minimum voltages on analog input signals.

Unipolar: A signal range that is always positive (e.g., 0 V to 10 V).

Update rate: The rate at which voltage values are generated per second.

EXERCISES

E9.1   Assume you are sampling a transducer that varies between 80 mV and 120 mV. Your device has a voltage range of 0 to +10 V, ±10 V, and ±5 V. What signal input range would you select for maximum precision if you use a DAQ device with 12-bit resolution? Use the signal input range values in Table 9.3 for your calculations.

E9.2 Open the Measurement & Automation Explorer. Locate your DAQ device under Devices and Interfaces≫NI-DAQmx Devices. Right-click on the device and select **Test Panels**. Run a test panel to observe the signals connected to your DAQ device. After running a test panel, go to **Data Neighborhood** and create a DAQmx Global Channel to measure one of your input signals. Test your DAQmx Global Channel.

E9.3 Open the NI Example Finder to Search and enter "function." Select the Function Generator with FM link. To open the **Find Examples**, select Find Examples in the **Help** pull-down menu. This VI is a good example of how virtual instruments emulate actual instruments. Run this VI and vary the inputs to obtain a sawtooth wave signal at 50 Hz and a sampling rate of 500 Hz.

E9.4 Construct a voltmeter VI that reads in a specified number of samples of an analog voltage and displays the input on a waveform graph on the front panel. Connect a sine wave to one of the analog input channels on your DAQ device and configure the DAQ Assistant to acquire *n* samples.

 If you do not have a DAQ device, use the Simulated DAQ as described in Section 9.7.3.

The front panel, illustrated in Figure E9.4, should have two numeric controls allowing the user to programmatically adjust the number of samples acquired and the sampling rate. The block diagram should utilize the DAQ Assistant. You can use the block diagram in Figure E9.4 as a guide to construct the VI. Experiment with different sampling rates and numbers of samples.

PROBLEMS

P9.1 Open a blank VI and place a DAQmx Task Name constant on the front panel. Right-click the constant and select New NI-DAQmx Channel≫Max. Using MAX, configure a voltage analog input named **MyVoltageChannel**. After configuring and testing your channel, use the Code Generation feature to generate code for your channel. To do this, right-click on the channel constant and select Generate Code≫Configuration and Example. Run the VI.

P9.2 Suppose that you want to design a VI to acquire analog data measuring the volume of a tank. You only want to take continual measurements when the machinery is on. Turning on the machinery can act as a trigger, specifically a digital trigger, to start measuring data. Construct a VI that acquires analog data only after a digital trigger has occurred.

P9.3 Consider a servo motor that requires a +5 V signal to operate. It is desired that the servo motor be under the control of an operator. Construct a VI such that the operator must set a Boolean toggle switch to True in order to send a +5 V signal

FIGURE E9.4
Voltmeter VI

to the motor to start the motor. When the operator sets the Boolean switch to False, the motor must stop. In this way, the operator can control the servo motor.

You can use either a digital output line or an analog output line.

P9.4   Construct a VI to monitor a security system that uses a light sensor to determine when a person has entered a restricted zone. When the sensor is activated, it returns a 0 V TTL signal. Write a VI to monitor the state of the light sensor and turn on a Boolean indicator when the sensor is activated.

You can use either a digital output line or an analog output line.

DESIGN PROBLEMS

D9.1 Design a VI which displays an analog voltage input on a numeric indicator with compensation for fluctuations caused by noise spikes. To reduce noise in the measurement, display the average of the 100 most recently acquired data points rather than simply displaying the most recently acquired sample. Your VI should have error handling capabilities.

D9.2 Construct a VI that outputs voltage from 10 V down to 0 V in 0.5 V steps using the DAQ Assistant. Display the output signal you send to the DAQ card on a meter indicator. To verify that the DAQ card is outputting the correct signal, you may wish to connect your DAQ board to an oscilloscope or digital multimeter, or change the settings on the voltmeter VI you created in Section 9.10 so that it can be used to monitor this output voltage.

You can perform both analog output and analog input tasks at the same time, but only one of each.

D9.3 Develop a VI that reads in an analog input voltage and then sets a Boolean output signal if the input voltage exceeds a limit entered by the user on a front panel control. Also display the input voltage on a meter. Use the DAQ Assistant to perform the I/O tasks.

CHAPTER 10

Strings and File I/O

This chapter introduces strings and file input/output (I/O). You have used strings in a limited fashion throughout the book, but here we discuss them more formally. In instrument control applications, numeric data are commonly passed as character strings, and LabVIEW has many built-in string functions that allow you to manipulate the string data. Writing to and reading data from files also utilizes strings. We will discuss how to use File I/O VIs to save data to and retrieve data from a disk file. We will also discuss the use of key Express VIs for string manipulation and file I/O.

GOALS

1. Practice creating string controls and indicators.

2. Be able to convert a number to a string, and vice versa.

3. Learn to use the File I/O VIs to write to and read data from a disk file.

4. Understand how to write data to a file in a format compatible with many common spreadsheet applications.

10.1   STRINGS

A **string** is a sequence of characters—these can be displayable or nondisplayable characters. Strings are often used in **ASCII** text messages. In LabVIEW, strings are also used in instrument control when numeric data is passed as character strings and subsequently converted back to numbers. Another situation requiring the use of strings is in storing numeric data, where numbers are first converted to strings before writing them to a file on disk. In this chapter, we will discuss strings and their various uses in file input/output (I/O).

String controls and indicators are in the **String & Path** subpalette of the **Controls≫Modern** palette, as illustrated in Figure 10.1. They can also be accessed on the **Text Controls** and **Text Indicators** subpalettes of the **Controls≫ Express** palette. As discussed in previous chapters (see Chapter 4 for example), you enter and change text inside a string control using the **Operating** tool or the **Labeling** tool. If there is not enough room to fit your text in the default size of the string control, you can enlarge the string controls and indicators by dragging a corner with the **Positioning** tool. If front panel space is limited, you can use a scrollbar to minimize the space that a front panel string control or indicator occupies, as illustrated in Figure 10.2. The Visible Items≫Vertical Scrollbar option is located on the string shortcut menu. If there is not enough room to place

FIGURE 10.1
(a) and (c) Accessing string controls and indicators using the **Express** palette. (b) Accessing string controls and indicators from the **String & Path** subpalette

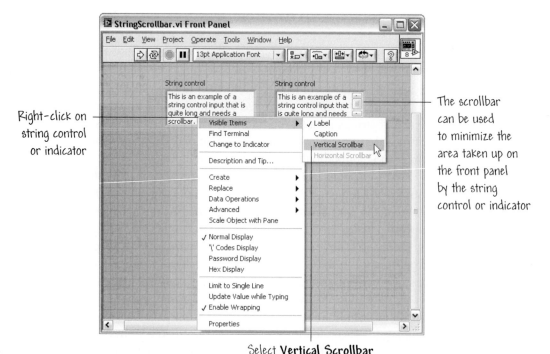

Right-click on string control or indicator

The scrollbar can be used to minimize the area taken up on the front panel by the string control or indicator

Select **Vertical Scrollbar**

FIGURE 10.2
The scrollbar can be used to minimize the size of string controls and indicators

the scrollbar within the string control or indicator, the scrollbar option will be dimmed. This indicates that you must increase the vertical size if you want a scrollbar. A horizontal scrollbar is also available.

You can configure string controls and indicators for different types of display, such as passwords, \ codes, and hex. String controls and indicators can display and accept characters that are usually **nondisplayable**—backspaces, carriage returns, tabs, and so on. Choose '\' **Codes Display** from the string shortcut menu to display these characters. In the '\' **Codes Display** mode, nondisplayable characters appear as a backslash followed by the appropriate code. A complete list of codes appears in Table 10.1.

To enter a nondisplayable character into a string control, press the appropriate key, like space or <tab>, or type the backslash character \, followed by the code for the character. As shown in Figure 10.3(a), after you type in the string and click the **Enter** button, any nondisplayable characters appear in backslash code format.

The characters in string controls and indicators are represented internally in ASCII format. You can view the ASCII codes in hex by choosing **Hex Display** from the string shortcut menu, as shown in Figure 10.3(b). You can also choose a password display by enabling the **Password Display** option from the string

TABLE 10.1   A List of Backslash Codes

Code	G Interpretation
\00 – \FF	Hex value of an 8-bit character; must be uppercase.
\b	Backspace (ASCII BS, equivalent to \08)
\f	Form feed (ASCII FF, equivalent to \0C)
\n	Line feed (ASCII LF, equivalent to \0A)
\r	Carriage return (ASCII CR, equivalent to \0D)
\t	Tab (ASCII HT, equivalent to \09)
\s	Space (equivalent to \20)
\\	Backslash (ASCII \, equivalent to \5C)

shortcut menu, as shown in Figure 10.3(c). With this option selected, only asterisks appear in the string front panel display, although on the block diagram the string data reflects the input string. This allows you to set up a security system requiring a password key before proper operation of the VI.

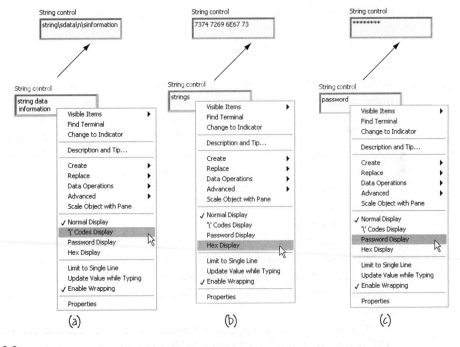

(a)   (b)   (c)

FIGURE 10.3
(a) Displaying characters that are usually nondisplayable (b) Choosing **Hex Display** to view the ASCII codes in hex (c) Choosing the **Password Display** option from the shortcut menu

**Practice
with
Manipulating
Strings**

In this exercise, you will practice with different ways to manipulate strings using three VIs:

- **Format Into String**: Concatenates and formats numbers and strings into a single output string.

- **Scan From String**: Scans a string and converts valid numeric characters (0 to 9, $+$, $-$, e, E, and period) to numbers.

- **Match Pattern**: Searches for an expression in a string, beginning at a specified offset, and if it finds a match, splits the string into three substrings—the substring before the matched substring, the matched substring itself, and the substring that follows the matched substring.

The VI that you will create is intended to simulate interaction with a digital multimeter (DMM). Open a new VI and construct a front panel using Figure 10.4 as a guide. Place two string controls, one numerical control, and one string indicator on the front panel. Switch to the block diagram. Select Format Into String in the **String** subpalette, as shown in Figure 10.4. In this exercise, this function converts the number you specify to a string and concatenates the inputs to form a command for the DMM. Using the **Positioning** tool, enlarge Format Into String so that two inputs appear in the lower left corner. Wire the numeric control **Number** to the first input argument (as shown in Figure 10.4). Wire the string control **Units** to the second input argument; the input type for Format Into String will automatically change from **DBL** to **abc** upon wiring. Wire the string control **DMM Command** to the **initial string** input at the top left. Finally, wire the output **resulting string** to the string indicator **Command sent to DMM**.

You can create strings according to a format specified using format strings. With format strings you can specify the format of arguments—the field width, base (hex, octal, and so on), and any text that separates the arguments. The format string can be seen in Figure 10.4 (wired at the top of the Format Into String function). A dialog box can be used to obtain the desired format string. On the block diagram, right-click on the Format Into String function and select **Edit Format String**, as illustrated in Figure 10.5. You can also double-click on the node to access the dialog box. Notice that the **Current format sequence** contains the argument types in the order that you wired them—**Format fractional number** and **Format string**. **Format fractional number** corresponds to the input value from the **Number** control and **Format string** corresponds to the input value from the **Units** control.

The only change we want to make is to set the precision of the numeric to 2. To do this, highlight **Format fractional number** in the **Current format sequence** list box, click the **Use specified precision** check box, and type in the number 2 in the box. When you are finished, press <Enter> (Windows) or

FIGURE 10.4
Constructing a VI using the Format Into String function to build a command to send to the DMM

<return> (Macintosh) and you should see the **Corresponding format string** (near the bottom of the dialog box). Press the **OK** button to automatically insert the correct format string information and wire the format string to the function, as shown in Figure 10.4.

Return to the front panel and type text inside the two string controls and a number inside the numeric control, and run the VI. You now have a VI that can concatenate strings and numbers to form a single string command that can be sent to an external instrument, such as the digital multimeter. Save the VI as **String.vi** in the **Users Stuff** folder.

To continue the exercise, we want to add the capability to scan a string and convert any valid numeric characters (0 to 9, +, −, e, E, and period) to num-

Highlight **Format fractional number** before changing precision

Change value to 2

Check here

The extra "2" appears to indicate precision

When finished, click here to create string

FIGURE 10.5
Setting the desired format string using the dialog box

bers. Use Figure 10.6 as a guide and add a string control and numeric indicator to the front panel. Then switch to the block diagram and wire the Scan From String function, as shown in Figure 10.6. The function itself is located on the **Programming≫String** palette, as illustrated in Figure 10.4. After the wiring is complete, double-click on the function (or right-click on the function and choose **Edit Format String**) to open the dialog box. The current scan sequence should indicate **Scan number** and the corresponding scan string should show %f. This is fine, so just select **OK** and make sure that the format string input (at the top of the Scan From String function) is properly wired (see Figure 10.6). To test the VI, type 5.56 V in the Present DMM Setting string control and run the VI. You should find that the VI extracts the numeric value 5.56 and displays that value in the numeric indicator DMM Setting - Numeric. Try other voltage values in the input.

Now, complete this VI using Figure 10.7 as a guide. The last addition to the VI uses the Match Pattern function in conjunction with the Scan From String function to detect and extract a series of DMM data points. The Match Pattern function detects the matched substring (in this case, a comma) and outputs the substring before the comma to the Scan From String function, which scans the substring and outputs a detected number. The While Loop iterates until the end

FIGURE 10.6
Constructing a VI using the Scan from String function to extract the present DMM setting from a string

of the string is detected—in other words, an empty string is found. You can find the empty string constant on the **Strings** palette, as illustrated in Figure 10.4.

When your VI has been wired properly, enter a few numbers in the string control **DMM Data Points**, such as 3.37, 4.56, 6.89, 5.67. Run the VI and verify that the output array **Data Points - Numeric** contains the four numbers.

A working version of String.vi *can be found in* Chapter 10 *of the* Learning *directory. Refer to the working version if your VI is not working as you think it should.* ◆

10.1.1   Converting Numeric Values to Strings with Build Text Express VI

The Build Text Express VI can be used to convert numeric values into strings. The Build Text Express VI is located on the **Functions≫Programming≫String** or from the **Functions≫Express≫Output** palette, as shown in Figure 10.8. If the input is not a string, this Express VI converts the input into a string based on the configuration of the Express VI.

FIGURE 10.7
Constructing a VI using the Match Pattern function to detect and extract a series of DMM data points

When you place the Build Text Express VI on the block diagram, the **Configure Build Text** dialog box appears. The dialog box shown in Figure 10.8 shows the Express VI configured to accept one input, **temperature**, and change it to a fractional. The input temperature concatenates on the end of the string **Temperature is**. A space has been added to the end of the **Temperature is** string.

A front panel and block diagram illustrating the use of the Build Text Express VI is shown in Figure 10.9. The Digital Thermometer.vi found in the Activity folder of the Learning directory generates the input temperature data. The output is directed to a string indicator. In the example shown, the displayed temperature is 79.59°.

The VI shown in Figure 10.9 is called Build Text Demo.vi and is located in Chapter 10 of the Learning directory.

Place Express VI on block diagram

Input text
with
Variable in %

Build Text Express VI

Choose variable properties

Click here when done

FIGURE 10.8
Using the Build Text Express VI

Build Text Express VI

Displayed
temperature

Simulated
temperature
data

String indicator

FIGURE 10.9
Converting temperature data to a string

10.2   FILE I/O

File I/O operations pass data to and from files. Use the File I/O VIs and functions on the **File I/O** palette to handle all aspects of file I/O, including the following:

- Opening and closing data files.
- Reading data from and writing data to files.
- Reading from and writing to spreadsheet-formatted files.
- Moving and renaming files and directories.
- Changing file characteristics.

The **File I/O** palette, shown in Figure 10.10, includes VIs and functions designed for common file I/O operations, such as writing to or reading from the following types of data:

- Numeric values to or from spreadsheet text files
- Characters (strings) to or from text files
- Lines from text files
- Data to or from binary files

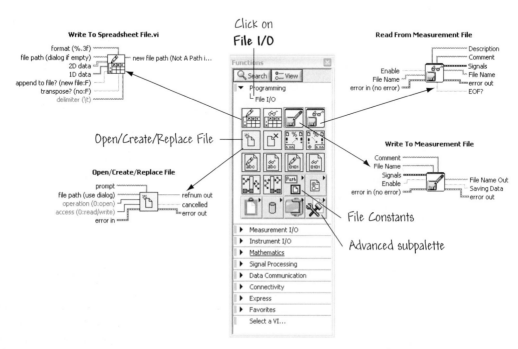

FIGURE 10.10
The **File I/O** palette

A typical file I/O operation involves the following process:

1. Create or open a file. Indicate where an existing file resides or where you want to create a new file by specifying a path or responding to a dialog box to direct LabVIEW to the file location. After the file opens, a reference number (refnum) represents the file.

2. Read from or write to the file.

3. Close the file.

The high-level File I/O VIs are located on the top row of the palette (see Figure 10.10) and can perform all three steps for common file I/O operations. They consist of the following VIs:

- **Write To Spreadsheet File**: Converts a 2D or 1D array of single-precision (SGL) numbers to a text string and writes the string to a new byte stream file or appends the string to an existing file.

- **Read From Spreadsheet File**: Reads a specified number of lines or rows from a numeric text file, beginning at a specified character offset, and converts the data to a 2D, single-precision array of numbers.

- **Write To Measurement File:** Writes data to a text-based measurement file (.lvm) or binary measurement file (.tdm).

- **Read From Measurement File:** Reads data from a text-based measurement file (.lvm) or binary measurement file (.tdm). You also can use this VI to read Multisim data if you have Multisim 9.0 or later installed.

The **File I/O** palette also includes functions to control each file I/O operation individually. Use these functions to create or open a file, read data from or write data to a file, and close a file. You also can use them to perform the following tasks:

- Create directories.

- Move, copy, or delete files.

- List directory contents.

- Change file characteristics.

- Manipulate paths.

For example, to create a new file or replace an existing file, you can use the intermediate File I/O function Open/Create/Replace File, as shown in Figure 10.10. To define a path to the file, right-click on the **file path (use dialog)** input on the left side of the function and select **Create Control** to create a File Path control. Once a file is selected and opened, a reference number (refnum) is created, and you can read from or write to the file. Remember to close an open file before your application completes execution. Many of LabVIEW's high-level

File I/O VIs automatically open or create new files, read from or write data to those files, and close the files upon completion.

The VIs and functions designed for multiple operations might not be as efficient as the functions configured or designed for individual operations.

You will find a selection of VIs on the **File I/O** palette that work with different data formats. You can read data from or write data to files in three formats—text, binary, and datalog. The format you use depends on the data you acquire or create and the applications that will access that data. Use the following basic guidelines to determine which format to use:

- If you want to make your data available to other applications (such as Microsoft Excel), use text files because they are the most common and the most portable.

- If you need to perform random access file reads or writes or if speed and disk space are crucial, use binary files because they are more efficient than text files in disk space usage and in speed.

- If you want to manipulate complex records of data or different data types in LabVIEW, use datalog files because they are the best way to store data if you intend to access the data only from LabVIEW and you need to store complex data structures.

When dealing with files, you will frequently see the terms **end-of-file**, **refnum**, **not-a-path**, and **not-a-refnum**. The end-of-file (EOF) is the character offset of the end of the file relative to the beginning of the file. Refnum is an identifier that LabVIEW associates with a file when opened. Not-a-path and not-a-refnum are predefined values that indicate that a path is invalid and that a refnum associated with an open file is invalid, respectively.

In the next three subsections we will cover the three most common file I/O operations, including writing to a file continuously, reading from a file, and writing to a spreadsheet file. In the last subsection, we will discuss the use of Express VIs for configuration-based file I/O functionality.

10.2.1   Writing Data to a File

In this section we will discuss writing data to a file using the Write To Text File function. This VI writes a string of characters or an array of strings as lines to a file. The Open/Create/Replace File function is used to create a new file, and the Close File function is used to close the file once all the data is written.

In this exercise, the objective is to create a VI to append temperature data to a file in ASCII format. This VI uses a For Loop to generate temperature values and store them in a file. During each iteration, the VI converts the temperature data from a numeric to a string, adds a comma as a delimiting character, and then appends the string to a file.

Open a new front panel and place the objects as shown in Figure 10.11. The front panel contains a numeric control and a waveform chart. The chart displays the temperature data. Right-click on the waveform chart and select Vis-ible Items≫Digital Display and deselect Visible Items≫Plot Legend. Also make sure that either the waveform chart has *y*-axis autoscaling enabled or that the maximum value of the *y*-axis is set to at least 90 degrees. The Number of Points control specifies how many temperature values to acquire and write to file. Right-click on Number of Points and choose Representation≫I32.

Switch to the block diagram and wire the code as shown in Figure 10.11. Add a For Loop and make it large enough to encompass the various components. Place the Digital Thermometer.vi on the block diagram—remember that this VI is located in the activity folder of the Learning directory, which you can select through **Functions≫Select a VI**, and it returns a simulated temperature measurement from a temperature sensor.

Add the Open/Create/Replace function outside the For Loop. Right-click on the input **operation** and choose Create≫Constant. From the list, choose **create**, meaning the VI will prompt the user to create a new file each time the VI is run. This VI then outputs a file refnum, which is passed to the Write To Text File function. Once the For Loop is done executing, the refnum is passed to the Close File function, which closes the file. If you do not close the file, the file stays in memory and is not accessible from other applications or to other users.

*When you wire the refnum out of the For Loop, auto-indexing is enabled by default. In order to sucessfully wire the refnum to the Close File function, right-click the tunnel and select **Disable Indexing**.*

The Format Into String function is used to convert the temperature mea-surement (a number) to a string and concatenates the comma that follows it. We need to modify the format string to change the precision to two digits af-ter the decimal point and to add the comma after each data point. The format string can be seen in Figure 10.11 (wired at the top of the Format Into String function). On the block diagram, right-click on the Format Into String function and select **Edit Format String** or double-click on the node to access the dia-log box. As shown in Figure 10.12(a), the **Current format sequence** contains **Format fractional number**. The first thing to do is set the precision to 2, as illustrated in Figure 10.12(b). Then select **Add New Operation** and in the **Se-lected operation** pull-down menu choose **Output exact string (abc)** as shown

FIGURE 10.11
Using the Write Temperature to File VI to write temperature data to a file

in Figure 10.12(c). Make sure that the exact string that you input is a comma (as shown in Figure 10.12(d)). When finished, select **OK** and verify that the string format is "%.2f,"—as desired.

Finish wiring the objects, return to the front panel, and run the VI with the number of points set to 20. A file dialog box prompts you for a file name—select a file name such as Test. When you enter a file name, the VI starts writing the temperature values to that file as each point is generated. Save the VI as Write Temperature to File.vi in the Users Stuff folder. Use any word processing software, such as Notepad for Windows or Teach Text for Macintosh, to open the data file and view the contents. You should get a file containing twenty data values (with a precision of two places after the decimal point) separated by commas.

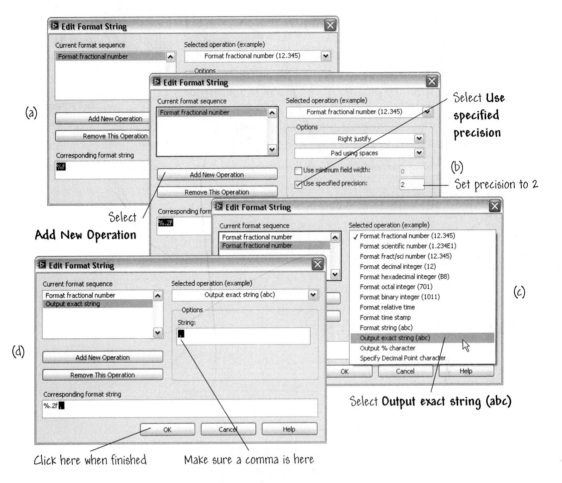

FIGURE 10.12
Setting the desired format string using the dialog box

A working version of Write Temperature to File.vi *can be found in the* Chapter 10 *folder in the* Learning *directory.* ◆

The VI architecture used in the previous exercise is an example of a disk-streaming operation. Disk streaming is a technique for keeping files open while you perform multiple write operations, for example, within a loop. Typical disk-streaming operations place the Open/Create/Replace File function before a loop, the Read or Write function in the loop, and the Close File function after the loop so continuous writing to a file can occur within the loop without the overhead associated with opening and closing the file in each iteration.

10.2.2   Reading Data from a File

As a natural follow-on to writing data to a file, in this section we discuss reading data from a file using the Read From Text File function. This function reads a specified number of characters or lines from a text byte stream file. By default, this function displays a file dialog box and prompts you to select a file and then closes it afterwards. In the following example, you will get a chance to construct a VI to read the temperature data that you wrote to a file in the previous exercise.

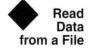

Read Data from a File

The goal here is to construct a VI that reads the temperature data file you wrote in the previous example and displays the data on a waveform graph. When reading from files be careful to read the data in the same data format in which you saved it. In this case, the data was originally saved in ASCII so it must be read in as string data.

Open a new front panel and build the front panel shown in Figure 10.13. You will need to place a string indicator and a waveform graph to display the temperature data that is read.

Located in **Strings.llb** found in **general** in the
examples folder of the **LabVIEW 8** directory

FIGURE 10.13
Using the Read From Text File function to read temperature data from a file

Build the block diagram as shown in Figure 10.13. The Read from Text File function reads the data from the file and outputs the information in a string. If no path name or refnum is provided, a file dialog prompts you to enter a file name.

The Extract Numbers VI is located in Strings.llb found in general in the examples folder of the LabVIEW8.0 directory. This VI takes an ASCII string containing numbers separated by commas, line feeds, or other non-numeric characters and converts them to an array of numbers. It uses the Match Pattern function to convert a spreadsheet string (that is, delimiter-separated columns with end-of-line characters between rows) into an array of numbers (by default) or strings. You can access the Extract Numbers VI using **Functions≫Select a VI**.

Complete the wiring of the block diagram, return to the front panel, and run the VI. When prompted for a file name, select the data file name that contains the temperature data from the previous exercise. You should see the same temperature data values displayed in the graph as you saw in the previous exercise. Save the VI as Read Temperature from File.vi in the Users Stuff folder and close the VI.

A working version of Read Temperature from File.vi *can be found in the* Chapter 10 *folder in the* Learning *directory.* ◆

10.2.3   Manipulating Spreadsheet Files

In many instances it is useful to be able to open saved data in a spreadsheet. In most spreadsheets, tabs separate columns and EOL (end-of-line) characters separate rows, as shown in Figure 10.14(a). Opening the file using a spreadsheet program yields the table shown in Figure 10.14(b).

We will concentrate in this section on writing data to a spreadsheet format, with the idea that you will be accessing that data using a spreadsheet application outside of LabVIEW. However, if you want to read spreadsheet data from within LabVIEW, it is possible to read data in text format from a spreadsheet using the Read From Spreadsheet File VI. This VI reads a specified number of lines or rows from a numeric text file beginning at a specified character offset and converts the data to a 2D, single-precision array of numbers. This is a high-level VI; hence it opens the file beforehand and closes it afterwards.

Write to a Spreadsheet File

The objective of this exercise is to construct a VI that will generate and save data to a new file in ASCII format. This data can then be accessed by a spreadsheet application.

Open a new VI and construct a front panel, as shown in Figure 10.15. This VI generates two data arrays, plots them on a graph, and writes them to a file where each column contains a data array. The front panel contains only one waveform graph.

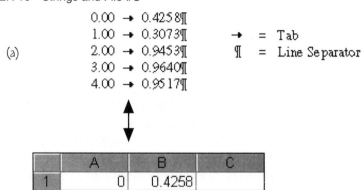

(a)

$\rightarrow$ = Tab

¶ = Line Separator

(b)

FIGURE 10.14
A common spreadsheet data format

FIGURE 10.15
Using the Write To Spreadsheet File VI to write data in ASCII format to a file

Open the block diagram and construct the VI by adding the block diagram functions shown in Figure 10.15. The Write To Spreadsheet File VI (for location of this VI see Figure 10.10) converts the 2D or 1D array of numbers to a text string and writes the string to a new text file or appends the string to an existing file. If you have not specified a path name, then a file dialog box appears and prompts you for a file name. You can write either a 1D or 2D array to file—in this exercise we have a 2D array of data, so the 1D input is not wired. With this VI, you can use spreadsheet delimiters such as tabs or commas in your data. The format string default is %.3f, which creates a string long enough to contain the number with three digits to the right of the decimal point.

This VI is a high-level VI that opens or creates the file before writing to it and closes it afterwards. You can use the Write To Spreadsheet File VI to create a text file readable by most spreadsheet applications.

The Boolean constant (see Figure 10.15) connected to the Write To Spreadsheet File VI controls whether or not the 2D array is transposed before writing it to file. To change the value of the Boolean constant, click on the constant with the **Operating** tool. Generally in LabVIEW, each row of a 2D array contains a data array (in this example, the first row of the 2D array is the square wave and the second row is the sine wave); thus in this case you want the data transposed because you want the two data arrays to be organized into columns in the spreadsheet.

After finishing up the wiring on the block diagram, return to the front panel and run the VI. After the data arrays have been generated, a file dialog box prompts you for the file name of the new file you are creating. Type in a file name and click on **OK**.

Do not attempt to write data in LLBs. Doing so may result in overwriting your library and losing your previous work.

Save the VI in Users Stuff and name it Spreadsheet.vi. You now can use a spreadsheet application or a text editor to open and view the file you just created.

In this example, the data was not converted or written to file until the fully data arrays had been collected. If you are acquiring large buffers of data or would like to write the data values to disk as they are being generated, then you must use a different File I/O function.

A working version of Spreadsheet.vi can be found in the Chapter 10 folder in the Learning directory. ◆

10.2.4   File I/O Express VIs

The two Express VIs that can be utilized for file I/O are the Read From Measurement File and the Write to Measurement File Express VIs. These two Express VIs can be found on the **Functions≫Programming≫File I/O** palette as illustrated in Figure 10.16(a). They can also be readily found on the **Functions≫Express** palettes. The Read From Measurement File Express VI is located on the **Input** palette and the Write To Measurement File Express VIs Express VI is on the **Output** palette, as shown in Figure 10.16(b). These Express VIs read and write .lvm files—LabVIEW measurement data files. The measurement data file (.lvm) is a tab-delimited text file that can be opened with a spreadsheet application (such as Microsoft Excel) or with a text-editing application (such as Notepad). In addi-

FIGURE 10.16
Locating the LabVIEW data-measurement Express VIs: (a) The **File I/O** palette (b) The **Express** palettes

tion to the data an Express VI generates, the .lvm file includes information about the data, such as the date and time the data was generated.

You may often find it necessary to permanently store measurement data acquired from your DAQ device. When planning to store data to a file, remember that not all data-logging applications use LabVIEW to process and analyze the stored data. Consider which applications might need to access the data, and write the data in an appropriate format—the data storage format defines which applications can read the file. Since LabVIEW contains standard file operation functions that exist in other languages, you have complete control over the data-logging process. The LabVIEW measurement file, or .lvm file, is an ASCII text file that is easy to create in LabVIEW and easy to read in other applications.

**Practice
with
Writing &
Reading
Measurement
Data Files**

In this example, we practice writing and reading LabVIEW measurement files. To begin, open the Write Measurement Data Demo.vi found in Chapter 10 of the Learning directory. The block diagram is shown in Figure 10.17. Central to the process of writing measurement data files is the Write To Measurement File Express VI, which includes open, write, close, and error-handling functions. It also handles formatting the string with either a tab or comma delimiter.

FIGURE 10.17
Writing measurement data to a file specified by the user

FIGURE 10.18
The Merge Signals and Split Signals functions

The Merge Signals function found in the Write Measurement Data Demo.vi combines the iteration count and the random number into the dynamic data type for use by the Express VI (see Chapter 2 for a review of the dynamic data type). The Merge Signals function is located on the **Functions≫Express≫ Signal Manipulation** palette, as shown in Figure 10.18.

When the Write To Measurement File Express VI is first placed on the block diagram, a dialog box opens automatically to configure the VI. The dialog box is shown in Figure 10.19. The configuration as shown requires the user to choose

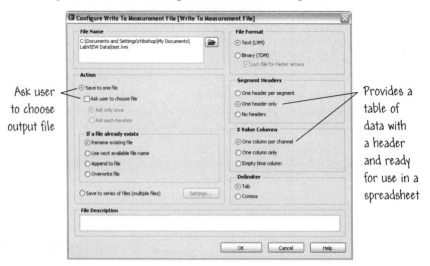

FIGURE 10.19
The Write To Measurement File Express VI dialog box

the output file name. When the VI is run, a dialog box will open, as illustrated in Figure 10.17, for specifying the output file name. This process will lead to the creation of a .lvm file that can be later opened in a spreadsheet or text editor application. Run the VI and save the results in a file named **test.lvm**.

You can double-click on the Write To Measurement File Express VI in the **Write Measurement Data Demo.vi** *to access the dialog box.*

Now we can use the Read From Measurement File Express VI as a central component of a VI that reads the .lvm file from the **Write Measurement Data Demo.vi** and outputs the results to a graph. Open the **Read Measurement Data Demo.vi** found in **Chapter 10** of the **Learning** directory. The block diagram is shown in Figure 10.20.

Double-click on the Read From Measurement File Express VI to access the configuration dialog box, shown in Figure 10.21. Notice that the following options have been set:

FIGURE 10.20
Reading measurement data from a file specified by the user

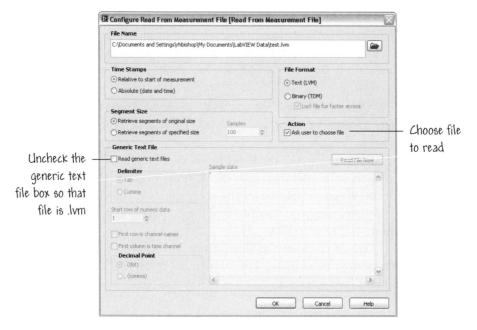

FIGURE 10.21
The Read From Measurement File Express VI dialog box

- In the **Action** section, a check mark is placed in the **Ask user to choose file** check box.

- Set the **Segment Size** to **Retrieve segments of original size** so that all the data stored in the file is retrieved.

- Set **Time Stamps** to **Relative to start of measurement**. Because the dynamic data type stores information about the signal timing, this setting aligns the data with the time of the measurement.

- In the **Generic Text File** section, remove the check mark from the **Read generic text files** check box because the data is stored in a LabVIEW measurement file.

Click the **OK** button to close the dialog box when you are finished. Display the front panel and run the VI. In the file name prompt that appears, select the test.lvm file that you created with the Write To Measurement File Express VI. The temperature data that was stored in the LabVIEW measurement file appears in the waveform chart. ◆

BUILDING BLOCK

10.3 BUILDING BLOCKS: PULSE WIDTH MODULATION

In this exercise, you will add the feature to the PWM with Chart VI (developed in Chapter 8) to write data to a file. To begin, navigate to the Users Stuff folder in the Learning directory and open PWM with Chart.vi. Once you have opened the PWM with Chart VI, save a new copy of the VI as PWM with File Output.vi.

A working version of PWM with Chart.vi can be found in the Building Blocks folder of the Learning directory. This working VI is provided in case you did not save your work in Chapter 8 to the Users Stuff folder.

To accomplish the objective of writing data to a file, use the Write to Measurement File Express VI found on the **Functions≫Express≫Output** palette. You might want to review the material in Section 10.2.4 on using the Express File I/O VIs.

Using Figure 10.22 as a guide, develop the code to create a LabVIEW measurement data file containing the array output by the While Loop. Configure the Express VI so that LabVIEW overwrites the same data file rather than creating a new one each time the VI runs. You may also wish to change the default file path and file name. Note that when you wire the array to the **Signals** input of the Write To Measurement File Express VI, LabVIEW automatically places a node to convert from numeric data type to **dynamic data type**, a data type used by Express VIs that carries additional information about a signal beyond its defining values, such as name and time of acquisition.

When you place the Write To Measurement File Express VI on the block diagram, the **Configure Write To Measurement File** dialog box will open automatically. When that occurs, make sure to select **Ask user to choose file**. Once you have configured the Write To Measurement File Express VI, add the Merge Error VI found on the **Programming≫Dialog & User Interface** palette. This VI alerts users of any errors that occur while writing to the measurement file. To the output of the Merge Error VI, wire a Simple Error Handler VI, also found on the **Programming≫Dialog & User Interface** palette. Follow the wiring shown in Figure 10.22.

After wiring the VI correctly, test it out. On the front panel, set the Duty Cycle to 0.5 and the Period to 100. Run the VI, and after it has completed executing, view the data file in Notepad (Windows) or SimpleText (Macintosh).

Close the VI when you are done experimenting and save your changes.

FIGURE 10.22
The PWM with File Output VI

   A working version of **PWM with File Output.vi** *can be found in the* **Building Blocks** *folder of the* **Learning** *directory.*

10.4   RELAXED READING: USING LABVIEW IN DRIVE-BY-WIRE TESTING

The traditional mechanical design of brakes in an automobile can lead to injury to the foot and lower leg of the driver in the event of a crash. Students from Luleå University of Technology and Stanford University were challenged to work together with the Volvo Car Corporation to develop new concepts for safer brake and accelerator pedals. The project was dubbed "virtual pedals."

The challenge was to develop a control system that incorporated continuous changes of key parameters using LabVIEW Real-Time in a PXI chassis with embedded controllers. This was to be implemented as part of a drive-by-wire system in a production car. The goal was to show significant performance improvement of the braking system during test driving in a real-time environment without delay or malfunction.

Since the newly designed brake system might replace the ordinary brake and accelerator system in a working car, it was necessary to incorporate redundancy into the design. The VOLVO S80 already has an electronically controlled accelerator system. Two separate signal circuits manage its redundancy: one analog signal and a digital PWM signal. The engine PU system compares these signals and if they differ by more than 2%, the control system shuts down the servo unit in the throttle valve system.

The student solution was to replace the brake pedal system with a DC engine controlled linear motion unit. A servo amplifier was employed to manage the high current in the DC. The hardware selected was the 1000B PXI chassis with a DC power supply, and the I/O unit selected was the NI PXI-6052E (333 kS/s) analog multifunction data acquisition card. The heart of the system was the NI PXI-8176 RT embedded controller with an Intel Pentium III processor. The controller communicates across a standard Ethernet with a Windows laptop running the LabVIEW Real-Time module.

The robust PXI chassis and the rest of the apparatus fit easily into the trunk of the car. For easy access to the driving parameters, a laptop provided access to the control panel of the system in the driving environment. The front panel of the system is shown in Figure 10.23.

LabVIEW Real-Time provided key advantages:

- It runs on its own real-time operating system, providing extremely good stability and reliability.

- The speed is outstanding because of its own operating system.

- You can modify and adjust control parameters during test driving and evaluation.

- You can easily add a CAN-bus system for communication with the vehicle onboard computer.

Another benefit of the flexible LabVIEW-based system was that it was easy to implement and test different subsystems, such as hill holding, which makes parking and starting on hills much easier for the driver, without difficult and time-consuming programming.

The students created a system to evaluate and test a modern production car braking system without loss of performance and reliability. The complex interfaces developed by the students would be difficult and very costly to design and produce utilizing typical microprocessing units and their software development

FIGURE 10.23
Student-designed VI for testing advanced drive-by-wire braking

environments. The successful student project provides the universities a great tool for use in future projects with the automotive industry. Due to the ease of adaptability and the user-friendly interface, incoming students can readily access the system and conduct their own experiments.

For more information about this project, contact:

Holger Andersson
Korngatan 3
943 31 Öjebyn
Sweden
ho.andersson@telia.com

10.5 SUMMARY

File input and output operations store information in and retrieve information from a disk file. LabVIEW supplies you with simple functions that take care of all aspects of file I/O. You can store data in or retrieve data from ASCII byte stream, binary byte stream, or datalog files. This allows you to interact with word-processing programs, spreadsheet programs, or with other VIs in LabVIEW.

KEY TERMS

ASCII: American Standard Code for Information Interchange.

Binary byte stream files: Files that store data as a sequence of bytes.

Datalog files: Files that store data as a sequence of records of a single, arbitrary data type that you specify when you create the file.

EOF: End-of-file. Character offset of the end of a file relative to the beginning of the file (that is, the EOF is the size of the file).

Hex: Hexadecimal. A base-16 number system.

Nondisplayable characters: ASCII characters that cannot be displayed, such as null space, backspace, and tab.

Not-a-path: A predefined value for the path control that means that a path is invalid.

Not-a-refnum: A predefined value that means the refnum associated with an open file is invalid.

Refnum: An identifier that LabVIEW associates with a file when you open it. You use the file refnum to indicate that you want a function or VI to perform an operation on the open file.

String: A sequence of displayable or nondisplayable ASCII characters.

Text (ASCII) byte stream files: Files that store data as a sequence of ASCII characters.

EXERCISES

E10.1 Develop a VI that writes data generated by the Generate Waveform VI to a spreadsheet. Use the block diagram in Figure E10.1 as a guide. Use the **Write to Spreadsheet File.vi** found on the **Functions≫Programming≫File I/O** palette. Wire the array of 10 data points from the For Loop to the **1D data** input of the Write To Spreadsheet File VI.

Run the VI, and when prompted, enter the file name to save the data. Upon completion of the VI execution, open the file in Notepad (Windows) or SimpleText (Macintosh) to view the data. Verify that 10 data points are in the file.

You can find the Generate Waveform VI in the activity *folder in the* Learning *directory.*

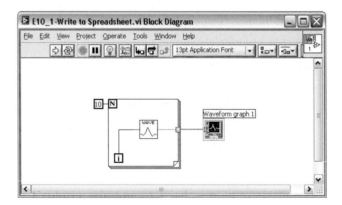

FIGURE E10.1
Use this block diagram as a guide and add the Write to Spreadsheet File VI to output the waveform data.

E10.2 Develop a VI that writes data generated by the Generate Waveform VI to a measurement file. Use the block diagram in Figure E10.1 as a guide. Use the Write to Measurement File Express VI found on the **Functions≫Express≫Output** palette. Wire the array of 10 data points from the For Loop to the **Signals** input of the Write to Measurement File VI.

Run the VI, and when prompted, enter the file name to save the data. Upon completion of the VI execution, open the file in Notepad (Windows) or SimpleText (Macintosh) to view the data. Verify that 10 data points are in the file.

E10.3 Create a new VI that acquires analog data and writes the data to a LabVIEW measurement file. Use the DAQ Assistant to configure the analog input operation. Right-click on the DAQ Assistant and select **Generate NI-DAQmx Code** to automatically create the code for this acquisition. Take 50 measurements and write the measurements using the Write to Measurement File VI. Refer back to Chapter 9 for more information on using the DAQ Assistant.

E10.4 Create a VI that reads the LabVIEW measurement file created in E10.3 and displays that information on the front panel.

E10.5 Modify the VI you created in E10.4 so that data is saved to a new file every time the VI runs. Also, make it so that LabVIEW appends the next sequential number to the file name.

E10.6 Open the VI you created in the Building Blocks exercise for this chapter, which you saved as **PWM with File Output.vi** in the **Users Stuff** folder in the **Learning** directory. On the front panel of the PWM with File Output VI, set the **Duty Cycle** to 0.5 and the **Period** to 100. Then run the VI and check to see that an output file was created at the expected location.

*In case you did not complete the Building Blocks exercise, you can find a working version of **PWM with File Output.vi** in the **Building Blocks** folder in the **Learning** directory.*

Construct a VI that prompts the user for the file name containing the measurement data created by the PWM with File Output VI and then displays the data on a waveform graph. Turn off autoscaling on the horizontal axis and make the scrollbar visible so that you can view large data sets more clearly. Run the VI and verify that it displays the same waveform obtained when running the PWM with File Output VI.

E10.7 Open **Build Text Demo.vi** found in **Chapter 10** of the **Learning** directory. This is the same example discussed in Section 10.1. Modify the VI so that a temperature reading can only be displayed if the user first enters a correct password.

Use the block diagram in Figure E10.7 as a guide. Place a string control on the front panel and right-click to change the display mode from **Normal** to **Password**. Place an LED next to the password control that lights when the correct password is entered. Using a Case structure, provide an empty output display and a message to the user with notification that an incorrect password was entered in the case that an invalid password is given. If the correct password is entered, display the temperature reading on the output indicator.

When you are finished, save the VI as **Password.vi** in the **Users Stuff** folder in the **Learning** directory.

PROBLEMS

P10.1 Although we did not cover the Binary File I/O functions in this chapter, you may find them useful for applications in which you wish to quickly store data in as little space as possible. In the NI Example Finder, find the Write Binary File VI by searching for the keywords "write binary." Once you have located the Write Binary File VI, open it and read the description of the VI and investigate the block diagram. Then run the VI and when prompted enter the file name **BinaryFile** for the saved data.

Return to the NI Example Finder and hunt for the Read Binary File VI by searching for the keywords "read binary." Once again, read the VI description and explore the block diagram. When you have done this, run the VI and when prompted, enter the same file name **BinaryFile** that you used to save the binary data created using the Write Binary File VI. You should recognize the data on the waveform graph as the same data that was graphed and saved with the Write Binary File VI.

FIGURE E10.7
The Password VI block diagram

Now return to the Write Binary File VI and modify this VI so that it creates a LabVIEW measurement file containing the same data. Use the Write to Measurement File Express VI. Use the Get File Size function found on the **Programming≫File I/O≫Advanced File Functions** palette to display the size of both the LabVIEW measurement file and the binary file on numeric indicators on the front panel.

Save the modified file as Write Two Files.vi in the Users Stuff folder in the Learning directory. Run the VI again and observe the size of each file created. Are the files the same size? Why or why not? List the different advantages of the two file formats.

P10.2 Open Basic Spectral Measurements.vi (located in the Express subfolder of the examples folder of the LabVIEW 8.0 directory. Using the Report Express VI found on the **Express≫Output** palette, modify this VI to create a report that displays the graphs of both signals and includes the report title, author, company name, and comments about the report. Give the user the option to enter additional comments. Save the report as an HTML file. Run the VI and view the HTML document that you created in a web browser or word-processing application.

P10.3 Create a VI that concatenates a message string, numeric, and unit string using the Build Text Express VI and writes the data to file. Create the input variables **Number** and **Units** and configure them in the Build Text Express VI. Wire a string control to the **Beginning Text** input of the Build Text Express VI. Write the resulting concatenated string to a datalog file. Then create a VI that reads the datalog file you just created and displays that information on the front panel. To find out more about datalog files, refer to the online help and examples.

P10.4 Construct a VI that converts a 2D array into a tab-delimited spreadsheet string. Then convert the tab-delimited spreadsheet string into a comma-delimited spreadsheet string and display both output strings in indicators on the front panel.

DESIGN PROBLEMS

D10.1 Construct a VI that performs number base conversions and then writes the resulting conversions to a text file. The front panel should have the following components:

- A numeric control into which the user will enter the decimal value to be converted. Change the representation of this control to **Unsigned Integer** (U64) so that the user will not enter any negative or non-integer values.

- A menu ring to select from **Binary**, **Octal**, or **Hexadecimal** conversion.

- A string indicator to display the result of the conversion.

- A file path control to enter the name of the text file to which the results are to be saved.

- A Boolean button which causes the VI to write the results of the most recent conversion to the designated file when true. You will need to change the mechanical action of this button so that it does not remain depressed after a write request is read by the VI.

- A stop button which ends VI execution. The program should perform as many new conversions and file writes as the user requests, until the stop button is pressed.

Each time a new file write is requested, concatenate a sentence to the end of the text file conveying the most recent results. For example, the sentence "Decimal 7 is 111 in binary" would be added to the file that converted the number 7 to binary. Be sure that your VI opens the text file once at the beginning of the VI and then closes it once just before stopping. You should not open and close the file with each loop iteration.

To convert a decimal number to a different base, use a While Loop that performs successive division by 2, 8, or 16 for binary, octal, and hexadecimal, respectively and outputs the remainders from each division. Concatenate each new digit to a result string as it is computed. You will need to use several string functions to convert the division results to characters and then manipulate the characters so that you can create a string that accurately represents the converted number.

Run the VI several times and verify that the correct results are being written to the output file.

D10.2   Create a VI which appends an array containing a time value and a numeric input to the end of a spreadsheet file, whose path is specified by the user. The time value should be recorded in seconds, relative to the time the VI began running. The VI should run continuously, saving the most recent input value to the spreadsheet file each time the user presses a Save button on the front panel. End execution when a stop button is pressed.

D10.3   Develop a VI to encode or decode words using a secret code. To encode a word in this code, first separate the word into two substrings – one with the first half of the original word, and one containing the second half. If there is an odd number of letters in the word, then split it so that the first substring has one more letter than the second one.

Create the encoded word by concatenating the first half of the word onto the end of the second half of the word, inserting the second letter of the original word in between these two halves. For example, the word "gravity" would be encoded as "ityrgrav" and the word "master" would be encoded as "teramas."

Place a string control on the front panel and a Boolean switch allowing the user to select whether the input is to be decoded or encoded. Display the output on a string indicator.

CHAPTER 11

Analysis

LabVIEW is an excellent environment for mathematical analysis and signal processing. The G programming language is ideally suited to developing programs for solving linear algebraic systems of equations, curve fitting, integrating ordinary differential equations, computing function zeroes, computing derivatives of functions, integrating functions, generating and analyzing signals, computing discrete Fourier transforms, and filtering signals. The Express VIs for curve fits, filters, signal generation, and spectral analysis provide the tools necessary to quickly prototype and solve problems. This chapter presents an overview of these topics.

GOALS

1. Introduce some of the LabVIEW capabilities for mathematical analysis.
2. Study some of the VIs available for analyzing signals and systems.

495

11.1   LINEAR ALGEBRA

LabVIEW provides a number of important VIs dedicated to solving systems of linear algebraic equations. These types of systems arise frequently in engineering and scientific applications. An entire branch of mathematics is dedicated to the study of *linear algebra*. We will only be able to touch on a few important subjects within the linear algebra arena in the space available to us here.

11.1.1   Review of Matrices

The basic element used in the computations is the *matrix*. A matrix is represented by an $m \times n$ array of numbers:

$$\mathbf{A} = \begin{bmatrix} a_{0,0} & a_{0,1} & \cdots & a_{0,n-1} \\ a_{1,0} & a_{1,1} & \cdots & a_{1,n-1} \\ \cdots & \cdots & \cdots & \cdots \\ a_{m-1,0} & a_{m-1,1} & \cdots & a_{m-1,n-1} \end{bmatrix},$$

where n is the number of columns, and m is the number of rows. When $m \neq n$, the matrix is called a *rectangular* matrix; conversely when $m = n$, the matrix is called a *square* matrix. An $m \times 1$ matrix is called a *column vector*, and a $1 \times n$ matrix is called a *row vector*. Other special forms of the matrix are the *diagonal* matrix, the *zero* matrix, and the *identity* matrix. Examples of these three types of matrices are

$$\mathbf{A} = \begin{bmatrix} 2 & 0 & 0 \\ 0 & 4 & 0 \\ 0 & 0 & -6 \end{bmatrix}, \quad \mathbf{0} = \begin{bmatrix} 0 & 0 & 0 \\ 0 & 0 & 0 \\ 0 & 0 & 0 \end{bmatrix}, \quad \mathbf{I} = \begin{bmatrix} 1 & 0 & 0 \\ 0 & 1 & 0 \\ 0 & 0 & 1 \end{bmatrix},$$

respectively. In LabVIEW, matrices can be created using the matrix control seen in Figure 11.1. The Real Matrix control can be found on the **Controls≫Modern ≫Array, Matrix & Cluster** palette.

The elements of a matrix, **A**, are denoted by $a_{i,j}$, where i indicates row number and j indicates column. Matrices can contain real numbers or complex numbers. If the elements of the matrix are complex numbers, the Complex Matrix control, found on the same palette as the Real Matrix control, should be used.

The unknowns are the variables x_1, x_2, and x_3. We can identify the two column vectors $\mathbf{x}$ and $\mathbf{b}$ as

$$\mathbf{x} = \begin{bmatrix} x_1 \\ x_2 \\ x_3 \end{bmatrix} \quad \text{and} \quad \mathbf{b} = \begin{bmatrix} 4 \\ 0 \\ 1 \end{bmatrix}.$$

Then we can write the system of algebraic equations as

$$\mathbf{A}\mathbf{x} = \mathbf{b},$$

where

$$\mathbf{A} = \begin{bmatrix} 4 & 6 & 1 \\ 1 & 2 & 3 \\ 0 & 5 & -1 \end{bmatrix}.$$

Thus, we have rewritten the problem in a compact matrix notation. Can we solve for the vector $\mathbf{x}$? In this case, the matrix $\mathbf{A}$ is invertible (i.e., the inverse exists), so the solution is readily obtained as

$$\mathbf{x} = \mathbf{A}^{-1}\mathbf{b} = \begin{bmatrix} 0.9123 \\ 0.1228 \\ -0.3860 \end{bmatrix}.$$

If the matrix $\mathbf{A}$ is singular, then the number of solutions depends on the vector $\mathbf{b}$. If a solution does exist in the singular case, it is not unique!

For any given $n \times n$ square matrix $\mathbf{A}$, we would like to know if there exists a scalar λ and a corresponding vector $\mathbf{v} \neq 0$ such that

$$\lambda\mathbf{v} = \mathbf{A}\mathbf{v}.$$

This scalar λ is called the *eigenvalue*, and the corresponding vector $\mathbf{v}$ is called the *eigenvector*. Rearranging yields

$$\lambda\mathbf{v} - \mathbf{A}\mathbf{v} = (\lambda\mathbf{I} - \mathbf{A})\,\mathbf{v} = 0.$$

Therefore, a solution exists (for $\mathbf{v} \neq 0$) if and only if

$$\det\,(\lambda\mathbf{I} - \mathbf{A}) = 0.$$

If $\mathbf{A}$ is an $n \times n$ matrix, then $\det\,(\lambda\mathbf{I} - \mathbf{A}) = 0$ is an nth-order polynomial (known as the *characteristic equation*) whose solutions are called the *characteristic roots* or eigenvalues. The eigenvalues of a square matrix are not necessarily unique and may be complex numbers (even for a real-valued matrix!). Given the eigenvalues of a matrix, we can compute the trace and determinant as

$$\text{tr } \mathbf{A} = \sum_{i=1}^{n} \lambda_i$$

$$\det \mathbf{A} = \prod_{i=1}^{n} \lambda_i$$

We see that if any eigenvalue is zero, then the determinant is zero, and thus it follows that the matrix is singular.

An interesting fact regarding eigenvalues is that if a matrix is real and symmetric, then its eigenvalues will be real. If we compute the "square" of a real matrix $\mathbf{A}$ according to $\mathbf{B} = \mathbf{A}\mathbf{A}^T$, then $\mathbf{B}$ is real and symmetric. In fact, the matrix $\mathbf{B}$ is nonnegative, in the sense that, for any nonzero column vector $\mathbf{v}$ it follows that the scalar value $\mathbf{v}^T \mathbf{B} \mathbf{v} \geq 0$. Now if we compute the eigenvalues of $\mathbf{B}$, we find that they are all nonnegative and real. Taking the square root of each eigenvalue yields quantities known as *singular values*, denoted here by β_i for $n = 1, 2, \ldots, n$. An $n \times n$ matrix has n nonnegative singular values. Two important singular values are the maximum and minimum values, $\beta_{\max}$ and $\beta_{\min}$, respectively. Singular values are important in computational linear algebra because they tell us something about how close a matrix is to being singular. We know that if the determinant of a matrix is zero, then it is a singular matrix. But what about if you compute the determinant numerically with the computer and determine that the determinant is 10^{-10}? Is this close enough to zero to say that the matrix is singular? The answer to this is provided by the **condition number** of a matrix. It can be defined in different ways, but it is commonly defined as

$$\text{cond } \mathbf{A} = \frac{\beta_{\max}}{\beta_{\min}}.$$

The condition number can vary from 0 to ∞. A matrix with a condition number near 1 is closer to being nonsingular than a matrix with a very large condition number. The condition number is useful in assessing the accuracy of solutions to systems of linear algebraic equations.

As a final practical note, it is not generally a good idea to explicitly compute the matrix inverse when solving systems of linear algebraic equations, because inaccuracies are associated with the numerical computations—especially when the condition number is high. The preferred solution technique involves using *matrix decompositions*. Popular techniques include

- Singular Value Decomposition (SVD)
- Cholesky decomposition (or QR)

The idea is to decompose the matrix into component matrices that have "nice" numerical properties.

For example, suppose we decompose the matrix

$$\mathbf{A} = \mathbf{QR},$$

where $\mathbf{Q}$ is an *orthogonal* matrix (that is, $\mathbf{Q}^T \mathbf{Q} = \mathbf{I}$), and $\mathbf{R}$ is upper triangular (all the elements below the diagonal are zero). Then,

$$\mathbf{Ax} = \mathbf{QRx} = \mathbf{b}.$$

Multiplying both sides by $\mathbf{Q}^T$ yields

$$\mathbf{Rx} = \mathbf{Q}^T \mathbf{b}.$$

Then you use the fact that $\mathbf{R}$ is upper triangular to *back-substitute* and solve for $\mathbf{x}$ without ever computing a matrix inverse.

11.1.3   Linear System VIs

LabVIEW comes with a complete set of VIs that can be used to perform all the matrix computations discussed previously (and much more!). You can find the VIs on the palette shown in Figure 11.4.

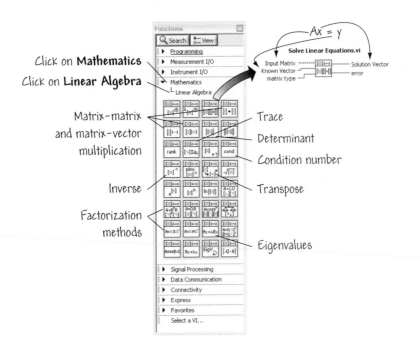

FIGURE 11.4
The linear algebra VIs

The Linear Algebra Calculator.vi is a useful Example VI that can be used to perform a variety of linear algebra operations on a matrix or to solve systems of linear algebraic equations. This VI can take data as input on the front panel or read the data from a file (that is, from a spreadsheet file). The next example gives you the opportunity to experiment with linear algebra computations.

Linear Algebra Calculator

Open Linear Algebra Calculator.vi using the NI Example Finder. In Lab-VIEW select **Find Examples** in the **Help** pull-down menu. When the NI Example Finder opens, browse by task to the Analyzing and Processing Signals folder and select Mathematics. Within the Mathematics folder you will find a variety of interesting examples, including the Linear Algebra Calculator VI. The front panel and block diagram are shown in Figure 11.5.

By default, the VI is set up to solve the system of linear algebraic equations

$$\mathbf{Ax} = \mathbf{b},$$

FIGURE 11.5
The linear algebra calculator

where

$$
\mathbf{A} = \begin{bmatrix} 4 & 2 & -1 \\ 1 & 4 & 1 \\ 0.1 & 1 & 2 \end{bmatrix}, \quad \text{and} \quad \mathbf{b} = \begin{bmatrix} 2 \\ 12 \\ 10 \end{bmatrix}.
$$

Choose **Solve Linear Equations** in the lower left side of the front panel and run the VI. Verify that you obtain the solution

$$
\mathbf{x} = \begin{bmatrix} 0.59 \\ 1.84 \\ 4.05 \end{bmatrix}.
$$

Use the calculator to compute the determinant of the matrix **A**. You should get the result det **A** = 23.6. Compute the condition number, the inverse, and the trace of the matrix **A** (condition number = 4.418 and trace = 10). The VI runs until the stop button is pressed.

Modify the input matrix to be

$$
\mathbf{A} = \begin{bmatrix} 4 & 2 & -1 \\ 1 & 4 & 1 \\ 2 & 8 & 2 \end{bmatrix}.
$$

Compute the condition number. You should find that the condition number is very, very large! Now compute the matrix inverse (by selecting **Inverse** from the menu). What happens? An error message appears that lets you know that the system of equations cannot be solved because the input matrix is singular. Vary the (3,3) element and observe the effect on the condition number. When the condition number reduces to less than 10, compute the matrix inverse again. Does it work in this case? It should!

Set the (3,3) term of the matrix to 2.01:

$$
\mathbf{A} = \begin{bmatrix} 4 & 2 & -1 \\ 1 & 4 & 1 \\ 2 & 8 & 2.01 \end{bmatrix}.
$$

Verify that the determinant is 0.14. Now compute the condition number—it should be quite high (above 2,500). What does this result lead you to conclude about the advisability of solving the system of linear algebraic equations by inverting the **A** matrix? Basically, when the condition number is high, it is not advisable to solve the system of equations by matrix inversion. ◆

11.2   STATISTICS AND CURVE FITTING

Statistics and curve fitting are common techniques used in science, engineering, business, medicine, and other fields in the analysis of data. Based on a set of assumptions, statistics attempt to determine trends in sets of related data. Statistics are commonly used every day in applications ranging from computing averages on test grades to predicting the weather. Curve fitting is another way to analyze related data, and involves extracting a set of curve parameters (or coefficients) from the data set to obtain a functional description of the data set. Using curve fitting, digital data can be represented by a continuous model. For example, you may want to fit the data with a straight-line model. The curve-fitting procedure would provide values for the linear curve fit in terms of slope and axis offset. In this section, we will cover curve fitting and discuss statistical methods for analyzing sets of data.

11.2.1   Curve Fits Based on Least Squares Methods

The main algorithm used in the curve-fitting process is known as the least squares method. Define the error as

$$e(\mathbf{a}) = [f(x, \mathbf{a}) - y(x)]^2,$$

where $e(\mathbf{a})$ is a measure of the difference between the actual data and the curve fit, $y(x)$ is the observed data set, $f(x, \mathbf{a})$ is the functional description of the data set (this is the curve-fitting function), and **a** is the set of curve coefficients that best describes the curve. For example, let $\mathbf{a} = (a_0, a_1)$. Then the functional description of a line is

$$f(x, \mathbf{a}) = a_0 + a_1 x.$$

The least squares algorithm finds **a** by solving the Jacobian system

$$\frac{\partial e(\mathbf{a})}{\partial \mathbf{a}} = 0.$$

LabVIEW's curve-fitting VIs solve the Jacobian system automatically and return the set of coefficients that best describes the input data set. The automatic nature of this process provides the opportunity to concentrate on the results of the curve fitting rather than dealing with the mechanics of obtaining the curve-fit parameters.

When we curve fit data, we generally have available two input sequences, Y and X. The sequence X is usually the independent variable (e.g., time) and the sequence Y is the actual data. A point in the data set is represented by (x_i, y_i), where x_i is the ith element of the sequence X, and y_i is the ith element of the sequence Y. Since we are dealing with samples at discrete points, statistics are appropriate for gathering additional information from the data. The VI calculates the mean square error (MSE), which is a relative measure of the residuals between the expected curve values and the actual observed values, using the formula

$$\text{MSE} = \frac{1}{n} \sum_{i=0}^{n-1} (f_i - y_i)^2,$$

where f_i is the sequence of fitted values, y_i is the sequence of observed values, and n is the number of input data points.

LabVIEW offers a number of curve-fitting algorithms, including:

- Linear Fit:

$$y_i = a_0 + a_1 x_i$$

- Exponential Fit:

$$y_i = a_0 e^{a_1 x_i}$$

- General Polynomial Fit:

$$y_i = a_0 + a_1 x_i + a_2 x_2 + \cdots$$

- General Linear Fit:

$$y_i = a_0 + a_1 f_1(x_i) + a_2 f_2(x_i) + \cdots$$

where y_i is a linear combination of the parameters $a_0, a_1, \ldots$. This type of curve fit provides user-selectable algorithms (including SVD, Householder, Givens, LU, and Cholesky) to help achieve the desired precision and accuracy.

- Nonlinear Levenberg-Marquardt Fit:

$$y_i = f(x_i, a_0, a_1, a_2, \ldots)$$

where $a_0, a_1, a_2, \ldots$ are the parameters. This method does not require y to have a linear relationship with $a_0, a_1, a_2, \ldots$. Although it can be used for linear curve fitting, the Levenberg-Marquardt algorithm is generally used for nonlinear curve fits.

The curve-fitting VIs can be found on the **Fitting** palette, as illustrated in Figure 11.6.

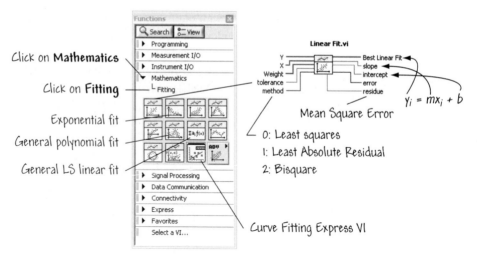

Click on **Mathematics**

Click on **Fitting**

Exponential fit

General polynomial fit

General LS linear fit

Curve Fitting Express VI

Mean Square Error

0: Least squares

1: Least Absolute Residual

2: Bisquare

$y_i = mx_i + b$

FIGURE 11.6
The curve-fitting VIs

Practicing with Curve Fitting

Open Regressions Demo.vi using the NI Example Finder. In LabVIEW select **Find Examples** in the **Help** pull-down menu. When the NI Example Finder opens, browse by task to the Analyzing and Processing Signals folder and select Curve Fitting and Interpolation. Within this folder you will find the Regressions Demo VI. The front panel is shown in Figure 11.7. The VI generates noisy data samples that are approximately linear, exponential, or polynomial and then uses the appropriate curve-fitting VIs to determine the best parameters to fit the given data. You can control the noise level with the knob on the front panel. You can also select an algorithm and the number of samples to fit. For the polynomial curve fits, the order of the polynomial can be varied via front panel input.

Select Linear in the Algorithm Selector control and set Noise Level to around 0.05. Run the VI and make a note of the computed error displayed in the mse indicator. Increase the noise level to 0.1 and again make a note of the computed error. Continue this process for noise levels of 0.15, 0.2, and 0.25. Did you detect any trends? You should have seen the MSE increase as the noise level increased.

Select Polynomial in the Algorithm Selector control and set Noise Level to around 0.05. Set the Order control to 2. Run the VI and make a note of the computed error. Follow the same procedure as above. Did you obtain the same trends as the noise increased? Reduce the noise level to 0.15. This time run an experiment with the noise level fixed but increase the polynomial order from 2 to 6. Did you detect any trends as the polynomial order increased? In this case, as you increase the polynomial order, the computed error fluctuates, but remains basically on the same order of magnitude.

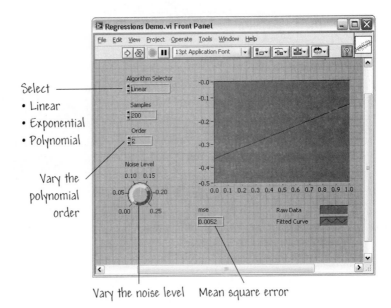

Select — Linear

• Linear
• Exponential
• Polynomial

Vary the polynomial order

Vary the noise level   Mean square error

FIGURE 11.7
A demo to investigate curve fitting with linear, exponential, and polynomial fits

With the **Algorithm Selector** control set to **Polynomial**, run the VI with the polynomial order set to 0. Then change the polynomial order to 1 and run the VI. With polynomial order equal to 0 and 1, the fitted curve is a horizontal line and a straight line with a (generally) nonzero slope, respectively. Experiment with the Regressions Demo VI and see if you can discover new trends. Consider comparing the linear fit with the exponential fit.   ◆

Open the NI Example Finder to search for examples that can be used to illustrate the curve fitting capability of LabVIEW. When the NI Example Finder opens, browse by task to the **Analyzing and Processing Signals** folder and select **Curve Fitting and Interpolation**. Within the **Curve Fitting and Interpolation** folder you will find a variety of interesting examples, including the Linear, Exp, and Power Fitting VI and the General LS Fitting VI.

11.2.2   Fitting a Curve to Data with Normal Distributions

Real-world data is very often normally (or Gaussian) distributed. The mathematical description of a normal distribution is

$$f(x) = \frac{1}{\sigma\sqrt{2\pi}} \exp\left[-\frac{1}{2}\left(\frac{x-m}{\sigma}\right)^2\right],$$

where m is the mean and σ is the standard deviation. Figure 11.8 shows the normal distribution with $m = 0$ and $\sigma = 0.5$, 1.0, and 2.0. As seen in the figure,

FIGURE 11.8
The normal distribution shape for three values of σ: 0.5, 1.0, and 2.0

the normal distribution is bell shaped and symmetric about the mean, m. The area under the bell-shaped curve is unity! The two parameters that completely describe the normally distributed data are the mean, m, and the standard deviation, σ. The peak of the bell-shaped curve occurs at m. The smaller the value of σ, the higher the peak at the mean and the narrower the curve.

The normal distribution is illustrated again in Figure 11.9. The standard deviation is an important parameter that defines the "bounds" within which a certain percentage of the data values is expected to occur. For example:

- About two-thirds of the values will lie between $m - \sigma$ and $m + \sigma$.

- About 95% of the values will lie between $m - 2\sigma$ and $m + 2\sigma$.

- About 99% of the values will lie between $m - 3\sigma$ and $m + 3\sigma$.

Therefore, an interpretation of these values is that the probability that a normally distributed random value lies outside $\pm 2\sigma$ is approximately 0.05 (or 5%).

Normal Distributions

In the folder Chapter 11 in the Learning directory you will find a VI called Normal (Gaussian) Fit.vi. This VI generates a random data set and then plots the distribution. The front panel is shown in Figure 11.10. If you want to experiment with normally distributed data, open and run the VI. You can vary the number of samples and the standard deviation (that is, the σ). Set $\sigma = 0.5$, 1.0, and 2.0, and experiment.

Run the VI and vary the number of samples from 10 to 10,000. You should notice that as the number of samples increases, the shape of the data distribution becomes more and more bell shaped. Does this exercise relate in any way to your experience with grade distributions in class? Would you expect to have a bell-shaped grade distribution in a class of 5 students? 100 students? 500 students? (Answers: no, maybe, yes)

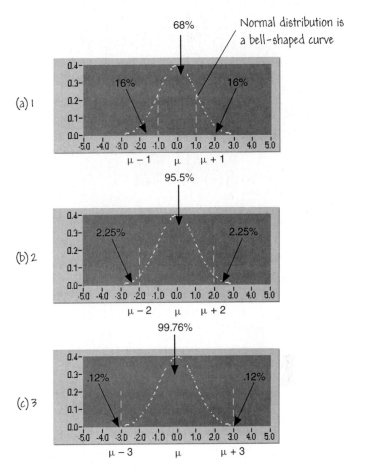

FIGURE 11.9
The normal distribution for 1σ, 2σ, and 3σ

Using Normal (Gaussian) Fit.vi as a starting point, construct a VI to curve-fit the normal distribution and to compute the mean and sigma (σ) from the curve-fit parameters.   ◆

11.2.3   The Curve Fitting Express VI

LabVIEW 8.0 provides a significant number of Express VIs designed for analysis purposes. The **Signal Analysis** palette is shown in Figure 11.11. We will discuss only a subset of the Express VIs: Curve Fitting, Spectral Measurements, Filter, and Simulate Signal. In this section, we present the Curve Fitting Express VI.

To use the Curve Fitting VI, you first place it on the block diagram. As illustrated in Figure 11.12, once the Curve Fitting Express VI is placed on the block diagram, a dialog box automatically appears to configure the VI. The curve fitting options include a linear fit, a quadratic fit, a spline fit, a polynomial fit (of

Vary the number of data points considered

Change this value to 0.5 and 2.0

As **samples** get larger, the data distribution will resemble a "bell shape" more closely

Found on **Express >> Signal Analysis** palette

FIGURE 11.10
Normally distributed data

Select **Express**

Select **Signal Analysis**

Spectral Measurements Express VI

Statistics

Create Histogram

Filter Express VI

Curve Fitting Express VI

Simulate Signal Express VI

FIGURE 11.11
The Signal Analysis Express VIs

FIGURE 11.12
Configuring the Curve Fitting Express VI

an order specified by the user), and a general least squares fit. In Figure 11.12, a fifth order polynomial fit is selected. Once the Curve Fitting VI is configured, click **OK** to return to the block diagram.

On the block diagram, the Express VI will expand to display the inputs and outputs. The key input is the signal, and the key output is the best fit signal. You can also access the curve fit residuals and mean squared error. These are useful in quantifying the accuracy of the curve fit itself. The Curve Fitting VI is now ready for inclusion in a VI for signal analysis.

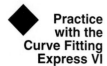

Consider the VI shown in Figure 11.13. Notice on the block diagram that two Express VIs—the Curve Fitting Express VI and the Simulate Signal Express VI—are utilized. This VI was developed to assist in investigating the impact of selecting various curve-fitting strategies to fit a noisy sine wave signal.

The Curve Fitting Express Demo.vi is located in Chapter 11 of the Learning directory.

There are four elements on the block diagram:

- The Curve Fitting Express VI—this can be configured by double-clicking on the icon to access the configuration dialog box.

FIGURE 11.13
The Curve Fitting Express Demo VI for investigating noisy sine wave signals

- The Simulate Signal Express VI—this VI will be described in Section 11.6 in more detail. It has been configured here to provide a sine wave signal at a frequency of 10.1 Hz, and normally distributed white noise is added to the signal.

- The Merge Signals function—this function was described in Section 10.2.4. Its purpose is to merge the raw signal and the best fit signal (after curve fitting) into a format acceptable to the waveform graph.

- The waveform graph—this is used to graphically display the smoothed signal and the raw signal.

Run Curve Fitting Express Demo.vi and observe the results on the waveform graph. Recall that the Curve Fitting Express VI was configured to use a fifth-order polynomial fit.

Double-click the Curve Fitting Express VI to access the configuration dialog box. Reduce the order of the polynomial fit to second-order. You should find that the curve fit is now much poorer than before with the fifth-order polynomial. Try the linear fit and see what happens. ◆

11.3 DIFFERENTIAL EQUATIONS

In LabVIEW, you can solve linear and nonlinear ordinary differential equations (ODEs) using one of seven VIs for solving first- and higher-order differential equations. Figure 11.14 shows the palette with the available ODE VIs.

The order of a differential equation is the order of the highest derivative in the differential equation.

FIGURE 11.14
Solving differential equations using LabVIEW VIs

Suppose that we have a set of first-order ordinary differential equations:

$$\dot{\mathbf{x}}(t) = \mathbf{f}(\mathbf{x}(t), \mathbf{u}(t)),$$

where $\mathbf{x}$ is the vector $(x_1, x_2, \cdots x_n)^T$, sometimes known as the *state vector*, and $\mathbf{u}$ is the vector $(u_1, u_2, \cdots u_m)^T$ of inputs to the system. When $\mathbf{u} = 0$, the differential equation is termed a *homogeneous* differential equation, and when

$$\mathbf{f}(\mathbf{x}(t), \mathbf{u}(t)) = \mathbf{A}\mathbf{x}(t) + \mathbf{B}\mathbf{u}(t),$$

the system is a system of *linear* ordinary differential equations.

An example of a *nonhomogeneous* system is

$$\dot{x}_1(t) = x_1^2(t) + \sin x_2(t) + u_1(t)$$
$$\dot{x}_2(t) = x_1(t) + x_2^3(t) + 3u_2(t),$$

where we write

$$\mathbf{x} = \begin{pmatrix} x_1 \\ x_2 \end{pmatrix} \qquad \mathbf{u} = \begin{pmatrix} u_1 \\ u_2 \end{pmatrix} \qquad \mathbf{f}(\mathbf{x}(t), \mathbf{u}(t)) = \begin{pmatrix} x_1^2(t) + \sin x_2(t) + u_1(t) \\ x_1(t) + x_2^3(t) + 3u_2(t) \end{pmatrix}.$$

To compute a solution we need to specify the initial conditions. For a system with n first-order differential equations, we need to specify n initial conditions, $x_1(0), x_2(0), \ldots, x_n(0)$.

We can also represent physical systems by higher-order differential equations. An example of a second-order mass-spring-damper system is

$$m\frac{d^2 y(t)}{dt^2} + b\frac{dy(t)}{dt} + ky(t) = g(t),$$

where the system parameters are $m = $ mass, $b = $ damping coefficient, $k = $ spring constant, and the input function is $g(t)$. When $g(t) = 0$, the system is homogeneous. To compute the solution, we need two initial conditions: $y(0)$ and $\dot{y}(0)$.

You can describe an nth-order differential equation equivalently by n first-order differential equations. Consider the second-order differential equation presented above and define

$$x_1 = y \quad \text{and} \quad x_2 = \dot{y} \quad \text{and} \quad u = g(t).$$

Taking time-derivatives of x_1 and x_2 yields

$$\dot{x}_1 = \dot{y} = x_2$$
$$\dot{x}_2 = \ddot{y} = -\frac{b}{m}\dot{y} - \frac{k}{m}y + \frac{1}{m}g(t) = -\frac{b}{m}x_2 - \frac{k}{m}x_1 + \frac{1}{m}u(t),$$

or,

$$\dot{\mathbf{x}}(t) = \mathbf{Ax}(t) + \mathbf{B}u(t) = \begin{bmatrix} 0 & 1 \\ -\dfrac{k}{m} & -\dfrac{b}{m} \end{bmatrix} \mathbf{x}(t) + \begin{bmatrix} 0 \\ \dfrac{1}{m} \end{bmatrix} u(t)$$

LabVIEW has eight VIs to solve sets of first-order differential equations, and two VIs to solve nth-order differential equations. Each VI uses a different numerical method for solving the differential equations. Each method has an associated *step size* that defines the time intervals between solution points. To solve differential equations of the form

$$\dot{\mathbf{x}}(t) = \mathbf{f}(\mathbf{x}(t), \mathbf{u}(t)),$$

use one of the following VIs:

- ODE Cash–Karp 5th-Order: Variable-step integrator that adjusts the step size internally to reduce numerical errors.

- ODE Euler Method: A very simple fixed-step integrator—that is, it executes fast, but the integration error associated with the Euler method is usually unacceptable in situations where solution precision is important.

- ODE Runge–Kutta 4th-Order: Fixed-step integrator that provides much more precise solutions than the Euler method.

To solve differential equations of the form

$$\dot{\mathbf{x}}(t) = \mathbf{Ax}(t)$$

where the coefficients of the matrix $\mathbf{A}$ are constant, use one of the following VIs:

- ODE Linear System Numeric: Generates a numerical solution of a homogeneous linear system of differential equations.

- ODE Linear System Symbolic: Generates a symbolic solution of a homogeneous linear system of differential equations.

By a symbolic solution, we mean that the output of the VI is actually a formula rather than an array of numbers. The solution is presented as a formula displayed on the VI front panel.

For solving homogeneous, higher-order differential equations of the form

$$a_0 \frac{d^n y}{dt^n} + a_1 \frac{d^{n-1} y}{dt^{n-1}} + \cdots + a_{n-1} y = 0,$$

LabVIEW has two VIs:

- ODE Linear nth-Order Numeric: Generates a numeric solution of a linear system of nth-order differential equations.

- ODE Linear nth-Order Symbolic: Generates a symbolic solution of a linear system of nth-order differential equations.

Which VI should you use? A few general guidelines follow:

- Use the ODE Euler Method VI sparingly and only for very simple ODEs.

- For most situations, choose the ODE Runge–Kutta 4th-Order or the ODE Cash–Karp 5th-Order VI.

  - If you need the solution at equal intervals, choose the ODE Runge–Kutta 4th-Order VI.

  - If you are interested in a global solution and fast computation, choose the ODE Cash–Karp 5th-Order VI.

The Pendulum

The objective of this exercise is to build a VI that solves a second-order differential equation that models the motion of a pendulum. Because the equations must be entered in symbolic form, you will need to incorporate the Substitution Variables VI, which allows you to vary the model parameters more easily. The pendulum model is given by

$$\frac{d^2\theta}{dt^2} + \frac{c}{ml}\frac{d\theta}{dt} + \frac{g}{l}\sin\theta = 0,$$

where m is the mass of the pendulum, l is the length of the rod, $g = 9.8$ is the acceleration due to gravity, and θ is the angle between the rod and a vertical line passing through the point where the rod is fixed (that is, the equilibrium position).

The pendulum model is a homogeneous, nonlinear equation (notice the $\sin\theta$ term!)—you have three choices of VIs:

- ODE Cash–Karp 5th-Order

- ODE Euler Method

- ODE Runge–Kutta 4th-Order

The first step is to formulate the pendulum model as two first-order differential equations. All three integration VIs listed above are for first-order systems. This is achieved by making the following substitution:

$$x_1 = \theta \quad \text{and} \quad x_2 = \dot{\theta}.$$

Go ahead and convert the second-order differential equation to two first-order differential equations. You should end with the resulting system:

$$\dot{x}_1 = x_2$$
$$\dot{x}_2 = -\frac{c}{ml}x_2 - \frac{g}{l}\sin x_1$$

Construct a VI to simulate the motion of the pendulum. Use the front panel of the VI shown in Figure 11.15 as a guide.

A block diagram is shown in Figure 11.16 that can be used as a guide in the VI development. To begin the development, use the ODE Euler Method VI. The inputs that you will need follow:

- **X:** An array of strings listing the dependent variables (x_1 and x_2).
- **time start:** The point in time at which to start the calculations ($t_0 = 0$).
- **time end:** The point in time at which to end the calculations ($t_f = 10$).
- **h (step rate):** The time increment for the calculations ($h = 0.01$).
- **X0:** The initial conditions ($x_1(0) = 1$, and $x_2(0) = 0$).

The pendulum differential equations are typed in the **F(x,t)** control. The Substitute Variables VI is used in this exercise as a vehicle for easily varying

FIGURE 11.15
Simulating the motion of a pendulum—front panel

Integration scheme: Euler
You can replace this with Runge-Kutta using the shortcut method

FIGURE 11.16
Simulating the motion of a pendulum—block diagram

specific pendulum parameters. This allows you to enter the pendulum model on the front panel and then to substitute numerical values for m, l, g, and c. The Substitute Variables VI can be found on the **Functions≫Mathematics≫Scripts & Formulas≫Formula Parsing** palette.

When the VI is ready to accept inputs, enter the right side of the pendulum model in the F(X,t) control. You will enter x_2 and $-(g/l)\sin x_1 - (c/ml)x_2$. For the substitution rules use the following:

- In element 0 of the Substitution Rules control, type in
  - parameter: m
  - value: 0.4
- In element 1 of the Substitution Rules control, type in
  - parameter: l
  - value: 0.3
- In element 2 of the Substitution Rules control, type in
  - parameter: g
  - value: 9.8
- In element 3 of the Substitution Rules control, type in
  - parameter: c
  - value: 0.2

Run the VI and observe the waveform on the graph display. You should see a nice, stable response damping out around 6 seconds.

On the front panel, change the h (step rate) from 0.01 to 0.1. Run the VI. Did you detect a problem? The system response is no longer stable! Nothing has changed with the physical model, so this must be due to the Euler integration scheme. Now, switch to the block diagram and right-click the ODE Euler Method VI and **Replace** it with the ODE Runge–Kutta 4th-Order VI. Both VIs have the same inputs and outputs. Run the VI. What happens? The expected smooth, stable response is obtained. This demonstrates the benefit of the Runge–Kutta method over the Euler method—you can run with larger time steps and obtain more accurate solutions.

Back on the front panel, switch the time step back to h (step rate) = 0.01. Run the VI and verify that the results remain essentially the same. Investigate the effect that varying the pendulum mass has on the response.

When you are finished, save the VI in User's Stuff and call it Pendulum.vi.

A working version of Pendulum.vi can be found in the Chapter 11 folder in the Learning directory. If you run this VI, make sure to verify that all the input parameters are correct! ◆

◆ The Pendulum with MathScript

The pendulum problem considered in the previous example can also be solved using MathScript (see Chapter 3). The ordinary differential equation solver in MathScript based on the Runge–Kutta method requires that the system of differential equations be described in a user-defined function in a predefined format. In this example we will employ the well-known ode45 function, which is a one-step solver implementing the Runge–Kutta (4,5) formula. For help with ode45 open the MathScript Interactive Window and enter help ode45 in the Command Window.

The ode45 function is used with the following syntax:

$$[t, y] = ode45(\text{`function'}, times, y0)$$

where the inputs and outputs are defined as follows:

- **Inputs**

    □ function - Specifies the name of the function that describes the ODE system. The function must have the following form:

    $$function\ dy = fun(times, y)$$

    where fun is a string.

    □ times - Specifies either the endpoints for the time span, such as [0 20], or the time points at which to approximate the y-values, such as 0:20. The input times must be a strictly monotonic, real, double-precision vector.

□ y0 - Specifies the *y*-value at the starting time. The input y0 is a real, double-precision vector.

- **Outputs**

    □ t - Returns the time values at which LabVIEW evaluates the *y*-values in a vector

    □ y - Returns the *y*-values that LabVIEW computes. The output y is a real matrix.

To begin, open the MathScript Interactive Window. Navigate to the Script Editor window where you will define the pendulum system using the same differential equations describing x_1 and x_2 considered in the previous example. Using the script in Figure 11.17 as a guide, enter the script in the Script Editor and when done **Save** the script to pendulum_system.m. See Section 3.5.1 for a review of user-defined functions.

It is suggested that you save pendulum_system.m in the LabVIEW Data folder so that MathScript can readily locate it. If needed, look back into Section 3.5 for a review of specifying path names. Since the ode45 solver calls the user-defined function pendulum_system.m as a source for the differential equations, it must be able to locate the file.

Again using Figure 11.17 as a guide, in the Script Editor window construct the calling script. The calling script calls the ode45 integrator which in turn uses the pendulum_system.m script. To visualize the pendulum motion, add a call

FIGURE 11.17
Using MathScript to simulate the motion of a pendulum

to the plot function in the script and graph the output vector y as a function of t. The output vector y contains the time-history of x_1 and x_2. Save the calling script as pendulum.m.

Run the pendulum.m script by selecting **Run** at the bottom of the Script Editor window. A plot will be produced as illustrated in Figure 11.17.

A working version of pendulum_system.m *and* pendulum.m *can be found in the* **Chapter 11** *folder in the* **Learning** *directory.*   ◆

11.4   FINDING ZEROES OF FUNCTIONS

LabVIEW provides seven VIs that can be used to compute zeroes of functions. The VIs can be used to determine the zeroes of general functions of the following form:

$$f(x, y) = 0$$
$$g(x, y) = 0$$

For example, you could use the symbolic VIs to find the zeroes of $\sin(x) + \cos(x)$ in the range $-10 \le x \le 10$.

Very often in the course of studying mathematics, engineering, business, and science it is necessary to compute the *zeroes* of a polynomial. An nth-order polynomial has the form

$$f(x) = x^n + a_{n-1}x^{n-1} + a_{n-2}x^{n-2} + \cdots + a_1x + a_0$$

The zeroes of the polynomial are the values of x such that $f(x) = 0$. The zeroes are also known as the *roots* of the polynomial. For example, we discussed in previous sections that the characteristic equation associated with an $n \times n$ matrix **A** is an nth-order polynomial, and the zeroes of the characteristic equation are the eigenvalues of the matrix. Eigenvalues can be real and imaginary. But in this discussion, when we talk about zeroes of a function, we mean the real roots.

LabVIEW VIs are shown in Figure 11.18. There are seven VIs:

- **Symbolic Formula**:
  - Find All Zeroes of $f(x)$: This VI determines all the zeroes of a 1D function in a specified interval.
  - Newton–Raphson Zero Finder: Uses derivatives to assist in determining a zero of a 1D function in a specified interval.
  - Ridders Zero Finder: This VI computes a zero of a function in a given interval, but the function must be continuous and when evaluated at the edges of the interval, the function must have different signs.

FIGURE 11.18
VIs for finding zeroes of functions

- **Nonlinear Systems**:
  - Nonlinear System Single Solution: Computes the zeroes of a nonlinear function, where an approximation is provided as input.
  - Nonlinear System Solver: Computes the zeroes of a set of nonlinear functions.
- **Polynomial**:
  - Polynomial Real Zero Counter: Determines the number of real zeroes of a polynomial in an interval without actually computing the zeroes.
  - Complex Polynomial Roots: Finds the complex roots of a complex polynomial.

Finding Roots of a Polynomial Function

The objective of this exercise is to build a VI to compute the roots of a polynomial function. Construct a VI using the front panel and block diagram shown in Figure 11.19 as a guide. Use Polynomial Roots.vi to compute the roots. By default, this VI uses the Simple Classification option (if not wired). This means that both real and complex roots are estimated and returned. If your polynomial has repeated roots, you may need to consider wiring the **option** input of the Polynomial Roots VI with a more advanced classification (see the **Context Help** for more information).

FIGURE 11.19
Computing the roots of a polynomial

The main input to the Polynomial Roots VI is a 1D array input composed of the coefficients of the polynomial. For example, if the polynomial whose roots you want to compute is given by

$$P(s) = 1 + 2x - 4x^2 + 9x^3 - x^4 = 0,$$

then the corresponding input array is

$$[1 \quad 2 \quad -4 \quad 9 \quad -1].$$

When your VI is ready to accept inputs, enter the following coefficients: [120 154 71 14 1]. Run the VI. Where are the roots? You should determine them to be $x = -5, -4, -3,$ and -2.

When you are finished, save the VI as roots.vi. in the Users Stuff folder in the Learning directory.

A working version of **roots.vi** *can be found in the* **Chapter 11** *folder in the* **Learning** *directory.*   ◆

11.5   INTEGRATION AND DIFFERENTIATION

LabVIEW VIs for integration and differentiation are shown in Figure 11.20. The figure shows that there are many other VIs available for working with 1D functions—unfortunately we cannot cover all the analysis capabilities of Lab-VIEW in this book.

FIGURE 11.20
VIs for integration and differentiation

A straightforward implementation of the integration and differentiation VIs is shown in Figure 11.21. You can find the VI shown in Figure 11.21 in the Chapter 11 folder within the Learning directory—it is called Derivative and Integration.vi. Open and run the VI. You will find that three plots appear on the graph: the function, the derivative of the function, and a plot of the integral of the function. The plot of the derivative of the sine function should appear as a cosine function. Use the graph palette zoom tool to zoom in on the derivative plot. If you run the VI in **Run Continuously** mode, you can vary the parameter # of sample and watch the sine function become smoother as the number increases and, conversely, become less smooth as the number of points decreases.

11.6   SIGNAL GENERATION

You can use LabVIEW to generate signals for testing algorithms and other purposes when real-world signals are not available. These signals can be generated using mathematical equations, arrays of data points, and by using signal generation VIs for common signals. To create signals that represent real-world signals you often need to accurately control the signal characteristics (such as magnitude, frequency and phase of periodic signals, and so on). In this section we discuss some of the possibilities for generating signals using VIs. The discussion begins by introducing normalized frequency, the discrete-time terminology used when generating periodic signals.

Select signal type

You may want to increase
this number to "smooth"
out the plots

Click on
**Signal
Processing**

Select
**Signal
Generation**

FIGURE 11.21
A VI for integration and differentiation of a 1D function

11.6.1   Normalized Frequency

In the analog or real world, signals vary continuously with time. When we analyze them in a computer we digitize them, changing them from a continuous signal to a sampled, discrete-time signal. For this reason, in the digital signal world (and with many Signal Generation VIs) we often use the so-called **digital frequency** or **normalized frequency** (in units of cycles/sample) to describe the frequency (1/period) of a signal. This is computed as

$$f = \text{digital frequency} = \frac{\text{analog frequency}}{\text{sampling frequency}}.$$

The analog frequency is generally measured in units of Hz (or cycles per second) and the sampling frequency in units of samples per second. The normalized frequency is assumed to range from 0.0 to 1.0 corresponding to a frequency range of 0 to the sampling frequency, denoted by f_s. The normalized frequency wraps around 1.0, so that a normalized frequency of 1.2 is equivalent to 0.2. As an example, a signal sampled at the **Nyquist frequency** (that is, at $f_s/2$) is sampled twice per cycle (that is, two samples/cycle). This corresponds to a normalized frequency of 1/2 cycles/sample = 0.5 cycles/sample. Therefore, we see that the reciprocal of the normalized frequency yields the number of times that the signal is sampled in one cycle (more on sampling in Section 11.7.4).

The following VIs utilize frequencies given in normalized units:

1. Sine Wave

2. Square Wave

3. Sawtooth Wave

4. Triangle Wave

5. Arbitrary Wave

6. Chirp Pattern

When using these VIs, you will need to convert the frequency units given in the problem to the normalized frequency units of cycles/sample. The VI depicted in Figure 11.22 illustrates how to generate two cycles of a sine wave and then convert cycles to cycles/sample.

In the example shown in Figure 11.22, the number of cycles (2) is divided by the number of samples (50), resulting in a normalized frequency of $f = 2/50$ cycles/sample. This implies that it takes 50 samples to generate two cycles of the sine wave. What if the problem specifies the frequency in units of Hz (cycles/sec)? In this case, if you divide the frequency in Hz (cycles/sec) by the sampling rate given in Hz (samples/sec), you obtain units of cycles/sample:

$$\frac{\text{cycles/sec}}{\text{samples/sec}} = \frac{\text{cycles}}{\text{sample}}.$$

The illustration in Figure 11.23 shows a VI used to generate a 60 Hz sine signal and to compute the normalized frequency when the input is in Hz. The normalized frequency is found by dividing the frequency of 60 Hz by the sampling rate of 1,000 Hz to get the normalized frequency of $f = 0.06$ cycles/sample:

$$f = \frac{60}{1,000} = 0.06 \, \frac{\text{cycles}}{\text{sample}}.$$

Therefore, we see that it takes almost 17 samples to generate one cycle of the sine wave. The number 17 comes from computing the reciprocal of $f = 0.06$.

FIGURE 11.22
Generating two cycles of a sine wave and converting cycles to cycles/sample

FIGURE 11.23
Generating a 60 Hz sine wave and computing normalized frequency

Normalized Frequencies

Open the VI called Normalized Frequency.vi located in the **Chapter 11** folder in the **Learning** directory. The front panel and block diagram are shown in Figure 11.24. You can use this VI to experiment with calculating the normalized frequency when the input is in cycles and in Hz. Make sure that the **f type** is selected as Hz and run the VI. You should find that the normalized frequency is $f = 0.01$ with the default VI input values.

Manually compute the normalized frequency for **f type** = Hz, **frequency** = 10, and **sampling rate (Hz)** = 1,000. Modify the VI input parameters accordingly and verify that you obtain the same answer as by hand. The answer should be $f = 0.01$, computed as

$$f = \frac{10}{1,000} = 0.01 \ \frac{\text{cycles}}{\text{sample}}.$$

FIGURE 11.24
A VI to compute the normalized frequency

You can also work in cycles (rather than Hz) by selecting **f type** to be cycles. Then, in this situation, the normalized frequency is computed as a ratio of **frequency** to **number of samples.**   ◆

11.6.2   Wave and Pattern VIs

The basic difference in the operation of the Wave or Pattern VIs is how signal timing is specified and tracked inside the VI. Wave VIs track phase internally generating continuous signals based on normalized frequency and on the number of samples. Pattern VIs generate signals based on cycles and do not track phase. You can distinguish between the two types of VIs by recognizing that the VI names contain either the word *wave* or *pattern*, as illustrated in Figure 11.25.

The Wave VIs operate with normalized frequencies in units of cycles/sample. The only Pattern VI that uses normalized units is the Chirp Pattern VI.

Since the Wave VIs can keep track of the phase internally, they allow the user to control the value of the initial phase. The **phase in** control specifies the initial phase (in degrees) of the first sample of the generated waveform and the **phase out** indicator specifies the phase of the next sample of the generated waveform. In addition, a **reset phase** control dictates whether or not the phase of the first sample generated when the wave VI is called is the phase specified at the **phase in** control, or whether it is the phase available at the **phase out** control when the VI last executed. A True value of reset phase sets the initial phase to **phase in**—a False value sets it to the value of **phase out** when the VI last executed.

FIGURE 11.25
Signal generation VIs

**Practice
with
Signal
Generation**

In this exercise you will construct a VI that uses the two types of signal generation VIs. The front panel and block diagram for the VI are shown in Figure 11.26. The two main VIs used on the block diagram are the Sine Pattern.vi and Sine Wave.vi. These VIs are located on the **Signal Generation** palette, as shown in Figure 11.25. The VI also calculates and displays the normalized frequency.

Construct the VI using Figure 11.26 as a guide and run the VI when it is ready. Vary the VI inputs, paying particular attention to the effect that changing

FIGURE 11.26
Using the Sine Wave and Sine Pattern VIs to construct a signal generation VI

the control **phase in (degrees)** has on the Sine Pattern waveform graph. You should notice that the waveform begins to shift left and right as you vary the phase up and down. When you set the **reset phase** switch to True (On), the initial phase is reset to the value specified by **phase in** each time the VI is called in the loop; otherwise, the initial phase is set to the previous phase output. While the VI is running, set the **reset phase** switch to the On position. The sine wave should be rendered stationary, whereas with the reset button set to the Off position, the sine wave varies with the varying phase.

Stop the execution of the VI using the **Stop** button located at the bottom left of the VI front panel in Figure 11.26. When you are finished experimenting with the VI, save it as **Signal Generation.vi** in the **Users Stuff** folder in the **Learning** directory.

*You can find a working version of **Signal Generation.vi** in the **Chapter 11** folder in the **Learning** directory.* ◆

11.6.3   The Simulate Signal Express VI

The Simulate Signal Express VI is located on the **Signal Analysis** palette shown in Figure 11.11. This VI simulates sine waves, square waves, triangular waves, sawtooth waves, and noise signals. In Section 11.2.3, we used the Simulate Signal Express VI to simulate a noisy sine wave.

As with all Express VIs, to use the Simulate Signal Express VI, you first place it on the block diagram. As illustrated in Figure 11.27, once the Simulate Signal Express VI is placed on the block diagram, a dialog box automatically appears to configure the VI. In Figure 11.27, a sawtooth wave is selected. The **Frequency (Hz)** is set to 10.1 Hz, although in the forthcoming example we will wire a control to the corresponding frequency input to programmatically vary the frequency from the front panel. We do the same with the **Amplitude**, set to 1 in Figure 11.27, but ultimately wired to a control to the corresponding input to programmatically vary the amplitude. We do the same as well for the **Offset** variable. Once the Simulate Signal Express VI is configured as desired, click **OK** to return to the block diagram.

Practice with the Simulate Signal Express VI

Consider the VI shown in Figure 11.28. Simulate Signal Express Demo.vi was developed to assist in investigating various signals generated by the Simulate Signal Express VI. Notice that the Simulate Signal Express VI is the central element on the block diagram in Figure 11.28. The Express VI is configured to generate a sawtooth signal. The remaining elements on the block diagram are placed there to permit easy access to key parameters of the sawtooth signal: fre-

FIGURE 11.27
Configuring the Simulate Signal Express VI

quency, amplitude, and offset. Noise has also been added to the signal and the control Noise Strength has been wired to the Noise Standard Deviation input.

 The Simulate Signal Express Demo.vi is located in Chapter 11 of the Learning directory.

Run the Simulate Signal Express Demo VI and observe the results on the waveform graph. Vary the frequency, amplitude, and offset and examine the resulting impact on the sawtooth signal.

Vary key parameters: frequency, amplitude, and offset

Noise strength

Sawtooth signal.

Terminate program execution

Simulate Signal Express VI

FIGURE 11.28
The Simulate Signal Express Demo VI for investigating signal generation

Double-click the Simulate Signal Express VI to access the configuration dialog box. Select a different signal, such as the DC signal, and repeat the above investigation. What happens? You should find that the Broken Run indicator appears and there are broken wires on the block diagram. Remember that when you change the signal type, the inputs also change. In the case of a DC signal, it makes no sense to input a frequency; hence, the wire to that input generates an error.

11.7   SIGNAL PROCESSING

In this section we discuss three main topics: the Fourier transform (including the discrete Fourier transform and the fast Fourier transform), smoothing windows, and a brief overview of filtering.

11.7.1   The Fourier Transform

In Chapter 9 we covered the subject of data acquisition, where we discussed the fact that the samples of a measured signal obtained from the DAQ system are a time-domain representation of the signal, giving the amplitudes of the sampled signal at the sampling times. A significant amount of information is coded into the time-domain representation of a signal—maximum amplitude, maximum overshoot, time to settle to steady-state, and so on. The signal contains other useful information that becomes evident when the signal is transformed into the frequency domain. In other words, you may want to know the frequency content of a signal rather than the amplitudes of the individual samples. The representation of a signal in terms of its individual frequency components is known as the **frequency-domain representation** of the signal.

A common practical algorithm for transforming sampled signals from the time domain into the frequency domain is known as the **discrete Fourier transform**, or DFT. The relationship between the samples of a signal in the time domain and their representation in the frequency domain is established by the DFT. This process is illustrated in Figure 11.29.

If you apply the DFT to a time-domain signal represented by N samples of the signal, you will obtain a frequency-domain representation of the signal of length N. We denote the individual components of the DFT by $X(i)$. If the signal is sampled at the rate f_s Hz, and if you collect N samples, then you can compute the frequency resolution as $\Delta f = f_s/N$. This implies that the ith sample of the DFT occurs at a frequency of $i \Delta f$ Hz. We let the pth element $X(p)$ correspond to the Nyquist frequency. Regardless of whether the input signal is real or complex, the frequency-domain representation is always complex and contains two pieces of information—the amplitude and the phase.

For real-valued time-domain signals (denoted here by $X(i)$), the DFT is symmetric about the index $N/2$ with the following properties:

$$|X(i)| = |X(N - i)| \quad \text{and} \quad \text{phase}(X(i)) = -\text{phase}(X(N - i)).$$

The magnitude of $X(i)$ is **even symmetric**, that is, symmetric about the vertical axis. The phase of $X(i)$ is **odd symmetric**, that is, symmetric about the origin.

Time-domain representation of x[n]   Frequency-domain representation

FIGURE 11.29
The DFT establishes the relationship between the samples of a signal in the time domain and their representation in the frequency domain

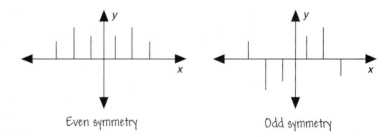

Even symmetry   Odd symmetry

FIGURE 11.30
Even and odd symmetric signals

This symmetry is illustrated in Figure 11.30. Since there is repetition of information contained in the N samples of the DFT (due to the symmetry properties), only half of the samples of the DFT need to be computed, since the other half can be obtained from symmetry.

Figure 11.31(a) depicts a *two-sided transform* for a complex sequence with $N = 8$ and $p = N/2 = 4$. Since $N/2$ is an integer, the DFT contains the Nyquist frequency. When N is odd, $N/2$ is not an integer, and thus there is no component at the Nyquist frequency. Figure 11.31(b) depicts a two-sided transform when $N = 7$.

The computationally intensive process of computing the DFT of a signal with N samples requires approximately N^2 complex operations. However, when N is a power of 2, that is, when

$$N = 2^m \quad \text{for} \quad m = 1 \text{ or } 2 \text{ or } 3 \text{ or } \cdots,$$

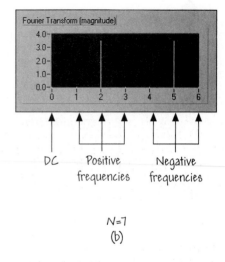

FIGURE 11.31
A two-sided transform representation of a complex sequence

FIGURE 11.32
The front panel of FFT_2sided.vi

you can implement the so-called **fast Fourier transforms** (FFTs), which require only approximately $N\log_2(N)$ operations. In other words, the FFT is an efficient algorithm for calculating the DFT when the number of samples (N) is a power of 2.

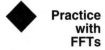

Practice with FFTs

In this example you will practice with FFTs by opening an existing VI and experimenting with the input parameters. Locate and open FFT_2sided.vi. You will find this VI in the Chapter 11 folder in the Learning directory. The front panel is shown in Figure 11.32.

The VI demonstrates how to use the FFT VI to analyze a sine wave of user-specified frequency. The VI block diagram, shown in Figure 11.33, contains three main VIs:

- FFT.vi: This VI computes the fast Fourier transform (FFT) or the discrete Fourier transform (DFT) of the input sequence. FFT.vi will execute FFT routines if the size of the input sequence is a power of 2. If the size of the input sequence is not a power of 2, then an efficient DFT routine is called.

- Sine Wave.vi: This VI generates an array containing a sine wave. The VI is located in **Functions≫Signal Processing≫Signal Generation** subpalette of the **Functions** palette.

FIGURE 11.33
The block diagram of FFT_2sided.vi

- **Complex To Polar.vi**: Separates a complex number into its polar components represented by magnitude and phase. The input can be a scalar number, a cluster of numbers, an array of numbers, or an array of clusters. In the case of Figure 11.33, the input is an array of numbers.

Run the VI and experiment with the input parameters. Run several numerical experiments using the **Run Continuously** mode. What happens to the spectrum when you vary the signal frequency? For example, set the input signal frequency to 10 Hz, the sampling frequency to 100 Hz, and the number of samples to 100. In this case, $\Delta f = 1$ Hz. The spectrum should have two corresponding peaks. Check this using the VI.

For another experiment, set the signal frequency to 50 Hz, the sampling frequency to 100 Hz, and the number of samples to 100, and run the VI. With the VI running in **Run Continuously** mode, set the sampling frequency to 101 and observe the effects on the time-domain sequence waveform and the corresponding spectrum. Now, slowly increase the sampling frequency and see what happens!

When you are finished, close the VI and do not save any changes.   ◆

11.7.2   Smoothing Windows

When using discrete Fourier transform methods to analyze a signal in the frequency domain, it is assumed that the available data of the time-domain signal represents at least a single period of a periodically repeating waveform. Unfortunately, in most realistic situations, the number of samples of a given time-domain signal available for DFT analysis is limited and this can sometimes lead to a phe-

FIGURE 11.34
A periodic waveform created from one period of a sampled waveform

nomenon known as **spectral leakage**. To see this, consider a periodic waveform created from one period of a sampled waveform, as illustrated in Figure 11.34.

The first period shown in Figure 11.34 is the sampled portion of the waveform. The sampled waveform is then repeated to produce the periodic waveform. Sampling a noninteger number of cycles of the waveform results in discontinuities between successive periods! These discontinuities induced by the process of creating a periodic waveform lead to very high frequencies (higher than the Nyquist frequency) in the spectrum of the signal—frequencies that were not present in the original signal. Therefore, the spectrum obtained with the DFT will not be the true spectrum of the original signal. In the frequency domain, it will appear as if the energy at one frequency has "leaked out" into all the other frequencies, leading to what is known as spectral leakage.

Figure 11.35 shows a sine wave and its corresponding Fourier transform. The sampled time-domain waveform is shown in Graph 1 in the upper left corner. In this case, the sampled waveform is an integer number of cycles of the original sine wave. The sampled waveform can be repeated in time, and a periodic version of the original waveform thereby constructed. The constructed periodic version of the waveform is depicted in Graph 2 in the upper middle section of Figure 11.35. The constructed periodic waveform does not have any discontinuities because the sampled waveform is an integer number of cycles of the original waveform. The corresponding spectral representation of the periodic waveform is shown in Graph 3. Because the time record in Graph 2 is periodic, contains no discontinuities, and is an accurate representation of the true waveform, the computed spectrum is correct.

In Figure 11.36 a spectral representation of another periodic waveform is shown. However, in this case a noninteger number of cycles of the original waveform is used to construct the periodic waveform, resulting in the discontinuities in the waveform shown in Graph 2. The corresponding spectrum is shown in Graph 3. The energy is now "spread" over a wide range of frequencies—compare this result to Graph 3 in Figure 11.35. The smearing of the energy is called spectral leakage, as mentioned earlier.

Original waveform

Periodic waveform constructed from a
complete cycle of the original waveform

Sine wave
frequency
spike

FIGURE 11.35
One complete period of a sine wave is repeated to obtain a periodic signal with no discontinuities; the corresponding Fourier transform shows no leakage

Discontinuity due to using a noninteger number of cycles
of the original waveform when constructiing the periodic waveform

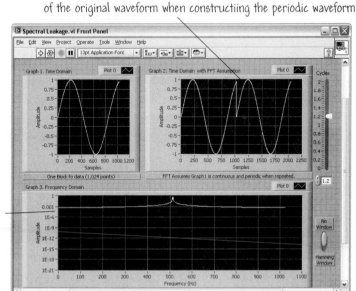

Leakage is evident
in this plot, where
the "clean" spike
shown in Figure 11.35
is now "washed" out

FIGURE 11.36
A portion of a sine wave period is repeated to obtain a periodic signal with discontinuities; the corresponding Fourier transform shows leakage

Leakage results from using only a finite time sample of the input signal. One (unpractical) solution to the leakage problem is to obtain an infinite time record, from $-\infty$ to $+\infty$, yielding an ideal FFT solution. In practice, however, we are limited to working with a finite time record. A practical approach to the problem of spectral leakage is the so-called **windowing** technique. Since the amount of spectral leakage depends on the amplitude of the discontinuity, the larger the discontinuity, the more the leakage. Windowing reduces the amplitude of the discontinuities at the boundaries of each period by multiplying the sampled original waveform by a finite length window whose amplitude varies smoothly and gradually towards zero at the edges.

One such windowing technique uses the *Hamming window*, as illustrated in Figure 11.37. The sinusoidal waveform of the windowed signal gradually tapers to zero at the ends—see the bottom graph in Figure 11.37. When computing the discrete Fourier transform on data of finite length, you can use the windowing technique applied to the sampled waveform to minimize the discontinuities of the constructed periodic waveform. This approach will minimize the spectral leakage.

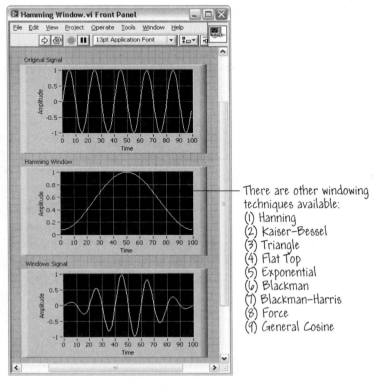

FIGURE 11.37
A sinusoidal signal windowed using a Hamming window

The Hamming Window

Open a new front panel and place four waveform graphs and one digital control, as shown in Figure 11.38.

Construct a block diagram using Figure 11.39 as a guide. In the block diagram, we use three main VIs:

- **Hamming Window.vi**: This VI applies the Hamming window to the input sequence. It is located in **Functions≫Signal Processing≫Windows**, as shown in Figure 11.39. If we denote the input sequence as **X** (with n elements) and the output sequence of the Hamming window as **Y**, then

$$\mathbf{Y}(i) = \mathbf{X}(i)\,[0.54 - 0.46\cos\omega]$$

where

$$\omega = \frac{2\pi i}{n}.$$

- The **Amplitude and Phase Spectrum.vi** computes the amplitude spectrum of the windowed and nonwindowed input waveforms. You can find this particular VI in the palette **Functions≫Signal Processing≫Spectral Analysis**, as shown in Figure 11.39.

Set value equal to noninteger value to examine the positive effects of the Hamming window

FIGURE 11.38
The front panel for a VI to investigate the use of windows

FIGURE 11.39
The block diagram for a VI to investigate the effect of windowing

- The Sine Pattern.vi generates a sine wave with the number of cycles specified in the control labeled **cycles**. It is located in **Functions≫Signal Processing≫Signal Generation**.

As interesting numerical experiment, make the following two runs:

- Set the control cycles to 10. Since this is an integer number, when you repeat the waveform to construct a periodic waveform, you will not have any discontinuities. You should observe that the spectrums of the windowed and the nonwindowed waveforms are both centered at 10, and that the spectrum of the original signal displays no spectral leakage.

- Set the control cycles to 10.25. Since this is not an integer, you should observe that the spectrums of the windowed and the nonwindowed waveforms are different. The nonwindowed spectrum should show distinct signs of spectral leakage due to the discontinuities when constructing the periodic waveform. The windowed waveform, while not a perfect spike centered at 10, displays significantly less leakage.

Save your VI as Windowed & Unwindowed Signal.vi in the Users Stuff folder in the Learning directory.

You can find a working version of Windowed & Unwindowed Signal.vi *in* Chapter 11 *of the* Learning *directory.*   ◆

11.7.3   The Spectral Measurements Express VI

The **Express»Signal Analysis** palette shown in Figure 11.11 contains the Spectral Measurements Express VI. This Express VI performs spectral measurements, such as spectral power density, on signals.

To use the Spectral Measurements VI, you first place it on the block diagram. As illustrated in Figure 11.40, once the Spectral Measurements Express VI is placed on the block diagram, a dialog box automatically appears to configure the VI. The spectral measurements options include choosing the measurement type, such as magnitude (peak or RMS), power spectrum, and power spectral density, and selecting the windowing technique from among Hanning, Hamming,

FIGURE 11.40
Configuring the Spectral Measurements Express VI

Blackman–Harris, Low Sidelobes, and more. In Figure 11.40, the magnitude RMS is selected as the measurement type and the Hanning window is selected for windowing. Once the Spectral Measurements VI is configured, click **OK** to return to the block diagram.

On the block diagram, you can expand the Express VI to display the inputs and outputs. The key input is the signal, and the key output is the spectral measurement, such as FFT-(RMS). The Spectral Measurements VI is now ready for inclusion in a VI for signal analysis.

Practice with the Spectral Measurements Express VI

Consider the VI shown in Figure 11.41. Notice on the block diagram that two Express VIs, the Spectral Measurements Express VI and the Simulate Signal Express VI, are utilized. This VI was developed to assist in investigating the power spectral density of a sine wave signal.

Original signal

Power spectral density of the windowed signal

Power spectral density of the sine wave

Generates a sine wave of frequency 10.25 Hz

Configured with a Hamming window

Configured with no windowing

FIGURE 11.41
The Spectral Express Demo VI for investigating the power spectral density of sine wave signals

Spectral Express Demo.vi is located in **Chapter 11** *of the* **Learning** *directory.*

There are four elements on the block diagram:

- Two Spectral Measurements Express VIs—these can be configured by double-clicking on their icons to access the configuration dialog boxes.

- The Simulate Signal Express VI—this VI was described in Section 11.6. It has been configured here to provide a sine wave signal at a frequency of 10.25 Hz.

- The waveform graphs—these are used to graphically display the power spectral density of the sine wave signal, both the windowed and unwindowed versions.

Run the Spectral Express Demo VI and observe the results on the waveform graphs. Double-click the Simulate Signal Express VI to access the configuration dialog box. Change the frequency of the sine wave signal to 50 Hz, run **Spectral Express Demo.vi**, and observe the effects on the power spectral density. You should observe the peak of the power spectral density shift to 50 Hz. Test the VI using different windowing methods. ◆

11.7.4   Filtering

There are two main types of filters: analog and digital. Analog filters can be built using analog components like resistors, capacitors, inductors, and amplifiers. Digital filters can be implemented in software, changed programmatically on the fly, are often more stable and predictable because they do not drift with changes in external environmental conditions, and generally have superior performance-to-cost ratios compared to their analog counterparts. In this section we will consider digital filters only.

LabVIEW can be used to implement digital filters and control digital filter parameters (such as filter order, cutoff frequency, stopband and passbands, amount of ripple, and stopband attenuation). You can envision a LabVIEW-based DAQ system wherein data is acquired from external sources (such as an accelerometer sensor) and filtered in the VI software. The results can be easily analyzed and studied using the graphics provided by LabVIEW and written to a spreadsheet. The key is that by using LabVIEW you can utilize digital filters, allowing the VIs to handle the design issues, computations, memory management, and the actual data filtering.

The theory of filters is a rich and interesting subject, and one that cannot be dealt with here in any depth. Please refer to other reference materials for in-depth coverage of filtering theory. A brief discussion of terms is needed, however, to give you a better understanding of the filter parameters and how they relate to the VI inputs.

The **sampling theorem** states that a continuous-time signal can be reconstructed from discrete, equally spaced samples if the sampling frequency is at least twice that of the highest frequency in the time signal. The sampling interval is often denoted by δt. The **sampling frequency** is computed as the inverse of the sampling interval:

$$f_s = \frac{1}{\delta t}.$$

Thus, according to the sampling theorem, the highest frequency that the filter can process—the **Nyquist frequency**—is

$$f_{nyq} = \frac{f_s}{2}.$$

As an example, suppose that you have a system with sampling interval $\delta t = 0.01$ second. Then the sampling frequency is $f_s = \frac{1}{0.01} = 100$ Hz. From the sampling theorem we find that the highest frequency that the system can process is $f_{nyq} = \frac{f_s}{2} = 50$ Hz. If we expect that the signals that we need to process have components at frequencies higher than 50 Hz, then we must upgrade the system to allow for shorter sampling intervals, say for example, $\delta t = 0.001$ second. What is f_{nyq} in this case?

One main use of filters is to remove unwanted noise from a signal—usually if the noise is at high frequencies. Depending on the frequency range of operation, filters either pass or attenuate input signal components. Filters can be classified into the following types:

1. A **lowpass filter** passes low frequencies and attenuates high frequencies. The ideal lowpass filter passes all frequencies below the *cutoff frequency* f_c.

2. A **highpass filter** passes high frequencies and attenuates low frequencies. The ideal highpass filter passes all frequencies above f_c.

3. A **bandpass filter** passes a specified band of frequencies. The ideal bandpass filter only passes all frequencies between f_{c1} and f_{c2}.

4. A **bandstop filter** attenuates a specified band of frequencies. The ideal bandstop filter attenuates all frequencies between f_{c1} and f_{c2}.

The ideal frequency response of these filters is illustrated in Figure 11.42.

The frequency points f_c, f_{c1}, and f_{c2} are known as the cutoff frequencies and can be viewed as filter design parameters. The range of frequencies that is passed through the filter is known as the **passband** of the filter. The gain of a filter at a specific frequency is typically represented in units of Decibels (dB), where dB is calculated as $10\,\mathrm{Log}_{10}\,(V_{out}/V_{in})$. The decibel is used to scale very large and very small numbers so that they can all be plotted on the same graph. For instance, when dB $= 0$ then $V_{out} = V_{in}$ indicating no gain. When dB $= 20$ then $V_{out} = V_{in} * 100$ indicating a gain of 100. An ideal

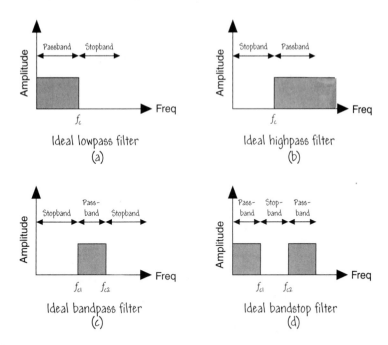

FIGURE 11.42
The ideal frequency response of common ideal filters

filter has a gain of one (0 dB) in the passband—that is, the amplitude of the output signal is the same as the amplitude of the input signal. Similarly, the ideal filter completely attenuates the signals in the stopband—that is, the stopband attenuation is $-\infty$ dB. The lowpass and highpass filters have one passband and one stopband. The range of frequencies that do not pass through the filter is known as the **stopband** of the filter. The stopband frequencies are rejected or attenuated by the filter. The passband(s) and the stopband(s) for the different types of filters are shown in Figure 11.42. The bandpass filter has one passband and two stopbands. Conversely, the bandstop filter has two passbands and one stopband.

Suppose you have a signal containing component frequencies of 1 Hz, 5 Hz, and 10 Hz. This input signal is passed separately through lowpass, high-pass, bandpass, and bandstop filters. The lowpass and highpass filters have a cutoff frequency of 3 Hz, and the bandpass and bandstop filters have cutoff frequencies of 3 Hz and 8 Hz. What frequency content of the signal can be expected? The output of the filter in each case is shown in Figure 11.43. The lowpass filter passes only the signal at 1 Hz because this is the only component of the input signal lower than the 3-Hz cutoff. Conversely, the highpass filter attenuates the 1-Hz component and passes a signal with components at 5 and 10 Hz. The bandpass filter passes only the signal component at 5 Hz, and the bandstop filter filters out the signal component at 5 Hz and passes a signal with frequency content at 1 and 10 Hz.

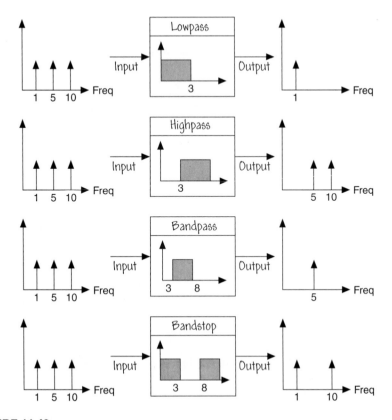

FIGURE 11.43
The output of the four filters in the case where the input signal contains component frequencies of 1 Hz, 5 Hz, and 10 Hz

Ideal filters are not achievable in practice. It is not possible to have a unit gain (0 dB) in the passband and a gain of zero ($-\infty$ dB) in the stopband—there is always a *transition region* between the passband and stopband. A more realistic filter will have the passband, stopband, and transition bands as depicted in Figure 11.44. As you can see, a real filter does not always have a unit gain (0 dB) across the passband and a gain of zero ($-\infty$ dB) across the stopband. The variation in the passband is called the *passband ripple*; see the bandpass filter in Figure 11.44. The *stopband attenuation*, also depicted in Figure 11.44, cannot be infinite (as it would be for an ideal filter).

If we view the filter as a linear system, then we can consider the response of the system (that is, the filter) to various types of inputs.[1] One interesting input is the *impulse*. If the input to the digital filter is the sequence $x(0), x(1), x(2), \ldots,$

---

1. A good source of information on systems and system response to various inputs is *Modern Control Systems*, by Richard C. Dorf and Robert H. Bishop, Prentice Hall, Upper Saddle River, NJ, 2005.

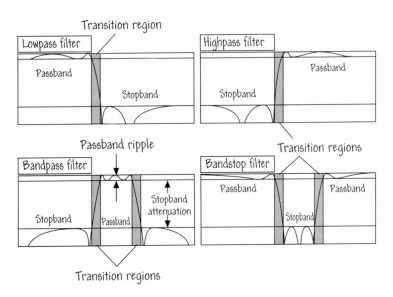

FIGURE 11.44
A realistic frequency response of a filter

then the impulse is given by $x(0) = 1, x(1) = x(2) = \cdots = 0$. The **impulse response** of a filter (that is, the output of the filter when the input is an impulse) provides another classification system for filters. The Fourier transform of the impulse response is known as the **frequency response**. The frequency response of a system provides a wealth of information about the system, including how it will respond to periodic inputs at different frequencies. Therefore, the frequency response will tell us about the filter characteristics: How does the filter respond in the passbands and stopbands? How accurately does the filter cut off and attenuate high-frequency components?

All digital filters can be classified into one of two types based on their impulse response, as either **finite impulse response** (FIR) filters or **infinite impulse response** (IIR) filter. For an IIR filter, the impulse response continues indefinitely (in theory), and the output depends on current and past values of the input signal and on past values of the output. In practical applications, the impulse response for stable IIR filters decays to near zero in a finite time. For an FIR filter the impulse response decays to zero in a finite time, and the output depends only on current and past values of the input signal. Take the example of processing noisy range measurements (that is, distance measurements) to a fixed target. Suppose that you want to determine the distance to the fixed object and you have available a ranging device (e.g., a laser ranging device) that is corrupted by random noise (as is the case for most realistic sensors!). One way to estimate the range is to take a series of range measurements $x(0), x(1), x(2), \ldots, x(k)$ and filter them by computing a running average:

$$x_{ave}(k) = \frac{1}{k} \sum_{i=1}^{k} x(i).$$

The output of the filter is $x_{ave}(k)$. This is an FIR filter because the output depends only on previous values of the input $(x(0), x(1), x(2), \ldots, x(k-1))$ and on the current value of the input $(x(k))$. Now, we can rewrite the filter as

$$x_{ave}(k) = \frac{k-1}{k} x_{ave}(k-1) + \frac{1}{k} x(k).$$

This is an IIR filter because the output depends on current and previous values of the input as well as on previous values of the output (that is, the $x_{ave}(k-1)$ term). Mathematically, the two filters provide the same output, but they are implemented differently. The FIR filter is sometimes referred to as a *nonrecursive* filter; the IIR filter is known as a *recursive* filter.

One disadvantage of IIR filters is that the phase response is nonlinear. You should use FIR filters for situations where a linear phase response is needed. A strong advantage of IIR filters is that they are recursive, thus reducing the memory storage requirements.

Some of the most common digital filters, such as Butterworth and Chebyshev, share their name with an analog counterpart because they share many of the same passband and stopband characteristics. Other filters are unique digital filters and are not typically used in an analog implementation. One such well-known filter, the Kalman filter, can be classified as an IIR filter (it is actually a bit more complicated to implement than an IIR filter) and was used successfully to filter navigation data acquired by the Apollo spacecraft to find its way to the moon and to rendezvous around the moon for the long journey home.[2] LabVIEW provides many different types of filters, as shown in Figure 11.45.

The IIR filter VIs available in LabVIEW include the following:

- Butterworth: Provides a smooth response at all frequencies and a mono-tonic decrease from the specified cutoff frequencies. Butterworth filters are maximally flat—the ideal response of unity in the passband and zero in the stopband—but do not always provide a good approximation of the ideal filter response because of the slow rolloff between the passband and the stopband.

- Chebyshev: Minimizes the peak error in the passband by accounting for the maximum absolute value of the difference between the ideal filter and the filter response you want (the maximum tolerable error in the passband). The frequency response characteristics of Chebyshev filters have an equiripple

---

FIGURE 11.45
IIR and FIR filter choices are located in the **Functions** palette

magnitude response in the passband, a monotonically decreasing magnitude response in the stopband, and a sharper rolloff than for Butterworth filters.

- Inverse Chebyshev: Also known as Chebyshev II filters. They are similar to Chebyshev filters, except that inverse Chebyshev filters distribute the error over the stopband (as opposed to the passband) and are maximally flat in the passband (as opposed to the stopband). Inverse Chebyshev filters minimize peak error in the stopband by accounting for the maximum absolute value of the difference between the ideal filter and the filter response you want. The frequency response characteristics have an equiripple magnitude response in the stopband, a monotonically decreasing magnitude response in the passband, and a rolloff sharper than for Butterworth filters.

- Elliptic: Minimize the peak error by distributing it over the passband and the stopband. Equiripples in the passband and the stopband characterize the magnitude response of elliptic filters. Compared with the same-order Butterworth or Chebyshev filters, the elliptic design provides the sharpest transition between the passband and the stopband. For this reason, elliptic filters are widely used.

- Bessel: Can be used to reduce nonlinear phase distortion inherent in all IIR filters. Bessel filters have maximally flat response in both magnitude and phase. Furthermore, the phase response in the passband of Bessel filters, which is the region of interest, is nearly linear. Like Butterworth filters, Bessel filters require high-order filters to minimize the error.

The FIR filter VIs available in LabVIEW include the following:

- Windowed: The simplest method for designing linear-phase FIR filters is the window design method. You select the type of windowed FIR filter you want—lowpass, highpass, bandpass, or bandstop—via input to the FIR Windowed Filter VI.

- Optimum filters based on the Parks–McClellan algorithm: Offers an optimum FIR filter design technique that attempts to design the best filter possible for a given filter complexity. Such a design reduces the adverse effects at the cutoff frequencies. It also offers more control over the approximation errors in different frequency bands—control that is not possible with the window method. The VIs available include

  - Equiripple Lowpass
  - Equiripple Highpass
  - Equiripple Bandpass
  - Equiripple Bandstop

Which filter is best suited for your application? Obviously, the choice of filter depends on the problem at hand. Figure 11.46 shows a flowchart that can serve as a guide for selecting the best filter for your needs. Keep in mind that you will probably use the flowchart to determine several candidate filters, and you will have to experiment to make the final choice.

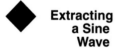
Extracting a Sine Wave

Open a new front panel and construct a front panel similar to the one shown in Figure 11.47. You will need to place one numeric control, two vertical slides, and two waveform graphs. Label them according to the scheme shown in the figure.

Construct a block diagram similar to the one shown in Figure 11.47. In the block diagram, we use three main VIs:

- Butterworth.vi: This VI is used to filter the noise. It is located in **Functions≫Signal Processing≫Filters**, as shown in Figure 11.45.

- Uniform White Noise.vi generates a white noise that is added to the sinusoidal signal. You can find this particular VI in the **Functions≫Signal Processing≫Signal Generation** palette.

- Sine Pattern.vi generates a sine wave of the desired frequency. It is located in the **Functions≫Signal Processing≫Signal Generation** palette.

With this VI you are generating 10 cycles of a sine wave (this value can be varied on the front panel), and there are 1,000 samples. Select a cutoff frequency of 25 Hz and a filter order of five. Note that we did not previously discuss filter order—it is a measure of filter complexity and is related to the number of terms retained in the filter. Run the VI. Vary the cutoff frequency and observe the

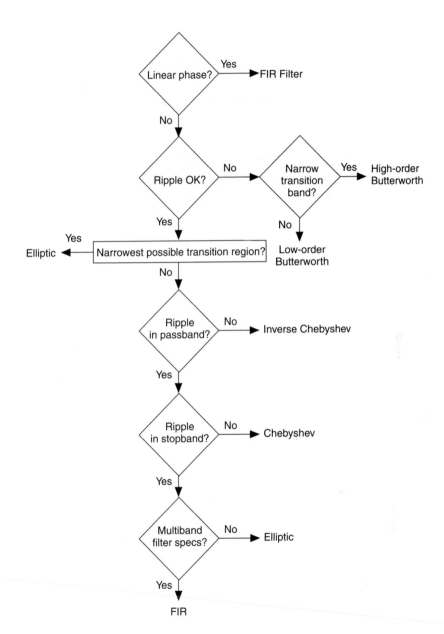

FIGURE 11.46
Flowchart that can serve as a guide for selecting the best filter

FIGURE 11.47
The front panel and block diagram for a VI to filter a noisy sine wave

effects. What happens when the cutoff frequency is set to 50? Does the filtered signal contain noise components? When you are finished exploring, save your VI as Extract the Sine Wave.vi in the Users Stuff folder.

You can find a working version of the VI in folder Chapter 11 in the Learning directory. It is called Extract the Sine Wave.vi.

◆

11.7.5   The Filter Express VI

The Filter Express VI is located on the **Express≫Signal Analysis** palette, shown in Figure 11.11. The Filter Express VI processes signals through filters and windows. The filter options include lowpass, highpass, bandpass, bandstop, and smoothing. To use the Filter Express VI, you first place it on the block diagram. As illustrated in Figure 11.48, once the Filter Express VI is placed on the block diagram, the **Configure Filter** dialog box automatically appears to configure the VI. Figure 11.48 shows a lowpass filter selected with a cutoff frequency of 100 Hz. Once the Filter Express VI is configured, click **OK** to return to the block diagram. The Filter Express VI is now ready for inclusion in a VI for signal analysis. On the block diagram, you can expand the Filter Express VI to display the inputs and outputs. The key input is the signal and the key output is the filtered signal.

FIGURE 11.48
Configuring the Filter Express VI

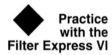
Practice with the Filter Express VI

Consider the VI shown in Figure 11.49. Notice on the block diagram that three Express VIs are used: the Filter Express VI, the Simulate Signal Express VI, and the Spectral Measurements Express VI. The Filter Express Demo VI was developed to assist in investigating the lowpass filtering of a sinusoidal signal.

FIGURE 11.49
The Filter Express Demo VI for investigating the filtering of a sinusoidal signal containing three main frequencies at 1, 10, and 100 Hz

Filter Express Demo.vi is located in Chapter 11 *of the* Learning *directory.*

The four Simulate Signal Express VIs have been configured to provide sine wave signals at frequencies of 1, 10, and 100 Hz and a DC signal of magnitude 10. The sinusoidal signal that is to be filtered is a sum of the four individual signals from the Simulate Signal Express VIs. The Filter Express VI has been configured to provide a lowpass filter with default cutoff frequency of 25 Hz. A numeric control is employed to vary the cutoff frequency directly from the front panel. The Spectral Measurements Express VIs appearing in Figure 11.49 have been configured to provide power spectral densities of the incoming signals. The waveform graphs display the power spectral density of the filtered and unfiltered sinusoidal signals.

Run the Filter Express Demo VI and observe the results on the waveform graphs. When the cutoff frequency is set to the default value of 25 Hz, you should observe that the filtered signal has a significantly reduced frequency component at 100 Hz. This is because the lowpass filter has essentially removed the 100 Hz component of the sinusoidal signal.

You can investigate the effect of varying the lowpass filter cutoff frequency on the filtering process. The numeric control located on the front panel will be employed to vary the lowpass filter cutoff frequency. Start the VI by pressing the **Run Continuously** button. Slowly reduce the cutoff frequency and observe the power spectral density graph of the filtered signal (lower right-hand side of the block diagram). Notice that as the cutoff frequency is reduced to a value under 10 Hz, the component of the filtered sinusoidal signal at 10 Hz is correspondingly reduced.

Experiment with increasing the lowpass filter cutoff frequency above 100 Hz. Can you find a value of the cutoff frequency for which the filtered signal power spectral density shows the 100 Hz component?   ◆

11.7.6   The LabVIEW Digital Filter Design Toolkit

The National Instruments Digital Filter Design Toolkit extends LabVIEW with functions (LabVIEW VIs that install into the palette) and interactive tools for design, analysis, and implementation of digital filters. Users that are new to digital filters can leverage interactive design tools to rapidly explore classical designs, while experienced users will find a breadth of algorithms, filter topologies, and analysis tools for design, modeling, and implementation of both fixed- and floating-point digital filters.

FIGURE 11.50
Digital Filter Design Utility

An extensive set of nearly 80 examples can serve as a starting point for your floating- or fixed-point designs. For fixed-point filters, you can model quantization effects and automatically generate ANSI C and LabVIEW/FPGA code for DSP or FPGA deployment.

Visit the Learning with LabVIEW companion website for a link to the free Interactive Filter Design Utility (see Figure 11.50) created in LabVIEW using the Digital Filter Design Toolkit.

BUILDING BLOCK

11.8 BUILDING BLOCKS: PULSE WIDTH MODULATION

LabVIEW's signal generation VIs can be used to simulate real-world signals (as described in Section 11.6). In this Building Block exercise, you will use the Square Wave VI to construct a VI that outputs a pulse width modulated waveform. Your VI will have a similar output to the pulse width modulation VIs in previous Building Blocks, but will be considerably less complicated in its implementation. As you become increasingly familiar with LabVIEW and its capabilities you will find simpler and more efficient ways to reach design objectives. Generally, the objectives of a VI can be met with a variety of solutions. Each solution comes with its own set of advantages and disadvantages.

Open a new VI and save it as **PWM with Square Wave VI.vi** in the **Users Stuff** folder in the **Learning** directory. The front panel and block diagram shown in Figure 11.51 can be used as a guide to developing your VI. The objective of the VI is to output a Square Wave using a vertical slide to adjust the duty cycle. The Square Wave VI is found on the **Signal Processing»Waveform Generation** palette. The Reciprocal function is found on the **Programming»Numeric** palette. One key difference between the PWM with Square Wave VI and the pulse width modulation VIs in previous Building Blocks is that with each loop iteration, this new VI generates an entire output cycle rather than one individual point.

Add the While Loop, Wait (ms) function, and other elements of the VI depicted in Figure 11.51. Once the VI is properly wired and ready to run, set the **Duty Cycle** = 50, and the **Period** = 100. Run the VI and observe the outputs displayed in the waveform chart. What happens when you set the **Period** = 10? When you are finished, push the **Stop** button.

FIGURE 11.51
The PMW with Square Wave VI

Using the Spectral Measurements Express VI and a waveform graph can you describe how the values of **Period** and **Duty Cycle** affect the frequency content of the signal? (Hint: The frequency is equal to 1 over the period.)

A working version of **PWM with Square Wave VI.vi** *is found in the* **Building Blocks** *folder of the* **Learning** *directory.*

11.9   RELAXED READING: MOTORCYCLE BRAKE TESTING

A tier-one supplier to large motorbike manufacturers in India wanted an on-road data acquisition system to test motorbike front disk brakes. The brakes were to be tested by measuring stopping distance during various road trials. The most straightforward way to determine stopping distance is to bring the motorbike to the required speed before reaching a certain predetermined marked point, and then to apply the brake from that point until the motorbike stops. Using a distance-measuring instrument, the stopping distance can be determined. This is a tedious and time-consuming process and not practical during long trials over different terrain. Additionally, in this method, other parameters relating to braking, such as the brake-pad temperature and hydraulic pressure, cannot be obtained.

An on-road brake test system was designed to measure the braking distance of a motorbike under different road conditions in real time. This data was then used offline to compute the performance parameters of the disc brakes. The test required a dedicated onboard computer to run the system and to acquire the test data. The test rig had to be compact enough to fit on the motorbike. A test system based on a small form factor, rugged FieldWorks computer with a National Instruments data acquisition card, signal-conditioning modules, and application software written in LabVIEW was constructed to meet the requirements.

The following sensors were used in the data acquisition system:

1. **Rotation encoder**—Measure the rotational displacement and the wheel speed.

2. **Thermocouple**—Measure the pad temperature during braking.

3. **Pressure sensor**—Measure the brake hydraulic line pressure.

4. **Accelerometer**—Measure the instantaneous acceleration of the vehicle.

For user inputs and displays, a small-sized integrated keyboard and LCD display was interfaced with the FieldWorks computer through the RS-232 port. Using this display, the motorbike rider could send commands, set parameters, and receive feedback from the system, such as the braking distance. The computer, user-interface kit, and signal-conditioning modules were all powered by the motorbike battery. The conditioned signals from the transducers connected to the National Instruments PCMCIA multipurpose high-speed data acquisition

FIGURE 11.52
On-Road Brake Test Review Data Screen

card installed in the FieldWorks computer. Three analog input channels and two counters/timers from the DAQ card were used for the data acquisition.

The rotational encoder and the accelerometer were both used to acquire the data used in calculating the braking distance. Two complementary and independent methods were used to calculate the stopping distance. The first method used the pulses per revolution output of the rotational encoder to measure both the speed and the displacement. The accelerometer data was used as supplementary data.

The data acquired by the various sensors was stored in the onboard computer. At the end of the test, the onboard computer could be connected to an Ethernet network and the data downloaded onto another PC for offline analysis. A software application was developed for the offline data review and analysis. This application included a review module that used to review the test data and view graphs that plot various parameters against time, such as pressure, acceleration, speed, and brake-pad temperature.

With the user-interface kit and integrated LCD display and keyboard, the user could enter various speeds at which to start braking, and the LabVIEW application monitored the motorbike speed and gave a signal to start braking at the desired speed range. When the bike came to a stop, the display immediately showed the braking distance. During the braking, the brake-pad temperature and acceleration were also acquired using the high-speed data acquisition card.

The on-road brake test system measured braking distance to an accuracy of 1% over a distance of 30 m. The small form factor of the NI PCMCIA data acquisition cards and the FieldWorks 2000 embedded computer made it easy to mount them on the motorbike. With the user interface, the user could set up different tests during the road trials and also see the results without having to connect a laptop to the embedded computer. The whole system was rugged enough to handle accidental falls during testing.

For more information, contact:

V. Arunachalam or Gokul DassT. V.
Soliton Automation Pvt. Ltd.
info@solitonautomation.com

11.10   SUMMARY

LabVIEW provides a great computational environment for analysis of signals and systems. In this chapter, we presented some applications of the many VIs available for analysis of signals, systems, functions, and systems of equations. The material was intended to motivate you to look further into the capabilities of LabVIEW in developing your own VIs for solving linear algebraic systems of equations, curve fitting, integrating ordinary differential equations, computing function zeroes, computing derivatives of functions, integrating functions, generating and analyzing signals, computing discrete Fourier transforms, and filtering signals.

KEY TERMS

Bandpass filter: A system that passes a specified band of frequencies.

Bandstop filter: A system that attenuates a specified band of frequencies.

Condition number: A quantity used in assessing the accuracy of solutions to systems of linear algebraic equations. The condition number can vary from 0 to ∞—a matrix with a condition number near 1 is closer to being nonsingular than a matrix with a very large condition number.

Determinant: A characteristic number associated with an $n \times n$ square matrix that is computed as a function of the minors and cofactors of the matrix. When the determinant is identically equal to zero, we say that the matrix is singular.

Digital frequency: Computed as the analog frequency divided by the sampling frequency. Also known as the **normalized frequency**.

Discrete Fourier transform (DFT): A common practical algorithm for transforming sampled signals from the time domain into the frequency domain.

Even symmetric signal: A signal symmetric about the y-axis.

Fast Fourier transform (FFT): The FFT is a fast algorithm for calculating the DFT when the number of samples (N) is a power of 2.

FIR filter: A finite impulse response filter in which the impulse response decays to zero in a finite time and the output depends only on current and past values of the input signal.

Frequency-domain representation: The representation of a signal in terms of its individual frequency components.

Frequency response: The Fourier transform of the impulse response.

Highpass filter: A system that passes high frequencies and attenuates low frequencies.

Homogeneous differential equations: A differential equation that has no input driving function (that is, $u(t) = 0$).

IIR filter: An infinite impulse response filter in which the impulse response continues indefinitely (in theory) and the output depends on current and past values of the input signal and on past values of the output.

Impulse response: The output of a system (e.g., a filter) when the input is an impulse.

Lowpass filter: A system that passes low frequencies and attenuates high frequencies.

Nyquist frequency: The highest frequency that a filter can process, according to the sampling theorem.

Odd symmetric signal: A signal symmetric about the origin.

Passband: The range of frequencies that is passed through a filter.

Sampling theorem: The statement that a continuous-time signal can be reconstructed from discrete, equally spaced samples if the sampling frequency is at least twice that of the highest frequency in the time signal.

Spectral leakage: A phenomenon that occurs when you sample a noninteger number of cycles, leading to "artificial" discontinuities in the signal that manifest themselves as very high frequencies in the DFT/FFT spectrum, appearing as if the energy at one frequency has "leaked out" into all the other frequencies.

Stopband: The range of frequencies that do not pass through a filter.

Trace: The sum of the diagonal elements of an $n \times n$ square matrix.

Windowing: A method used to reduce the amplitude of sampled signal discontinuities by multiplying the sampled original waveform by a finite-length window whose amplitude varies smoothly and graduates towards zero at the edges.

EXERCISES

E11.1   (a) Open a new VI and place a While Loop on the block diagram by going to **Functions≫Programming≫Structures**.

(b) Within the While Loop, place the Simulate Signal Express VI on the block diagram by navigating to **Functions≫ProgrammingExpress≫Input**. The **Configuration Window** will open in LabVIEW by default, but if it does not, double-click on the Simulate Signal Express VI to view it.

(c) In the **Configuration Window**, alter the **Signal Type** (i.e., Sine, Square, Triangle, Sawtooth, DC) and the additional inputs associated with the signal (i.e., Frequency, Amplitude, Phase, etc.). Notice that the **Result Preview** graph changes according to these controls.

(d) Add noise to the signal by adding a check mark to the box **Add noise** and view how the various types of noise affect the signal.

(e) Feel free to experiment with the other options in the **Configuration Window** to create the simulated signal of your choice. Press the **OK** button to return to the block diagram of your VI.

(f) Right-click on the signal output of your Simulate Signal Express VI and select Create≫Graph Indicator. Wire a boolean data source (such as a button) to the Conditional Terminal of the While Loop.

(g) Run the VI and confirm that the signal that you configured in the Simulate Signal Express VI matches the signal displayed on the front panel of your VI.

(h) Stop the VI and go back to the **Configuration Window** to make changes to the simulated signal in order to observe other types of signals that can be generated.

E11.2 Open a new VI and go to Help≫Find Examples. . . .. Click on the **Search** tab, type "express" into the string labeled **Enter keyword(s)**, and press the **Search** button.

(a) Open Express Comparison.vi, which shows how to compare values using the Express VI. Using the Value A and Value B controls, you can modify the signals. If Value A is greater than Value B, the result will be indicated by a 1 on the graph. If Value A is less than Value B, the result will be indicated by a 0.

(b) Open Express Vibration Lab.vi, which simulates a vibratory system of two masses and springs. The parameters of the system, such as the mass and rigidity of the springs, can be changed. Refer to **Context Help** for a more detailed explanation of this example VI, and explore the block diagram.

E11.3 Navigate to the windxmpl.llb in the analysis subfolder in the examples folder in the LabVIEW 8 directory and open the Window Comparison.vi. Observe the effects of the native LabVIEW window algorithms, including Hanning, Hamming, Triangle, and describe in your own words your observations.

E11.4 Navigate to the fltrxmpl.llb in the analysis subfolder in the examples folder in the LabVIEW 8 directory and open the IIR Filter Design.vi. Observe the effects of the native LabVIEW IIR filters, including Elliptic, Bessel, Butterworth, and describe in your own words your observations.

PROBLEMS

P11.1 Create a VI that generates a sine wave, which is displayed on a waveform graph. The user should have the ability to change the frequency and the amplitude of this sine wave programmatically. A power spectrum analysis should be performed on the sine wave and displayed on a waveform graph.

P11.2 Continue your analysis on the sine wave that is being generated from your solution to Problem 11.1. Determine the positive peak value of the signal and the root mean square value of the signal, and display these numeric values on the front panel.

Use a Hanning window and a Flat Top window to perform the spectral analysis and display both on the same waveform graph.

Hint: Explore the Express VIs by navigating to **Functions≫Express≫Signal Analysis**.

P11.3   Continue your analysis on the sine wave that is being generated from your solution to Problem 11.2. Add uniform white noise with noise amplitude of 3 to the sine wave being generated. Filter the noisy sine wave through a lowpass Butterworth filter (Order = 4), and display the filtered sine wave to compare the effect the filter has on the noise.

DESIGN PROBLEMS

D11.1   Consider the following linear, time-invariant system:

$$\dot{\mathbf{x}} = \mathbf{Ax} + \mathbf{Bu}$$

where

$$\mathbf{A} = \begin{bmatrix} k_0 & 1 \\ -1 & -1 \end{bmatrix} \text{ and } \mathbf{B} = \begin{bmatrix} 0 \\ 1 \end{bmatrix}.$$

The vector $\mathbf{x} = [\ x_1 \quad x_2\ ]^T$ is known as the state vector. The parameter k_0 is a constant. The input to the system is $\mathbf{u}$. Develop a script using MathScript to perform the following computations:

(a) For $0 \le k_0 \le 5$, compute the eigenvalues of $\mathbf{A}$. The eigenvalues can be complex (imaginary) numbers. Generate a plot of the real part of the eigenvalues versus the imaginary part.

(b) With the input $\mathbf{u} = 2$ for $t \ge 0$ and $k = 0.5$, find the solution using numerical integration and plot $\mathbf{x}$ versus t for $0 \le t \le 20$. Use the initial conditions $\mathbf{x}(0) = [\ 1 \quad 0\ ]^T$.

D11.2   Design a VI to serve as a filter for sinusoidal input signals. The VI should separate the low frequency and high frequency components of the input signal. Make the filter cutoff frequency an input to the VI via a front panel control. Plot the original signal and its components on a waveform graph. Use Simulate Signal Express VIs to generate the input signals for testing.

APPENDIX A

Instrument Control

This appendix introduces the concept of communicating with and controlling external instruments. In this appendix we focus on the GPIB (General Purpose Interface Bus) and RS-232 (a serial interface bus standard). A brief introduction to the main components of an instrument control system is given. The application of the **Measurement & Automation Explorer (MAX)** in detecting instruments and installing instrument drivers is discussed. The notion of an instrument driver is an important topic woven throughout the appendix. The Instrument I/O Assistant is also introduced.

GOALS

1. Learn about GPIB and serial instrument control.

2. Understand how to interact with your instruments using MAX.

3. Gain some experience with instrument drivers (for the AGILENT 34401 multimeter).

4. Gain experience with the Instrument I/O Assistant.

A.1   COMPONENTS OF AN INSTRUMENT CONTROL SYSTEM

LabVIEW communicates with and controls external instruments (such as oscil-loscopes and digital multimeters) using GPIB (General Purpose Interface Bus), RS-232 (a serial interface bus standard), VXI (VME eXtensions for Instrumentation), Ethernet, USB, and other hardware standards. In this appendix we focus on GPIB and serial communication.

A.1.1   What Is GPIB?

Hewlett Packard developed the General Purpose Interface Bus (or **GPIB**) standard in the late 1960s to interconnect and control its line of programmable instruments. The interface bus was originally called HP-IB. In this context, a **bus** is the means by which computers and instruments transfer data. National Instruments made GPIB available to users of non-Hewlett-Packard equipment.

At the time it was developed, GPIB provided a much-needed specification and protocol to govern this communication. Figure A.1 shows a typical GPIB system. While the GPIB is one way to bring data into a computer, it is fundamentally different from data acquisition with boards that plug into the computer. Using a special protocol, GPIB brings data that has been acquired by another computer or instrument into the computer using a "handshake," while data acquisition involves connecting a signal directly to a DAQ device in the computer.

The original purpose of GPIB was to provide computer control of test and measurement instruments. GPIB was soon applied to intercomputer communication and control of scanners, film recorders, and other peripherals because of its 1 Mbyte/sec maximum data transfer rates. The Institute of Electrical and Electronic Engineers (**IEEE**) standardized the GPIB in 1975, and it became accepted as IEEE Standard 488-1975. The standard has since evolved into ANSI/IEEE Standard 488.2-1987. The GPIB functions for LabVIEW follow the IEEE 488.2 specification. The terms GPIB, HP-IB, and IEEE 488 are synonymous.

FIGURE A.1
A typical GPIB system

A.1.2   GPIB Messages

GPIB carries two types of messages:

- **Device-dependent messages** contain device-specific information such as programming instructions, measurement results, machine status, and data files. These are often called **data messages**.

- **Interface messages** manage the bus itself, and perform such tasks as initializing the bus, addressing and unaddressing devices, and setting device modes for remote or local programming. These are often called **command messages**.

Physically, GPIB is a digital, 24-conductor, parallel bus. It comprises 16 signal lines and 8 ground-return lines. The GPIB connector is depicted in Figure A.2. The 16 signal lines are divided into three groups:

- **Eight data lines**: The eight data lines (denoted DIO1 through DIO8) carry both data and command messages. GPIB uses an eight-bit parallel, asynchronous data transfer scheme where whole bytes are sequentially handshaked across the bus at a speed determined by the slowest participant in the transfer. Because the GPIB sends data in bytes (1 byte = 8 bits), the messages transferred are frequently encoded as ASCII character strings. All commands and most data use the 7-bit ASCII or International Standards Organization (ISO) code set, in which case the eighth bit, DIO8, is unused or is used for parity.

- **Three handshake lines**: The three handshake lines asynchronously control the transfer of message bytes among devices. These lines guarantee that message bytes on the data lines are sent and received without transmission error.

- **Five interface management lines**: The five interface management lines manage the flow of information across the interface from device to computer.

 *The next generation in high-speed GPIB, proposed by National Instruments, is known as **HS488**. Speed increases up to 8 Mbyte/sec have been achieved by removing the propagation delays associated with the three handshake lines.*

A.1.3   GPIB Devices and Configurations

You can have many GPIB devices (that is, many instruments and computers) connected to the same GPIB. Typical GPIB linear and star configurations are illustrated in Figure A.3. You can even have more than one GPIB device in your

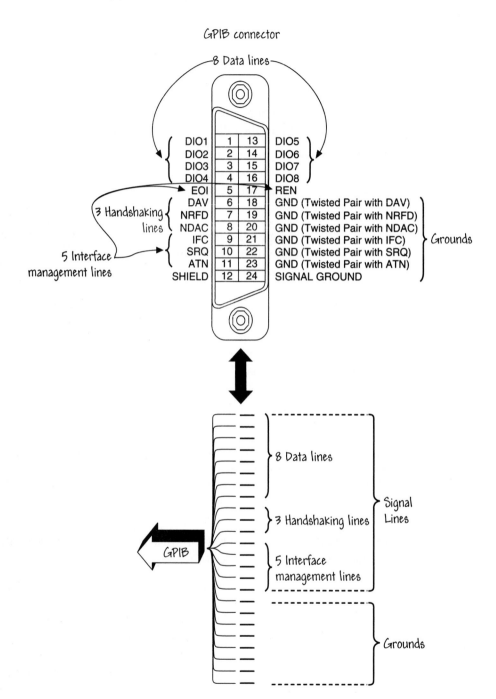

FIGURE A.2
The GPIB connector

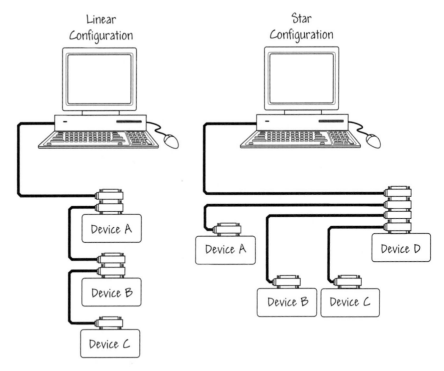

FIGURE A.3
Typical GPIB configurations

computer. A typical multiboard GPIB configuration is illustrated in Figure A.4. GPIB devices are grouped in three categories:

- **Talkers**: send data messages to one or more listeners.
- **Listeners**: receive data messages from the talker.
- **Controllers**: manage the flow of information on the GPIB by sending commands to all devices.

GPIB devices can fall into multiple categories. For example, a digital voltmeter can be both a talker and a listener.

The GPIB has one controller (usually a computer) that controls the bus. The role of the GPIB controller is similar to the role of a central processing unit in a computer. A good analogy is the switching center of a city telephone system. The communications network (the GPIB) is managed by the switching center (the controller). When a party (a GPIB device) wants to make a call (that is, to send a data message), the switching center (the controller) connects the caller (the talker) to the receiver (the listener). To transfer instrument commands and data on the bus, the controller addresses one talker and one or more listeners. The controller must address the talker and a listener before the talker can send its message to the listener. The data strings are then handshaked across the bus from

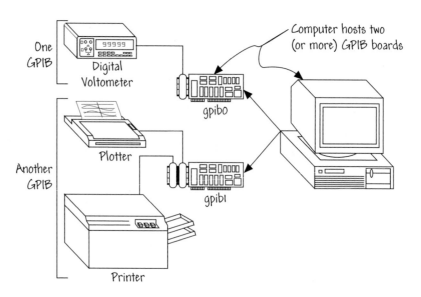

One
GPIB

Digital
Voltometer

Another
GPIB

Plotter

Printer

gpib0

gpib1

Computer hosts two
(or more) GPIB boards

FIGURE A.4
A typical multiboard GPIB configuration

the talker to the listener(s). After the talker transmits the message, the controller may unaddress both devices. LabVIEW provides VIs that automatically handle these GPIB functions.

Some bus configurations do not require a controller. For example, one device may always be a talker (called a talk-only device) and there may be one or more listen-only devices. A controller is necessary when you must change the active or addressed talker or listener. A computer usually handles the controller function. With the GPIB board and its software, the personal computer plays all three roles: controller, talker, and listener.

There can be multiple controllers on the GPIB but only one controller at a time is active. The active controller is called the **controller-in-charge** (CIC). Active control can be passed from the current CIC to an idle controller. Only the system controller (usually the GPIB board) can make itself the CIC.

With GPIB, the physical distance separating the GPIB devices matters. To achieve high-rate data transfers, we must accept a number of physical restrictions. Typical numbers are as follows:

- A maximum separation of 4 meters between any two devices and an average separation of 2 meters over the entire bus. For high-speed applications, you should have at least one device per meter of cable.

- A maximum total cable length of 20 meters. A maximum of 15 meters is desirable for high-speed applications.

- A maximum of 15 devices connected to each bus, with at least two-thirds of the devices powered on. For high-speed applications, all devices should be powered on.

Bus extenders and expanders can be used to increase the maximum length of the bus and the number of devices that can be connected to the bus. You can also communicate with GPIB instruments through a TCP/IP network. For more information, refer to the National Instruments website.

A.1.4   Serial Port Communication

Serial communication is another popular means of transmitting data between a computer and another computer, or between a computer and a peripheral device. However, unlike GPIB, a serial port can communicate with only one device, which can be limiting for some applications. Serial port communication is also very slow.

Most computers and many instruments have built-in serial port(s). Since serial communication uses the built-in serial port in your computer, you can send and receive data without buying any special hardware. Serial communication uses a transmitter to send data—one bit at a time—over a single communication line to a receiver. This method works well when sending or receiving data over long distances and when data transfer rates are low. It is slower and less reliable than GPIB, but you do not need a special board in your computer, and your instrument does not need to conform to the GPIB standard. Figure A.5 shows a typical serial communication system.

There are different serial port communication standards. Developed by the Electronic Industries Association (EIA) to specify the serial interface between equipment known as Data Terminal Equipment (DTE), such as modems and plotters, and Data Communications Equipment (DCE), such as computers and terminals, the RS-232 standard includes signal and voltage characteristics, connector characteristics, individual signal functions, and recipes for terminal-to-modem connections. The most common revision to this standard is the RS-232C used in connections between computers, printers, and modems.

FIGURE A.5
A typical serial communication system

FIGURE A.6
The RS-232 serial port connectors

The RS-232 serial port connectors come in two varieties: the 25-pin connector and the 9-pin connector. Both connectors are depicted in Figure A.6. The 9-pin connector has two data lines (denoted TxD and RxD in the figure) and five handshake lines (denoted RTS, CTS, DSR, DCD, and DTR in the figure). This compact connector is occasionally found on smaller RS-232 laboratory equipment and has enough pins for the "core" set used for most RS-232 interfaces. The 25-pin connector is the "standard" RS-232 connector with enough pins to cover all the signals specified in the RS-232 standard. Only the "core" set of pins is labeled in Figure A.6.

Serial communication requires that you specify four parameters: the baud rate of the transmission, the number of data bits encoding a character, the sense of the optional parity bit, and the number of stop bits. Each transmitted character is packaged in a character frame that consists of a single start bit followed by the data bits, the optional parity bit, and the stop bit or bits. The baud rate is a measure of how fast data is moving between instruments that use serial communication.

A.2 DETECTING AND CONFIGURING INSTRUMENTS

In this section NI-GPIB configuration for Windows, Mac OS X, and Linux will be discussed. National Instruments provides a wizard for helping you configure your NI-GPIB device.

A.2.1   Windows

With the **Measurement & Automation Explorer (MAX)** you can automatically detect connected instruments, install the required instrument drivers, and manage the instrument drivers already installed. The software architecture for GPIB instrument control using LabVIEW is similar to the architecture for DAQ (see Chapter 9.1). Figure A.7 shows the software architecture on the Windows platforms. Instrument drivers are LabVIEW applications written to control a specific instrument. More on instrument drivers in the next section. You can use the instrument drivers or parts of the instrument drivers to develop your own application quickly. You can also set naming aliases for your instruments for easier instrument access.

FIGURE A.7
Software architecture for Windows platforms

A.2.2   Macintosh OS X and Linux

If you are using a Macintosh, the NI-488.2 Configuration Utility is called the GPIB Explorer. To start the GPIB Explorer from the Finder, double-click on Applications≫National Instruments≫NI-488.2≫Explore GPIB. Macintosh OS automatically recognizes plug-and-play GPIB devices. You can view or modify the default configuration settings using the GBIB Explorer.

**Practice
with the
Instrument
Wizard**

In this example, you will use MAX to configure and test the GPIB interface. MAX interacts with the various diagnostic and configuration tools installed with the driver and also with the Windows Registry and Device Manager (see Figure A.7). The driver-level software is in the form of a dynamically linked library (DLL) and contains all the functions that directly communicate with the GPIB board. The LabVIEW Instrument I/O VIs and functions directly call the driver software.

You configure an object listed in MAX by right-clicking on the item and making a selection from the shortcut menu. Figure A.8 shows the GPIB interface board in the MAX utility and the results of pressing the **Scan For Instruments** button at the top of the window.

 The configuration utilities and hierarchy described in this example are specific to Windows platforms. If you are using a Macintosh or other operating system, refer to the manuals that came with your GPIB interface board for the appropriate information for configuring and testing that board.

As discussed in Chapter 9, MAX is a configuration utility for your software and hardware. MAX executes system diagnostics, adds new channels, interfaces, and tasks, and views devices and instruments connected to your system. Open MAX by double-clicking on its icon on the desktop or by selecting **Measurement & Automation Explorer** from the **Tools** menu of LabVIEW.

Select this option to enable the search for instruments

FIGURE A.8
Using MAX to search for instruments

Four possible selections in MAX are:

- **Data Neighborhood**—Use this selection to create virtual channels, aliases, and tags to your channels or measurements configured in Devices and Interfaces as you did in the DAQ chapter.

- **Devices and Interfaces**—Use this selection to configure resources and other physical properties of your devices and interfaces. Using this selection, you can view attributes of one or more devices, such as serial numbers.

- **Scales**—Use this selection to set up simple operations to perform on your data, such as scaling.

- **Software**—Use this selection to determine which drivers and application software are installed and their version numbers.

If you do not have an instrument attached to your computer, MAX will obviously not find any instruments. In the example in Figure A.9, the attached instrument is a National Instruments GPIB Device Simulator, denoted by **Instrument0** in MAX.

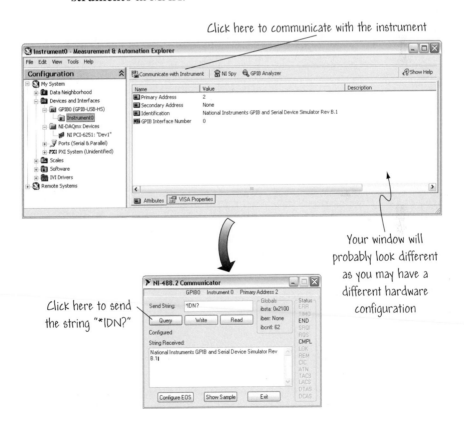

FIGURE A.9
A National Instruments GPIB device simulator is detected

Click **Communicate with Instrument** to communicate with the instrument, as shown in Figure A.9. By default, the string *IDN? appears in the command to send to this GPIB device. The string *IDN? is an IEEE 488.2 standard identification request to the instrument. Click **Query** to send the string. In this example the string is sent to the NI GPIB Device Simulator. The instrument replies with the string "National Instruments GPIB and Serial Device Simulator Rev B.1." The next time you launch MAX, it will automatically contain your stored instrument configuration.

You can also examine the properties of your GPIB board. In MAX, highlight the GPIB hardware and then choose **Properties**, as illustrated in Figure A.10. A window will open that provides all the information about your GPIB board.

FIGURE A.10
Examining the properties of the GPIB devices

A.3 USING THE INSTRUMENT I/O ASSISTANT

The Instrument I/O Assistant can be employed to establish communication with message-based instruments. You can communicate with an instrument that uses a serial, Ethernet, or GPIB interface. The Instrument I/O Assistant organizes instrument communication into ordered steps using a dialog box associated with

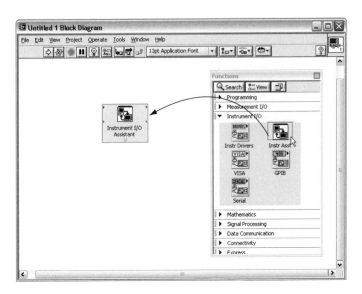

FIGURE A.11
Placing the Instrument I/O Assistant on the block diagram

the I/O Assistant. Like the DAQ Assistant (see Chapter 9), the Instrument I/O Assistant is readily configured to provide a communication link between the instrument and the computer. The Instrument I/O Assistant can be used whenever an instrument driver is not available.

To launch the Instrument I/O Assistant, place the Instrument I/O Assistant on the block diagram, as illustrated in Figure A.11. The Instrument I/O Assistant is located on the **Functions≫Instrument I/O** and **Functions≫Express≫Input** palettes. Once the Instrument I/O Assistant is placed on the block diagram, the Instrument I/O Assistant configuration dialog box will automatically appear. The dialog box is shown in Figure A.12.

If the Instrument I/O Assistant configuration dialog box does not appear, double-click the Instrument I/O Assistant icon.

The desired communication steps in the instrument communication process are specified in the Instrument I/O Assistant configuration dialog box. Once the configuration dialog box appears, the general procedure for configuring the Instrument I/O Assistant is as follows:

1. **Select an instrument.**
   Instruments that have been configured in MAX appear in the **Select an instrument** pull-down menu. In Figure A.12, we see the National Instruments GPIB Device Simulator that was configured in MAX.

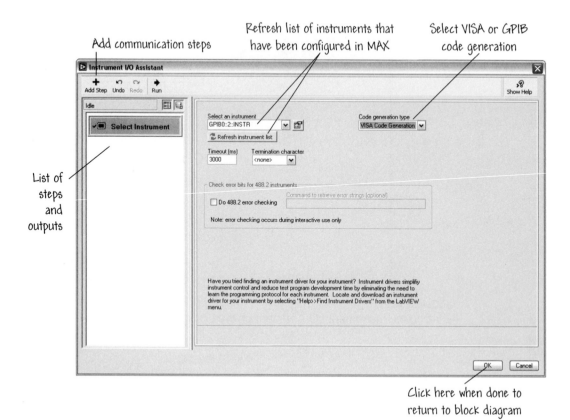

FIGURE A.12
The Instrument I/O Assistant dialog box

2. **Choose a Code generation type.**
   VISA code generation allows for more flexibility and modularity than GPIB code generation.

3. **Specify the communication steps.**
   Select from the following communication steps using the **Add Step** button:

   (a) **Query and Parse**—Sends a query to the instrument and parses the returned string.

   (b) **Write**—Sends a command to the instrument.

   (c) **Read and Parse**—Reads and parses data from the instrument.

   To use the Instrument I/O Assistant, you place the communication steps into a sequence. As you add steps to the sequence, they appear in the **Step Sequence** window (see Figure A.12).

4. **Test the communication sequence.**
   After adding the desired number of steps, click the **Run** button to test the sequence of communication that you have configured for the Express VI.

5. **Return to the block diagram and complete the VI.**

   Click the **OK** button to exit the Instrument I/O Assistant configuration dialog box. LabVIEW adds input and output terminals to the Instrument I/O Assistant on the block diagram that correspond to the data you will receive from the instrument.

Practice with the Instrument I/O Assistant

In this example, we will construct a VI that uses the Instrument I/O Assistant to communicate with a National Instruments GPIB Device Simulator. This device uses a GPIB interface.

To begin the VI development, open a blank VI and place the Instrument I/O Assistant (located on the **Functions≫Instrument I/O** palette) on the block diagram. The dialog box will appear to allow you to configure the Instrument I/O Assistant.

As previously discussed, the first step is to select an instrument. In the dialog box, make sure that the National Instruments GPIB Device Simulator is selected (denoted by GPIB0::2::INSTR in Figure A.12). You can click on the **Refresh instrument list** button to make sure all the available instruments are listed.

The second step is to choose a code generation type. Since VISA code generation allows for more flexibility and modularity than GPIB code generation, we will choose **VISA Code Generation** from the **Code generation type** pull-down menu.

We will add two communication steps using the **Query and Parse** sequence. To accomplish this, click the **Add Step** button, and in the dialog box that appears, select **Query and Parse** to write and read from the NI GPIB Device Simulator, as illustrated in Figure A.13. The Step Sequence window will update automatically to show the Query and Parse step. To test the communication link, type *IDN? as the command, and click the **Run this step** button. If no error warning appears in the lower half of the dialog box, this step has successfully completed. To parse the data received, click the **Auto parse** button. Notice that "**Token**" now appears in the **Outputs** pane on the left side of the dialog box. This value represents the string returned from the identification query—the name of the instrument.

Now, repeat the previous configuration to include a second communication step, as illustrated in Figure A.14. Click the **Add Step** button and select **Query and Parse**. Enter the command MEAS:DC? and click the **Run this step** button to test the communication. As before, to parse the data received, click the **Auto parse** button. The data returned is a random numeric value, which in this case is 4.8442 as shown in Figure A.14. This is a simulated voltage reading. The name "**Token2**" now appears in the **Outputs** pane on the left side of the dialog box.

Generally we should rename the outputs to more accurately represent the variables. Right-click on the output **Token** and select **Rename** to change the name of the output, as illustrated in Figure A.15. Rename **Token** to Voltage and rename **Token2** to ID String.

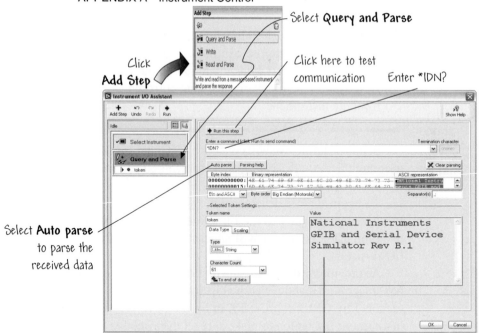

FIGURE A.13
Configuring the Instrument I/O Assistant

FIGURE A.14
Adding a second communication step

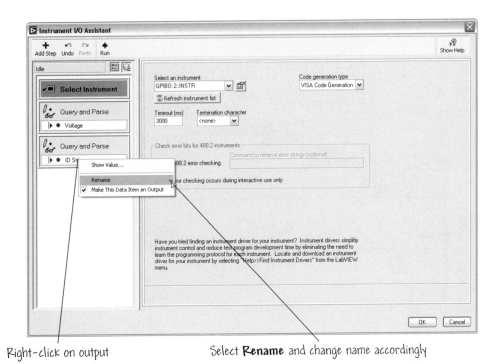

Right-click on output   Select **Rename** and change name accordingly

FIGURE A.15
Renaming the outputs

Click the **OK** button to exit the Instrument I/O Assistant and return to the block diagram. The configuration of the Instrument I/O Assistant is complete.

To view the code generated by the Instrument I/O Assistant, right-click the Instrument I/O Assistant icon and select **Open Front Panel** from the shortcut menu. When asked if you want to convert to a subVI select **Convert**. This converts the Express VI to a subVI. Switch to the block diagram to see the code generated. Figure A.16 shows the code generated by the Instrument I/O Assistant after completing the above steps.

Once an Express VI has been converted to a subVI, it cannot be converted back. If you changed an Express VI to a subVI without making changes, you can **Undo** *the conversion.*

The final step in the development of the VI is to wire appropriate controls and indicators on the block diagram. In our case, we need to wire an indicator to the **ID String** and another to the **Voltage** output. Right-click the **ID String** output and select Create≫Indicator from the shortcut menu. Similarly, right-click the **Voltage** output and select Create≫Indicator from the shortcut menu. The appropriate indicators will automatically wire to the outputs of the Instrument I/O Assistant, as illustrated in Figure A.17.

FIGURE A.16
Viewing the code generated by the Instrument I/O Assistant

FIGURE A.17
Completing the instrument VI

Now we finally test the VI. Display the front panel and select **Run**. You will most likely need to resize the string indicator to accommodate the instrument identification. As seen in Figure A.17, the NI GPIB Device Simulator returns the identification National Instruments GPIB and Serial Device Simulator Rev B.1, and on this test it returned a voltage reading of 9.907.

Save the VI as IO Assistant Demo.vi in the Users Stuff folder in the Learning directory. ◆

A.4 INSTRUMENT DRIVERS

An **instrument driver** is a piece of software (that is, a VI) that controls a particular instrument. Instrument drivers eliminate the need to learn the complex, low-level programming commands for each instrument. LabVIEW is ideally suited for creating instrument drivers since the VI front panel can simulate the operation of an instrument front panel. The block diagram sends the necessary commands to the instrument to perform the various operations specified on the front panel. When using an instrument driver, you do not need to remember the commands necessary to control the instrument—this is specified via input on the front panel.

LabVIEW provides many VIs that can be used in the development of an instrument driver for your hardware. These VIs can be grouped into the following categories:

- Standard VISA I/O functions

- Traditional GPIB functions and added capability via the GPIB 488.2

- Serial port communication functions

VISA stands for Virtual Instrument Software Architecture. In essence, VISA is a VI library for controlling GPIB, serial, Ethernet, USB, or VXI instruments and making the appropriate calls depending on the type of instrument. VISA by itself does not provide instrumentation programming capability—it is a high-level application programming interface (API) that calls lower-level code to control the hardware. Each VISA instrument driver VI corresponds to a programmatic operation, such as configuring, reading from, writing to, and triggering an instrument. The GPIB and serial port functions provide similar capabilities.

Two questions come to mind:

1. When should you attempt to develop your own instrument driver from "scratch" using the VISA, GPIB, or serial port functions?

And if you must develop your own instrument driver,

2. When should you use the VISA functions, and when should you use the GPIB functions?

The answer to the first question is simple—students new to LabVIEW should not attempt to develop their own instrument drivers! They should use the instrument drivers developed by National Instruments, rather than attempt to develop drivers from "scratch." Instrument drivers can be downloaded from the National Instruments website using the Instrument Driver Network. To access the driver network, connect to

http://www.ni.com/idnet

directly or use Tools≫Instrumentation≫Visit Instrument Driver Network.

In LabVIEW 8, instrument drivers can be found and installed using the Find Instrument Drivers wizard found in Tools≫Instrumentation≫Find Instrument Drivers. This new wizard allows you to automatically connect to the network, search for instrument drivers, view their rating, and install them, as illustrated in Figure A.18. If an instrument driver is not available, use the Instrument I/O Assistant.

FIGURE A.18
Link automatically to the National Instruments, Instrument Driver Network

The LabVIEW instrument driver library contains instrument drivers for a variety of programmable instruments that use the GPIB, serial, Ethernet, USB, or VXI interfaces. You can use an instrument driver from the library to control your instrument, or you can customize the instrument drivers, since they are distributed with their block diagram source code. Once you have properly installed the appropriate instrument drivers using MAX, you can access the instrument drivers on the **Functions** palette, as shown in Figure A.19.

If you have a web browser installed, you can link automatically to the National Instruments Instrument Driver Network, as shown in Figure A.18. The Instrument Driver Network provides the complete library of available instrument drivers for LabVIEW. On this web page, you can search for your instrument driver among over 4000 drivers with free source code available from more than 200 vendors. You then can download the instrument driver you need and install it into the **instr.lib** *folder in the LabVIEW root directory.*

FIGURE A.19
Locating the installed instrument drivers on the **Functions** palette

If you decide to build your own instrument driver, VISA is the standard API throughout the instrumentation industry. In addition, the VISA functions can control a suite of instruments of different types, including GPIB, serial, Ethernet, USB, or VXI. In other words, VISA provides interface independence. Students that need to program instruments for different interfaces only need to learn one API.

The Agilent 34401 Instrument Driver

In this example we will take a look at the Agilent 34401 instrument driver. The Agilent 34401 digital multimeter instrument driver is installed automatically as part of the *LabVIEW Student Edition*, from the Device Driver CD. Begin by first opening a new VI. Use Figure A.19 as a guide to locating the Agilent 34401 VI Tree. Drop the VI on the block diagram. Then double-click on the VI, which should open up the front panel. The value of this VI can be found on the block diagram, illustrated in Figure A.20. The VIs are organized in a Tree depicting the key VIs for performing various functions. A typical instrument driver application will include steps for initializing communication, configuring the instrument, passing data, and closing communication.

Open the NI Example Finder to see an example using these VIs. Navigate to the Example Finder by clicking Help≫Find Examples, then navi-

FIGURE A.20
The Agilent 34401 VI Tree

FIGURE A.21
The Agilent 34401 Read Single Measurement Vi front panel

gate through the folder structure to Hardware Input and Output≫Instrument Drivers≫LabVIEW Plug and Play and select Agilent 34401 Read Single Measurement.vi. The Instrument Driver Example VIs are used to verify communication with your instrument and test a typical programmatic instrument operation. The Agilent 34401 Read Single Measurement VI front panel is shown in Figure A.21.

With the exception of the address field, the defaults for most controls on the front panel will be sufficient for your first run. You will need to set the VISA resource name appropriately. Typically this is accomplished by clicking the down arrow on the control and selecting your device from the list. MAX populates the VISA resource name list box. If you do not know the VISA resource name, or it is not present in the list, refer to MAX for help. After running the VI, check to see that reasonable data was returned and an error was not reported in the error cluster (bottom right corner of the front panel).

If your Agilent 34401 Read Single Measurement VI does not work, you need to check that

- NI-VISA is installed. If you did not choose this as an option during your LabVIEW installation, you will need to install it from the Device Driver CD before rerunning your Agilent 34401 Read Single Measurement VI.

For more information on installing Device Drivers, visit the LabVIEW Student Edition Companion page at www.prenhall.com/bishop.

- The instrument VISA resource name is correct.
- The selected instrument driver supports the exact instrument model you are using.

Once you have verified basic communication with your instrument using the Agilent 34401 Read Single Measurement VI, you may want to customize instrument control for your particular needs. If your application needs are similar to the Agilent 34401 Read Single Measurement VI, the simplest means of creating a customized VI is to save a copy of the Agilent 34401 Read Single Measurement VI by selecting **Save As** from the **File** menu. You can change the default values on the front panel by selecting **Make Current Values Default** from the **Operate** menu.

The block diagram of an instrument driver generally consists of three VIs: the Initialize VI, an application function VI, and the Close VI. A simple instrument driver can be assembled from the VIs provided by LabVIEW for the Agilent 34401. The three VIs we need are Agilent 34401 Initialize.vi, Agilent 34401 Read Multiple Points.vi, and Agilent 34401 Close.vi. A straightforward instrument driver comprised of these three basic VIs is shown in Figure A.22. Build the block diagram shown in Figure A.22, if you have the Agilent 34401 or do not have an instrument. If you are using another instrument driver, build a VI similar to Figure A.22 using the Initialize, Read Points, and Close VIs for that instrument.

FIGURE A.22
A simple Agilent 34401 instrument driver

A working version of **Figure A.22.vi** *can be found in the* **Appendix A** *folder of the* **Learning** *directory.*

◆

A.4.1   Developing Your Own Instrument Driver

Make sure to check thoroughly for existing instrument drivers for your instrument before starting to build your own from scratch. You should check first at the National Instruments website as discussed in the previous sections. If that search fails, you should then search the website of the manufacturer of your equipment—they may have developed the instrument driver already. During your website searches, be on the lookout for instrument drivers that support a similar instrument, since instruments from the same model series often have similar command sets. If you find any such drivers, download them and assess the similarity of their command sets to that of your instrument—they may work directly or with minor modifications. For instruments from the same model series, you might need to contact the manufacturer and ask for details on the differences between the command sets.

If an instrument driver for your particular instrument does not exist, you can do one of several things (in rank order):

1. Use a driver for a similar instrument. Often similar instruments from the same manufacturer have similar if not identical command sets. The degree to which an instrument driver will need to be changed will depend on how similar the instruments and their command sets are. If the command sets are very different, you may be better off starting from scratch.

2. Use the Instrument I/O Assistant.

3. Develop a complete, fully functional instrument driver. To develop a National Instruments quality driver, you can download the **Application Note AN006** *Developing LabVIEW Plug and Play Instrument Drivers* from the National Instruments website at

http://www.ni.com/idnet

This application note will help you to develop a complete instrument driver. To aid in the development of your instrument driver, National Instruments created an Instrument Driver Project Wizard to help you implement standards for instrument driver structure, device management, instrument I/O, and error reporting. The Instrument Driver Project Wizard can be accessed by navigating to Tools≫Instrumentation≫Create Instrument Driver Project in LabVIEW. The **Application Note AN006** describes these standards, as well as the purpose of a LabVIEW instrument driver, its components, and the integration of these components. In addition, this application note suggests a process for developing useful instrument drivers.

As to the question of whether to use VISA, GPIB, or serial functions, the recommendation is to rely on VISA. There are four major reasons for using VISA VIs over GPIB VIs. VISA

- is the industry standard,
- provides interface independence,
- provides platform independence, and
- will be easily adaptable in future instrumentation control applications.

VISA uses the same operations to communicate with instruments regardless of the interface type. VISA is designed so that programs written using VISA function calls are easily portable from one interface type to another like GPIB, serial, Ethernet, USB, or VXI. To ensure platform independence, VISA strictly defines its own data types. The VISA function calls and their associated parameters are uniform across all platforms, so that software can be ported to other platforms and then recompiled. A LabVIEW program using VISA can be ported to any platform supporting LabVIEW. A final advantage of using VISA is that it is an object-oriented API that will easily adapt to new instrumentation interfaces as they are developed in the future, making application migration to the new interfaces easy.

 If you must develop your own instrument driver, you should use VISA functions rather than GPIB functions because of the versatility of VISA.

A.5   FUTURE OF INSTRUMENT DRIVERS AND INSTRUMENT CONTROL

Current instrument drivers use VISA, which is a common software interface for controlling GPIB, serial, Ethernet, USB, or VXI instruments. At the present time and into the future, there will be one instrument driver for all oscilloscopes, for instance, no matter the manufacturer, model, or hardware interface (such as GPIB). These new instrument drivers are called Interchangeable Virtual Instruments (IVI) drivers and are supported by the IVI Foundation. The IVI Foundation is comprised of end-user test engineers and system integrators with many years of experience building GPIB- and VXI-based test systems. By defining a standard instrument driver model that enables engineers to swap instruments without requiring software changes, the IVI Foundation members believe that significant savings in time and money will result because of:

- Software that does not change when instruments become obsolete.

- A single software application that can be used on a system with different instrument hardware, which maximizes existing resources.

- Portable code that can be developed in test labs and hosted on different instruments in the production environment.

The following instrument types or classes have been defined by the IVI Foundation: oscilloscope, DMM, arbitrary waveform generator, switch, and power supply. More instrument types will be defined in the future. For more information on the IVI drivers and the IVI Foundation, refer to the **Application Note AN121** *Using IVI Drivers to Build Hardware-Independent Test Systems with LabVIEW and LabWindows/CVI* on the National Instruments website, www.ni.com/appnotes.nsf. You may also want to visit the IVI Foundation website at

http://www.ivifoundation.org/

A.6 SUMMARY

As we already know from the chapter on data acquisition, LabVIEW can communicate with external devices. As discussed in Chapter 9, you can use DAQ devices in conjunction with LabVIEW to read and generate analog input, analog output, and digital signals. In this appendix we learned that LabVIEW can also control external instruments (such as digital voltmeters and oscilloscopes) over the GPIB bus or through a serial port. MAX is used to detect instruments and to install instrument drivers. An instrument driver is a VI that controls a particular instrument. Students should use the instrument drivers developed by National Instruments rather than attempt to develop drivers from "scratch." The Instrument I/O Assistant provides a relatively straightforward way to build instrument communication VIs. Instrument drivers can be downloaded from the National Instruments website using the Instrument Driver Network.

KEY TERMS

Bus: The means by which computers and instruments transfer data.

Controller: A GPIB device that manages the flow of information on the GPIB by sending commands to all devices. A computer usually handles the controller function.

Controller-in-charge: The active controller in a GPIB system.

Device-dependent messages: Messages that contain device-specific information such as programming instructions, measurement results, machine status, and data files. These are often called **data messages**.

GPIB: General Purpose Interface Bus is the common name for the communications interface system defined in ANSI/IEEE Standard 488-1975 and 488.2-1987.

IEEE: Institute for Electrical and Electronic Engineers.

Instrument driver: A set of LabVIEW VIs that communicates with an instrument using standard VISA I/O functions.

Interface messages: Messages that manage the bus itself and perform such tasks as initializing the bus, addressing and unaddressing devices, and setting device modes for remote or local programming. These are often called **command messages**.

Listener: A GPIB device that receives data messages from the talker.

MAX: A utility that provides information about connected instruments and installs and manages instrument drivers.

Serial communication: A popular means of transmitting data between a computer and a peripheral device by sending data one bit at a time over a single communication line.

Talker: A GPIB device that sends data messages to one or more listeners.

VISA: Virtual Instrument Software Architecture. VISA is a VI library for controlling GPIB, serial, Ethernet, USB, or VXI instruments.

APPENDIX B

Other LabVIEW Applications

LabVIEW 8.0 features and applications—additional toolkits and modules that expand the core functionality of LabVIEW—are presented in this appendix. The expanded functionality may be of interest after you master the core LabVIEW programming elements introduced earlier in the book. LabVIEW has evolved from a tool used primarily for computer-based data acquisition and instrument control to a powerful graphical system design platform for designing, simulating, prototyping, and deploying complex solutions to perform a variety of applications in areas such as signal and image processing, embedded design, instrumentation, communications, circuit design, control, and mechatronics. The new core features described in this chapter include improved support for built-in sound cards, a discussion of the LabVIEW Project, simplified distributed application development with the shared variable, toolkits, and modules.

GOALS

1. Understand how to acquire data with the sound card and when it is useful.

2. Understand when you might want to use a LabVIEW Project.

3. Know how shared variables can simplify network communication.

4. Know where to find the right toolkit for your application.

B.1   ACQUIRING DATA WITH THE SOUND CARD

Sound cards are standard equipment on most personal computers. The sound card, which is similar to an NI DAQ device, usually has one or two channels of analog-to-digital conversion and digital-to-analog conversion for acquiring signals from a microphone or sending signals to a speaker, respectively. Express VIs for sound card communication are one of the many improvements of the new sound card API. The new VIs utilize Microsoft ActiveX technology, giving users of LabVIEW 8 for Windows added functionality. The new LabVIEW sound VIs treat the sound card like a standard data acquisition device by passing data into and out of VIs as waveforms, increasing the number of possible sampling rates, and adding the ability to continuously read and output buffered signals.

When using the sound card on your personal computer, it is important to keep the following challenges in mind:

- Sound cards usually have a bandpass anti-aliasing filter that attenuates frequencies lower than 100 Hz and higher than 22 kHz.

- Sound card data is not read or generated in engineering units (volts). Instead, the data will be read/written as integers representing the binary result of the A/D or D/A conversion.

- Sound cards are commonly not calibrated, are prone to noise, and do not support timing or triggering functions, features that are all common to data acquisition devices.

- Sound cards are designed for microphones only; attaching other devices to the sound card can damage the sound card and your computer.

The LabVIEW **Sound Card API** can be found on the **Functions≫Programming≫Graphics & Sound≫Sound** palette as shown in Figure B.1.

For simple sound generation, use the Play Waveform Express VI found on the **Functions≫Programming≫Graphics & Sound≫Sound≫Output** palette; for simple sound acquisition, use the Acquire Sound Express VI, found on the **Functions≫Programming≫Graphics & Sound≫Sound≫Input** palette. Both Express VIs allow you to choose the sound card (if you have more than one installed on your PC) and allow you to test to make sure the device is working properly.

For more information and examples using the Sound Card API, navigate to the Sound folder of the NI Example Finder as illustrated in Figure B.2.

B.2   MANAGING PROGRAMS WITH THE LABVIEW PROJECT

The **LabVIEW Project** is introduced with LabVIEW 8 to help organize application development and is especially attractive for large application development with multiple developers. LabVIEW Project libraries are collections of VIs, type

FIGURE B.1
The sound card API is found on the **Functions≫Programming≫Graphics & Sound≫Sound** palette

FIGURE B.2
Examples using the sound card API in the NI Example Finder

definitions, shared variables, palette menu files, DAQ Task configurations, and other files, including other project libraries. LabVIEW Projects also help when managing multiple VIs running on multiple targets within one application.

Project libraries in LabVIEW have the file extension .lvlib, which includes the properties of the project library and references to the files that the project library owns. Application development features of the project library include organizing files into a single hierarchy of items, avoiding potential VI name

FIGURE B.3
Accessing the **Project Explorer** window

duplication, limiting public access to certain files, limiting editing permission for a collection of files, and setting a default palette menu for a group of VIs. You can view the structure of a project library from the **Project Explorer** window or in a stand-alone project library window. The **Project Explorer** window is depicted in Figure B.3.

Consider using the LabVIEW Project any time you need more than a single VI to complete your application. Projects are great for organizing subVIs, custom controls, and other LabVIEW native files, as well as third-party information like your custom .dll files, data sheets, or CAD files. For additional information on the LabVIEW Project, click on the LabVIEW Projects link on the Getting Started window of LabVIEW to access the LabVIEW Help files for the LabVIEW Project, as illustrated in Figure B.3.

B.3   SHARING DATA WITH SHARED VARIABLES

Shared variables are introduced in LabVIEW 8 to simplify complex applications that share data across a network. Sharing information is the fundamental, and often a critical, component of many applications. The shared variable solves many problems by creating a shared FIFO buffer striving for both data integrity and deterministic behavior. LabVIEW users can now harness the power of a network more easily when creating solutions. The shared variable is a part of a LabVIEW project and looks similar to a local or global variable on the block diagram.

To create a shared variable, follow these steps:

1. In the **Project Explorer** window (see Figure B.3), right-click a project library or a folder under a project library and select New≫Variable to display the Shared Variable Properties dialog box.

If you right-click a target or a folder that is not inside a project library and select New≫ Variable from the shortcut menu to create a shared variable, LabVIEW creates a new project library and places the shared variable inside.

2. Enter a name for the shared variable in the **Name** text box.

3. Select the data type for the shared variable from the **Data Type** pull-down list. If you do not see the data type you want to use, select **Custom** to use the data type of a custom control.

4. Select the type of shared variable you want to create from the **Variable Type** pull-down list. Select **Network-Published** from the Variable Type list to create shared variables that you want to read and write on remote computers and targets on the same network. Select **Single-Process** from the Variable Type list to create shared variables that you want to read and write on a single computer. Configuration options vary with the variable type you select. LabVIEW modules and toolkits you have installed might provide additional shared variable types, configuration options, and limitations. Refer to the documentation for the LabVIEW module or toolkit for more information about the shared variable types, configuration options, and limitations that the product adds to shared variables.

5. If you select Network-Published from the Variable Type list, place a check mark in the Use Buffering check box to enable client-side buffering for the shared variable data. Depending on the data type you select from the Data Type pull-down list, you might need to enter values in additional fields in the Use Buffering section.

You can configure buffering only for network-published shared variables. If a buffer for a network-published shared variable overflows, the data is lost, and LabVIEW does not return an error.

6. Place a check mark in the **Single Writer** check box to accept changes in value from only one target at a time.

7. Click the **OK** button to create the shared variable.

For more information on shared variables, search for the term "shared variables" in the LabVIEW Help, as shown in Figure B.4.

FIGURE B.4
Using LabVIEW Help to access information about shared variables

B.4   LABVIEW TOOLKITS

Toolkits are libraries of functions, virtual instruments, interactive wizards, examples, utilities, and documentation that fully integrate into your LabVIEW development environment and applications. These functions are designed and optimized for specific needs, ranging from advanced analysis and signal-processing algorithms to communication with databases and external applications.

You can reduce your development time for specialized application needs with LabVIEW add-on software. Toolset components are available for sound and vibration measurement, transient signal analysis, motion control, image processing, and digital filter design. You can also take advantage of the development of applications that use database communications, creating professional reports with standardized tools such as Microsoft Word and Excel, and implementing monitoring and message notification of remote applications using internet-based e-mail or browser technology.

For more information on available LabVIEW toolkits, visit http://www.ni.com /toolkits.

B.5   PROTOTYPING A MODEL-BASED CONTROL SOLUTION

The LabVIEW System Identification Toolkit, LabVIEW Control Design Toolkit, and LabVIEW Simulation Module integrate a suite of dynamic system and control tools into the LabVIEW environment. You can model linear, nonlinear, discrete, and continuous plant or control systems in block diagram form and create and simulate modern and classical control systems. The interactive nature of the LabVIEW tools allows you to modify parameters throughout the design process. Models can be built, simulated, tested, and seamlessly downloaded to a real-time target with the LabVIEW Real-Time Module for control prototyping and hardware-in-the-loop (HIL) simulation.

Model-based control design involves the following four phases:

1. Developing and analyzing a model to describe a plant,
2. Designing and analyzing a controller for the dynamic system,
3. Simulating the dynamic system, and
4. Deploying the controller.

Because **model-based control design** involves iterations, you might need to repeat one or more of these phases before the design is complete. Figure B.5 illustrates the LabVIEW tools for each of these phases. National Instruments also provides products for I/O and signal conditioning that you can use to gather and process data. Using these tools, which are built on the LabVIEW platform, you can experiment with different approaches at each phase in model-based control design and quickly identify the optimal design solution for a control system.

B.5.1   Developing a Plant Model

The first phase of model-based control design involves developing and analyzing a mathematical model of the plant you want to control. This process can be done mathematically by analyzing the dynamics of the system or through other mathematical properties. In some cases you can use a process called system

FIGURE B.5
The four phases of model-based control design using LabVIEW

identification to obtain and analyze this model. The system identification process involves acquiring data from a plant and then numerically analyzing stimulus and response data to estimate the parameters and order of the model.

The system identification process requires a combination of the following components:

- **Signal generation and data acquisition**—NI provides software and hardware that you can use to stimulate and measure the response of the plant.

- **Mathematical tools to model a dynamic system**—The LabVIEW System Identification Toolkit contains VIs to help you estimate and create accurate mathematical models of dynamic systems. You can use this toolkit to create discrete linear models of systems based on measured stimulus and response data.

B.5.2   Control Design

The second phase of model-based control design is the analysis of the plant model and the design of the controller. Control design is a process that involves developing mathematical models that describe a physical system, analyzing the models to learn about their dynamic characteristics, and creating a controller to achieve certain dynamic characteristics. Control systems contain components that direct, command, and regulate the physical system, also known as the plant. In this setting, the control system refers to the sensors, the controller, and the actuators. The reference input refers to a condition of the system that you specify. Figure B.6 illustrates the feedback-controlled system, including the controller, actuators, sensors, and the physical system.

The controlled system in Figure B.6 is known as a closed-loop system (also referred to as a feedback system). In closed-loop systems, the control system monitors the outputs of the plant (employing a variety of sensors) and adjusts the inputs to the plant (using actuators) to make the actual response (known as the output) of the system track the desired output (represented by the reference input).

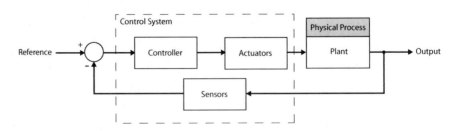

FIGURE B.6
Feedback-controlled system showing controller, actuators, sensors, and the physical system

An example of a closed-loop system is a heating and cooling system that regulates room temperature. The reference input is the desired room temperature. The sensor is a temperature-sensitive device that provides a measure of the actual room temperature. Based on the reference input, the thermostat activates the heating/cooling unit (the actuator) to produce either warm or cool air. In this example, the physical system is the room and the underlying physical processes that affect the temperature of the room, the thermometer (or other temperature device) is the sensor, the thermostat is the controller, and the heating/cooling unit is the actuator.

Other common examples of control systems include the following applications:

- Automobile cruise control systems

- Robots in manufacturing

- Wind gust control of modern aircraft

- Hard-drive head control systems

The LabVIEW Control Design toolkit adds functions to LabVIEW that are ideal for analyzing a plant model and designing the controller. It allows you to create simple and advanced transfer functions and place them in series or in parallel, or establish feedback. In addition to the Bode Plot, the toolkit supports other visualization mechanisms, such as the Root Locus, Nyquist, Interactive Pole-Zero, Interactive PID, and more.

B.5.3   Simulation

The third phase of model-based control design involves validating the controller design. Often times this verification and validation is performed by simulating the closed-loop dynamic system. Simulation is a process that involves using software to re-create and analyze the behavior of dynamic systems. You use the simulation process to lower product development costs by testing algorithms and ideas, thereby accelerating the design process. You also use the simulation process to provide insight into the behavior of dynamic systems you cannot replicate conveniently in the laboratory. For example, simulating a jet engine saves time, labor, and money compared with building, tweaking, testing, and rebuilding an actual jet engine each time you want to try something different.

The LabVIEW Simulation Module can be used to simulate the dynamic system or a component of a dynamic system. For example, you can simulate only the plant while using hardware for the controller, actuators, and sensors. Figure B.7 shows an image of a dynamic system simulation using the LabVIEW Simulation Loop.

FIGURE B.7
The LabVIEW Simulation Loop

*The **simulationexample.vi** depicted in Figure B.7 can be found in the folder **Appendix B** in the **Learning** directory.*

B.5.4   Deployment

The fourth phase of model-based control design involves deploying the actual controller, which is often done using a real-time (RT) target, which allows for deterministic execution. One nice feature of the LabVIEW Simulation Module is that it allows you to use the same model and LabVIEW block diagram used to simulate the system on your PC, add the appropriate data acquisition and control VIs using NI-DAQmx, and deploy the system to a LabVIEW Real-Time target, greatly simplifying the process of implementing control systems by using a single tool chain.

*For more information on deploying control systems using LabVIEW, visit the website **www.ni.com/academic/controls**.*

B.6   USING MULTISIM FOR INTEGRATED CIRCUIT SIMULATION AND VERIFICATION

Multisim is an industry standard circuit design and simulation tool. It is now part of the National Instruments family of design tools. The combination of LabVIEW 8.0 and Multisim 9 offers an effective means to share data between environments that is particularly useful for design prototyping and verification.

As an example, consider the following process of analyzing circuit simulation results using Multisim 9 and LabVIEW 8.0. Following the steps below, you can open and run a sample program in Multisim, and then you can open data files within LabVIEW produced by Multisim.

To open and run a sample program in Multisim:

1. Open Multisim.

2. Navigate to File≫Open Samples.

3. Open the AdjFreqAmp.ms9 sample program. This program demonstrates the use of an active adjustable frequency divider.

4. Open the oscilloscope on the top right. The oscilloscope will allow you to see both the input and the results of the frequency divider.

5. Press the **Simulate** button (lightning bolt) on the palette to begin simulation.

6. After a few seconds, stop the simulation by pressing the **Simulate** button again.

7. Click the **Save** button in the lower left of the oscilloscope window.

8. In the **Save Scope Data** window, change the **Save As Type** to **LabVIEW Measurement File** (*.lvm).

LabVIEW measurement files contain a short header and tab-delimited data. They can be easily imported into LabVIEW and Microsoft Excel.

Once the Multisim data is saved, you can open the Multisim data within LabVIEW:

1. Open LabVIEW.

2. Create a new VI.

3. Navigate to the Read from Measurement File Express VI found on the **Functions** palette under **Programming≫File IO**.

4. After placing the Read From Measurement File Express VI on your block diagram, the Configuration Window will appear.

5. Click on the file folder button to the right to select the file path of the *.lvm file you just saved from Multisim.

6. Select **OK**.

At this point you can access the Multisim data in LabVIEW. You can create a graph to verify that the data has been imported into LabVIEW.

7. Right-click the **Signals** output on the Read From Measurement File Express VI and select Create≫Graph Indicator.

8. Switch to the front panel and press the **Run** button to execute your VI. You may want to customize the graph by increasing the length and width and increasing the size of the graph legend to see the key for both plots. The Read LVM VI is shown in Figure B.8.

FIGURE B.8
Using Multisim to demonstrate the use of an active adjustable frequency divider and then loading the Multisim data into LabVIEW for graphing purposes

Now that you have successfully imported the Multisim simulation data, Lab-VIEW can be used for further analysis or design verification from actual hardware.

The Read LVM.vi in Figure B.8 can be found in the folder **Appendix B** *in the* **Learning** *directory.*

B.7 SUMMARY

New LabVIEW 8.0 features and applications that may be of interest to you as you begin to master the core LabVIEW programming elements were covered in this appendix. The material was intended to stimulate your curiosity about advanced LabVIEW programming features. From its inception 20 years ago, LabVIEW has evolved from a tool used primarily for computer-based data acquisition and and instrument control into a powerful graphical system design platform for designing, simulating, prototyping, and deploying complex solutions to perform a variety of applications in areas such as signal and image processing, embedded design, instrumentation, communications, circuit design, control, and mechatronics.

The new core features covered here include improved support for built-in sound cards, a discussion of the LabVIEW Project, and simplified distributed application development with the shared variable. LabVIEW toolkits and modules add further functionality to LabVIEW for specific areas or applications. For example, the LabVIEW Control Design Toolkit and Simulation Module add suites of tools for graphically modeling and designing continuous- and discrete-time systems. LabVIEW brings all of these capabilities together into a single graphical programming environment, giving you the ability to solve more advanced problems in a single environment. The appendix was intended to provide a brief introduction to these topics. You should consider other information sources from which you can continue to discover additional information. A good repository of information regarding advanced LabVIEW features is the NI website at www.ni.com.

KEY TERMS

LabVIEW Project: Collections of VIs, type definitions, shared variables, palette menu files, DAQ Task configurations, and other files, including other project libraries that help organize application developments.

Model-based control design: A control system design methodology that involves four phases: developing and analyzing a model to describe a plant, designing and analyzing a controller for the dynamic system, simulating the dynamic system, and deploying the controller.

Multisim: An industry standard circuit design and simulation tool that is now part of the National Instruments family of design tools.

Project Explorer window: A window used to create and edit projects.

Shared variables: Used to simplify complex applications that share data across the network and look similar to local or global variables on the block diagram.

Sound card API: Sound VIs that treat the sound card like a standard data acquisition device by passing data into and out of VIs as waveforms, increasing the number of possible sampling rates, and adding the ability to continuously read and output buffered signals.

Toolkits: Libraries of functions, virtual instruments, interactive wizards, examples, utilities, and documentation designed and optimized for specific needs.

INDEX